Encyclopaedia of Mathematical Sciences
Volume 2

Editor-in-Chief: R. V. Gamkrelidze

Ya. G. Sinai (Ed.)

Dynamical Systems II

Ergodic Theory
with Applications to Dynamical Systems
and Statistical Mechanics

With 25 Figures

Springer-Verlag
Berlin Heidelberg New York
London Paris Tokyo
Hong Kong

Consulting Editors of the Series: N.M. Ostianu, L.S. Pontryagin
Scientific Editors of the Series:
A.A. Agrachev, Z.A. Izmailova, V.V. Nikulin, V.P. Sakharova
Scientific Adviser: M.I. Levshtein

Title of the Russian edition:
Itogi nauki i tekhniki, Sovremennye problemy matematiki,
Fundamental'nye napravleniya, Vol. 2, Dinamicheskie sistemy 2
Publisher VINITI, Moscow 1985

QA
805
.D5613
1988
v.2

Mathematics Subject Classification (1980):
28Dxx, 34C35, 58Fxx

ISBN 3-540-17001-4 Springer-Verlag Berlin Heidelberg New York
ISBN 0-387-17001-4 Springer-Verlag New York Berlin Heidelberg

Library of Congress Cataloging-in-Publication Data
Dynamical systems.
(Encyclopaedia of mathematical sciences ; v.   )
Translation of: Dinamicheskie sistemy, issued as part of the series:
Itogi nauki i tekhniki. Seriia Sovremennye problemy matematiki.
Bibliography: p. Includes index.
1. Mechanics, Analytic. 2. Mechanics, Celestial.
I. Arnol'd, V. I. (Vladimir Igorevich), 1937–.
II. Iacob, A. III. Series: Encyclopaedia of mathematical sciences ; v.
QA805.D5613   1989   531   87-20655
0-387-17002-2 (v. 3)

This work is subject to copyright. All rights are reserved, whether the whole or part of the material is concerned, specifically the rights of translation, reprinting, reuse of illustrations, recitation, broadcasting, reproduction on microfilms or in other ways, and storage in data banks. Duplication of this publication or parts thereof is only permitted under the provisions of the German Copyright Law of September 9, 1965, in its version of June 24, 1985, and a copyright fee must always be paid. Violations fall under the prosecution act of the German Copyright Law.
© Springer-Verlag Berlin Heidelberg 1989
Printed in the United States of America
Typesetting: Asco Trade Typesetting Ltd., Hong Kong
2141/3140-543210 – Printed on acid-free paper

Preface

Each author who took part in the creation of this issue intended, according to the idea of the whole edition, to present his understanding and impressions of the corresponding part of ergodic theory or its applications. Therefore the reader has an opportunity to get both concrete information concerning this quickly developing branch of mathematics and an impression about the variety of styles and tastes of workers in this field.

Ya. G. Sinai

List of Editors, Contributors and Translators

Editor-in-Chief

R.V. Gamkrelidze, Academy of Sciences of the USSR, Steklov Mathematical Institute, ul. Vavilova 42, 117966 Moscow; Institute for Scientific Information (VINITI), Baltiiskaya ul. 14, 125219 Moscow, USSR

Consulting Editor

Ya.G. Sinai, Landau Institute of Theoretical Physics, ul. Kosygina 2, Moscow V-334, USSR

Contributors

L.A. Bunimovich, Institute of Oceanology of the Academy of Sciences of the USSR, ul. Krasikova 23, 117218 Moscow, USSR

I.P. Cornfeld, Central All-Union Institute of Complex Automation, Olkhovskaya 25, Moscow, USSR

R.L. Dobrushin, Institute of Information Transmission of the Academy of Sciences of the USSR, ul. Ermolovoj 19, 101447 Moscow GSP-4, USSR

M.V. Jakobson, Department of Mathematics, University of Maryland, College Park, Maryland, MD 20742, USA

N.B. Maslova, Leningrad Branch of the Institute of Oceanology of the Academy of Sciences of the USSR, 191028 Leningrad, USSR

Ya.B. Pesin, All-Union Extramural Engineering Construction Institute, ul. Sr. Kalitnikovskaya 30, Moscow, USSR

Ya.G. Sinai, Landau Institute of Theoretical Physics, ul. Kosygina 2, Moscow V-334, USSR

Yu.M. Sukhov, Institute of Information Transmission of the Academy of Sciences of the USSR, ul. Ermolovoj 19, 101447 Moscow GSP-4, USSR

A.M. Vershik, Mathematics Department, Leningrad State University, 198904 Leningrad, USSR

Translators

L.A. Bunimovich, Institute of Oceanology of the Academy of Sciences of the USSR, ul. Krasikova 23, 117218 Moscow, USSR

I.P. Cornfeld, Central All-Union Institute of Complex Automation, Olkhovskaya 25, Moscow, USSR

M.V. Jakobson, Department of Mathematics, University of Maryland, College Park, Maryland, MD 20742, USA

Yu.M. Sukhov, Institute of Information Transmission of the Academy of Sciences of the USSR, ul. Ermolovoj 19, 101447 Moscow GSP-4, USSR

Contents

I. General Ergodic Theory of Groups of Measure Preserving Transformations
1

II. Ergodic Theory of Smooth Dynamical Systems
99

III. Dynamical Systems of Statistical Mechanics and Kinetic Equations
207

Subject Index
279

I. General Ergodic Theory of Groups of Measure Preserving Transformations

Contents

Chapter 1. Basic Notions of Ergodic Theory and Examples of
Dynamical Systems (*I.P. Cornfeld, Ya.G. Sinai*) 2
§ 1. Dynamical Systems with Invariant Measures 2
§ 2. First Corollaries of the Existence of Invariant Measures. Ergodic
    Theorems 11
§ 3. Ergodicity. Decomposition into Ergodic Components. Various
    Mixing Conditions 17
§ 4. General Constructions 22
    4.1. Direct Products of Dynamical Systems 22
    4.2. Skew Products of Dynamical Systems 23
    4.3. Factor-Systems 24
    4.4. Integral and Induced Automorphisms 24
    4.5. Special Flows and Special Representations of Flows 25
    4.6. Natural Extensions of Endomorphisms 27
Chapter 2. Spectral Theory of Dynamical Systems
(*I.P. Cornfeld, Ya.G. Sinai*) 28
§ 1. Groups of Unitary Operators and Semigroups of Isometric Operators
    Adjoint to Dynamical Systems 28
§ 2. The Structure of the Dynamical Systems with Pure Point and
    Quasidiscrete Spectra 30
§ 3. Examples of Spectral Analysis of Dynamical Systems 33
§ 4. Spectral Analysis of Gauss Dynamical Systems 34
Chapter 3. Entropy Theory of Dynamical Systems
(*I.P. Cornfeld, Ya.G. Sinai*) 36
§ 1. Entropy and Conditional Entropy of a Partition 36
§ 2. Entropy of a Dynamical System 38
§ 3. The Structure of Dynamical Systems of Positive Entropy 41
§ 4. The Isomorphy Problem for Bernoulli Automorphisms and
    K-Systems 43

§ 5. Equivalence of Dynamical Systems in the Sense of Kakutani 51
§ 6. Shifts in the Spaces of Sequences and Gibbs Measures 55
Chapter 4. Periodic Approximations and Their Applications. Ergodic Theorems, Spectral and Entropy Theory for the General Group Actions (*I.P. Cornfeld, A.M. Vershik*) 59
§ 1. Approximation Theory of Dynamical Systems by Periodic Ones. Flows on the Two-Dimensional Torus....................... 59
§ 2. Flows on the Surfaces of Genus $p \geq 1$ and Interval Exchange Transformations 64
§ 3. General Group Actions 67
    3.1. Introduction 67
    3.2. General Definition of the Actions of Locally Compact Groups on Lebesgue Spaces 68
    3.3. Ergodic Theorems...................................... 69
    3.4. Spectral Theory... 71
§ 4. Entropy Theory for the Actions of General Groups 73
Chapter 5. Trajectory Theory (*A.M. Vershik*) 77
§ 1. Statements of Main Results 77
§ 2. Sketch of the Proof. Tame Partitions 81
§ 3. Trajectory Theory for Amenable Groups 86
§ 4. Trajectory Theory for Non-Amenable Groups. Rigidity 88
§ 5. Concluding Remarks. Relationship Between Trajectory Theory and Operator Algebras 91
Bibliography... 93

Chapter 1
Basic Notions of Ergodic Theory and Examples of Dynamical Systems

I.P. Cornfeld, Ya.G. Sinai

§ 1. Dynamical Systems with Invariant Measures

Abstract ergodic theory deals with the measurable actions of groups and semigroups of transformations. This means, from the point of view of applications, that the functions defining such transformations need not satisfy any smoothness conditions and should be only measurable.

A pair $(M, \mathcal{M})$ where M is an abstract set and $\mathcal{M}$ is some σ-algebra of subsets of M, is called a measurable space. In the sequel M will be the phase space of a

Chapter 1. Basic Notions of Ergodic Theory

dynamical system. The choice of $\mathscr{M}$ will always be clear from the context. We shall make use of the notions of the direct product of measurable spaces and of $\mathscr{M}$-measurable functions.

Definition 1.1. A transformation $T: M \to M$ is measurable if $T^{-1}C \in \mathscr{M}$ for any $C \in \mathscr{M}$.

A measurable transformation T is also called *an endomorphism of the measurable space* $(M, \mathscr{M})$. Any endomorphism generates a cyclic semigroup $\{T^n\}$ of endomorphisms ($n = 0, 1, 2, \ldots$).

If T is invertible and T^{-1} (as well as T) is measurable, then T is said to be *an automorphism of the measurable space* $(M, \mathscr{M})$. Any automorphism generates the cyclic group $\{T^n\}$ of automorphisms, $-\infty < n < \infty$.

A natural generalization of the above notions can be achieved by considering an arbitrary countable group or semigroup G and by fixing for each $g \in G$ a measurable transformation T_g such that $T_{g_1} \cdot T_{g_2} = T_{g_1 g_2}$ for all $g_1, g_2 \in G$, $T_e = \text{id}$.

Definition 1.2. The family $\{T_g\}$, $g \in G$, is said to be *a measurable action* of the countable group (semigroup) G. The simplest example is as follows. Suppose that $(X, \mathscr{X})$ is a measurable space and M is the space of all X-valued functions on G, i.e. any $x \in M$ is a sequence $\{x_g\}$, $x_g \in X$, $g \in G$. For any $g_0 \in G$ define the transformation $T_{g_0}: M \to M$ by the formula $T_{g_0} x = x'$, where $x'_g = x_{g_0 g}$. In this case $\{T_g\}$ is called a group (semigroup) of shifts. In particular,

1) if G is the semigroup $\mathbb{Z}_+^1 = \{n: n \geq 0, n \text{ is an integer}\}$, then M is the space of all 1-sided X-valued sequences, i.e. the points $x \in M$ are of the form $x = \{x_n\}$, $x_n \in X$, $n \geq 0$, and $T_m x = \{x_{n+m}\}$, $m \in \mathbb{Z}_+^1$. T_1 is called a 1-sided shift.

2) if G is the group $\mathbb{Z}^1 = \{n: -\infty < n < \infty, n \text{ is an integer}\}$, then M is the space of all 2-sided sequences $x = \{x_n\}$, $x_n \in X$, $-\infty < n < \infty$, and $T_m x = \{x_{n+m}\}$, $m \in \mathbb{Z}^1$. T_1 is called a 2-sided shift, or, simply, a shift.

3) if $G = \mathbb{Z}^d = \{(n_1, n_2, \ldots, n_d): n_i \in \mathbb{Z}^1, 1 \leq i \leq d\}, d \geq 1$, then M is the space of all sequences x of the form $x = \{x_n\} = \{x_{n_1, \ldots, n_d}\}$, while $T^m x = \{x_{n+m}\}$, $m = \{m_1, \ldots, m_d\} \in \mathbb{Z}^d$.

The above examples arise naturally in probability theory, where the role of M is played by the space of all realizations of d-dimensional random field.

Now suppose G is an arbitrary group or semigroup endowed with the structure of measurable space $(G, \mathscr{G})$ compatible with its group structure, i.e. all transformations $T_{g_0}: g \mapsto g_0 g$ ($g, g_0 \in G$) are measurable.

Definition 1.3. The family $\{T_g: M \to M\}$, $g \in G$, where G is a measurable group is called a *measurable action of the group G* (or a G-flow) if

1) $T_{g_1} \cdot T_{g_2} = T_{g_1 g_2}$ for all $g_1, g_2 \in G$;
2) for any $\mathscr{M}$-measurable function $f: M \to \mathbb{R}^1$ the function $f(T_g x)$ considered as a function on the direct product $(M, \mathscr{M}) \times (G, \mathscr{G})$ is also measurable.

Our main example is $G = \mathbb{R}^1$ with the Borel σ-algebra of subsets of $\mathbb{R}^1$ as $\mathscr{G}$. There also exist natural examples with $G = \mathbb{R}^d$, $d > 1$ (cf Chap. 10).

Let now $G = \mathbb{R}^1$. If T^t is the transformation in $\mathbb{R}^1$-flow corresponding to a $t \in \mathbb{R}^1$, then we have $T^{t_1} \cdot T^{t_2} = T^{t_1+t_2}$. We will describe a natural situation in which the actions of $\mathbb{R}^1$ arise.

Suppose M is a smooth compact manifold and α is a smooth vector field on M. Consider the transformation T^t sending each point $x \in M$ to the point $T^t x$ which can be obtained from x by moving x along the trajectory of α for the period of time t (T^t is well defined because of compactness of M). Then $T^{t_1+t_2} = T^{t_1} \cdot T^{t_2}$ and T^t is a measurable action of $\mathbb{R}^1$.

Measurable actions of $\mathbb{R}^1$ are usually called *flows*, and those of $\mathbb{R}^1_+$ — *semiflows*. The cyclic groups and semigroups of measurable transformations are also known as *dynamical systems with discrete time*, while flows and semiflows are known as *dynamical systems with continuous time*.

Now, let $(M, \mathcal{M}, \mu)$ be a measure space (probability space), i.e. $(M, \mathcal{M})$ is a measurable space and μ is a nonnegative normalized ($\mu(M) = 1$) measure on $\mathcal{M}$. Consider a measure ν on $\mathcal{M}$ given by $\nu(C) = \mu(T^{-1}C)$, $C \in \mathcal{M}$. This measure is said to be the image of the measure μ under T (notation: $\nu = T\mu$).

Definition 1.4. A measure μ is *invariant* under a measurable transformation $T: M \to M$ if $T\mu = \mu$.

If μ is invariant under T, then T is called an *endomorphism of the measure space* $(M, \mathcal{M}, \mu)$. If, in addition, T is invertible, it is called an *automorphism* of $(M, \mathcal{M}, \mu)$. If $\{T^t\}$ is a measurable action of $\mathbb{R}^1$ and each T^t, $-\infty < t < \infty$, preserves the measure μ, then $\{T^t\}$ is called a *flow on the measure space* $(M, \mathcal{M}, \mu)$.

Now consider the general case.

Definition 1.5. Let $\{T_g\}$ be a measurable action of a measurable group $(G, \mathcal{G})$ on the space $(M, \mathcal{M})$. A measure μ on $\mathcal{M}$ is called *invariant* under this action if, for any $g \in G$, μ is invariant under T_g.

We now introduce the general notion of metric isomorphism of dynamical systems which allows us to identify systems having similar metric properties.

Definition 1.6. Suppose $(G, \mathcal{G})$ is a measurable group and $\{T_g^{(1)}\}$, $\{T_g^{(2)}\}$ are two G-flows acting on $(M_1, \mathcal{M}_1)$, $(M_2, \mathcal{M}_2)$ respectively and having invariant measures μ_1, μ_2. Such flows are said to be *metrically isomorphic* if there exist G-invariant subsets $M_1' \subset M_1$, $M_2' \subset M_2$, $\mu_1(M_1') = \mu_2(M_2') = 1$, as well as an isomorphism $\varphi: (M_1', \mathcal{M}_1, \mu_1) \to (M_2', \mathcal{M}_2, \mu_2)$ of measure spaces M_1', M_2' such that $T_g^{(2)} \varphi x^{(1)} = \varphi T_g^{(1)} x^{(1)}$ for all $g \in G$, $x^{(1)} \in M_1'$.

Ergodic theory also studies measurable actions of groups on the space $(M, \mathcal{M}, \mu)$ which are not necessarily measure-preserving.

Definition 1.7. Suppose $\{T_g\}$ is a measurable action of a measurable group $(G, \mathcal{G})$ on $(M, \mathcal{M})$. The measure μ on $\mathcal{M}$ is said to be *quasi-invariant* under this action if for any $g \in G$ the measure $\mu_g \stackrel{\text{def}}{=} T_g \mu$, i.e. the image of μ under T_g, is equivalent to μ. In other words, μ and $T_g \mu$ have the same sets of zero measure.

Chapter 1. Basic Notions of Ergodic Theory

In the case of quasi-invariant measure the notions of endomorphisms, automorphisms, G-flows can also be introduced in a natural way. The metric isomorphism of groups of transformations with quasi-invariant measure is naturally defined "up to a transformation with quasi-invariant measure", i.e. our requirement imposed on the transformation φ in Definition 1.6 is that the measure $\varphi\mu_1$ should be equivalent but not necessarily equal to μ_2.

The basic properties of transformations with invariant measure will be discussed in Section 2, and now we turn to the problem of existence of such measures.

Suppose α is a smooth vector field on a m-dimensional manifold M, $\{T^t\}$ is the corresponding group of shifts along the trajectories of α, and μ is an absolutely continuous measure, i.e. in any system of local coordinates $(x_1, \ldots, x_m)$ μ is given by its density: $d\mu = \rho(x_1, \ldots, x_m) dx_1 \ldots dx_m$.

The well-known Liouville's theorem says that μ is invariant under $\{T^t\}$ if ρ satisfies the Liouville equation: $\text{div}(\rho\alpha) = 0$. Such a measure μ is known as a Liouville measure, or integral invariant of the dynamical system $\{T^t\}$. We will give now some applications of Liouville's theorem.

1. Let M be a $2m$-dimensional symplectic manifold and let the vector field α be given by a Hamilton function H not depending on time. In a local system of coordinates $(q^1, \ldots, q^m, p_1, \ldots, p_m)$ such that the symplectic form ω can be written as $\omega = \sum_{i=1}^{m} dq^i \wedge dp_i$, the vector field α is given by the Hamilton system of equations

$$\frac{dq^i}{dt} = \frac{\partial H}{\partial p_i}, \qquad \frac{dp_i}{dt} = -\frac{\partial H}{\partial q^i} \tag{1.1}$$

The measure μ with density $\rho(q, p) = 1$ is invariant.

Dynamical systems given by the equations (1.1) are called *Hamiltonian dynamical systems*. The class of such systems includes, in particular, geodesic flows which frequently appear in applications. They can be introduced by the following construction.

Let Q be a smooth compact m-dimensional Riemannian manifold. Denote by $\mathcal{T}_q(\mathcal{T}_q^*)$ the tangent (cotangent) space to Q at a point $q \in Q$ and consider the unit tangent bundle $M = \{(q, v): q \in Q, v \in \mathcal{T}_q, \|v\| = 1\}$ whose points are called linear elements on Q.

The geodesic flow on Q is a group $\{T^t\}$ of transformations of M such that a specific transformation T^t consists in moving a linear element $x = (q, v)$ along the geodesic line which it determines, by a distance t. The measure μ on M with $d\mu = d\sigma(q) d\omega_q$, where $d\sigma(q)$ is the element of the Riemann volume, and ω_q is the Lebesgue measure on the unit sphere S^{m-1} in $\mathcal{T}_q$, is invariant under $\{T^t\}$.

Another way to introduce the geodesic flow is as follows:

The tangent bundle $\mathcal{T}_q = \{(q, v): q \in Q, v \in \mathcal{T}_q\}$ may be naturally identified with the cotangent bundle $\mathcal{T}^*Q = \{(q, p): q \in Q, p \in \mathcal{T}_q^*\}$. Let $(q^1, \ldots, q^m)$ be the local coordinates at the point $q \in Q$. Each point $p \in \mathcal{T}_q^*$ is uniquely determined by its components $(p_1, \ldots, p_m)$. The nondegenerate differential 2-form $\omega = \sum_{i=1}^{m} dq^i \wedge dp_i$ induces the symplectic structure on $\mathcal{T}^*Q$, and the geodesic flow $\{T^t\}$ which we have just introduced is naturally isomorphic to the restriction to the unit tangent bundle of the Hamiltonian dynamical system with Hamiltonian $H(q, p) = \frac{1}{2}\|p\|^2$. Ergodic properties of the geodesic flow are uniquely determined by the Riemannian structure on Q.

2. Let $M = \mathbb{R}^m \times \mathbb{R}^m$ be the tangent bundle over $\mathbb{R}^m$, $m \geq 1$, and the vector field $\alpha = (\alpha_1, \ldots, \alpha_m)$ be given by

$$\frac{dx_i}{dt} = y_i, \qquad \frac{dy_i}{dt} = \alpha_i(x_1, \ldots, x_m), \qquad 1 \leq i \leq m, \tag{1.2}$$

or

$$\frac{d^2 x_i}{dt^2} = \alpha_i(x_1, \ldots, x_m).$$

The measure μ with $d\mu = dx_1 \ldots dx_m d\dot{x}_1 \ldots d\dot{x}_m$ is invariant under the group $\{T^t\}$ of translations along the solutions of the system (1.2).

3. $M = \mathbb{R}^m$, $m \geq 1$, and the vector field α is given by

$$\frac{dx_i}{dt} = \alpha_i(x_1, \ldots, x_m), \qquad 1 \leq i \leq m,$$

and $\operatorname{div} \alpha = \sum_{i=1}^{m} \frac{\partial \alpha_i}{\partial x_i} = 0$. Then the measure μ with $d\mu = dx_1 \ldots dx_m$ will be invariant under the action of the group $\{T^t\}$ corresponding to the vector field α. The important case is $m = 3$. The trajectories of the vector field are in this case the magnetic flux of the stationary magnetic field with tension α. The equality $\operatorname{div} \alpha = 0$ is one of the Maxwell equations.

Remark. It was mentioned above that the invariant measure given by Liouville's theorem is usually infinite. If the system under consideration has a prime integral $H(x)$ with compact "level surfaces" $M_c = \{x \in M : H(x) = c\}$, $c \in \mathbb{R}^1$, a finite invariant measure can be constructed out of this infinite measure. In this case any trajectory lies on a single manifold M_c. If $\rho(x)$ is the density of the Liouville measure, the measure μ_c concentrated on M_c with $d\mu_c = \frac{1}{|\nabla H|} \rho \, d\sigma$, where $d\sigma$ is the element of Riemann volume, will also be invariant. This measure may turn out to be finite and, therefore, can be normalized. The main example: for a Hamiltonian system (cf Example 1) the Hamilton function H itself is a prime integral. The measure μ_c induced by the Liouville measure on the surface $H = c$ is called a microcanonical distribution.

4. *Billiards.* Suppose Q_0 is a closed m-dimensional Riemannian manifold of class C^∞, and Q is a subset of Q_0 given by the system of inequalities of the form $f_i(q) \geq 0$, $q \in Q_0$, $f_i \in C^\infty(Q_0)$, $1 \leq i \leq r < \infty$. The phase space of the billiards in

Q is the set M whose points are the linear elements $x = (q, v)$, $x \in \text{Int } Q$, $v \in S^{m-1}$, as well as those $x = (q, v)$ for which $x \in \partial Q$, $v \in S^{m-1}$ and v is directed inside Q. The motion of a point $x = (q, v)$ under the billiard flow is the motion with unit speed along the trajectory of the geodesic flow until the boundary ∂Q is reached. At such moments the point reflects from the boundary according to the "incidence angle equals reflection angle" rule and then continues its motion. The measure μ on M such that $d\mu = d\rho(q)\, d\omega_q$, where $d\rho(q)$ is the element of Riemann volume, ω_q is the Lebesgue measure on S^{m-1}, is invariant under $\{T^t\}$. A more detailed construction of billiards flow see in [CFS].

5. M is a commutative compact group, μ is the Haar measure on M. If T is a group translation on M, i.e. the transformation of the form $Tx = x + g$, $(x, g \in M)$ then μ is invariant under T. This fact follows immediately from the definition of Haar measure.

Now let T be a group automorphism of the group M, i.e. T is a continuous and one-to-one mapping of M onto itself such that $T(x_1 + x_2) = Tx_1 + Tx_2$ for all $x_1, x_2 \in M$. The invariance of μ under T follows in this case from the uniqueness of the Haar measure.

6. Suppose $(X, \mathscr{X}, \lambda)$ is a probability space, M is the space of all sequences of the form $x = \{x_n\}$ where $x_n \in X$, $n \in \mathbb{Z}^1$ or $n \in \mathbb{Z}^1_+$, T is the shift on M, i.e. $Tx = x'$, $x'_n = x_{n+1}$. Let a measure μ on M be the product-measure of measure λ. In other words, the random variables x_n are mutually independent and have the distribution λ. The shift T is called in this case a *Bernoulli automorphism* (*endomorphism*). It gives us one of the most important examples of automorphisms (endomorphisms) in ergodic theory. The space $(X, \mathscr{X}, \lambda)$ is called *the state space* of Bernoulli automorphism (endomorphism).

7. The above example can be generalized as follows. Suppose $(Y, \mathscr{Y}, \lambda)$ is the probability space and we are given the transition operator $P_y(\cdot)$. This means that for any $y \in Y$ there is a probability measure P_y on $\mathscr{Y}$ and the family of all P_y is measurable in the sense that for any $\mathscr{Y}$-measurable function f the integral $\int_Y f(z)\, dP_y(z)$ is also a $\mathscr{Y}$-measurable function on Y. Assume further that λ is an invariant measure for $P_y(\cdot)$, i.e.

$$\lambda(C) = \int_Y P_y(C)\, d\lambda(y). \tag{1.3}$$

The space M of all sequences $x = \{y_n\}$, $y_n \in Y$, $n \in \mathbb{Z}^1$ or $n \in \mathbb{Z}^1_+$, and the shift T on M are just as in the previous example, but the construction of the measure μ differs from it. The measure is now given by

$$\mu(\{y: y_i \in C_0, y_{i+1} \in C_1, \ldots, y_{i+k} \in C_k\})$$
$$= \int_{C_0 \times C_1 \times \ldots C_{k-1}} d\lambda(y_i)\, dP_{y_i}(y_{i+1}) \cdot \ldots \cdot dP_{y_{i+k-2}}(y_{i+k-1}) \cdot P_{y_{i+k-1}}(C_k),$$

where $C_0, C_1, \ldots, C_k \in \mathscr{Y}$, $1 \leq k < \infty$.

It follows from (1.3) that this measure is invariant under T. The transformation T is called in this case *a Markov automorphism (endomorphism)*.

There are many groups and semigroups of transformations for which the existence of at least one invariant measure is not self-evident. We will present now the general approach to this problem due to N.N. Bogolyubov and N.M. Krylov.

Suppose M is a compact metric space, $\mathcal{M}$ is its Borel σ-algebra and $T: M \to M$ is a continuous mapping.

Theorem 1.1 (N.N. Bogolyubov, N.M. Krylov [BK]). *There exists at least one normalized Borel measure invariant under T.*

Indeed, let μ be an arbitrary normalized Borel measure on $\mathcal{M}$. For $n = 1, 2, \ldots$ consider the measures μ_n given by $\mu_n(C) = \mu_0(T^{-n}C)$, $C \in \mathcal{M}$, as well as $\mu^{(n)} = \sum_{k=0}^{n-1} \mu_k$. The compactness of M implies that the space of all normalized Borel measures on M is weakly compact. Therefore the sequence $\{n_s\}$ of integers, $n_s \to \infty$ as $s \to \infty$, exists such that $\mu^{(n_s)}$ weakly converges as $s \to \infty$ to some measure μ. This limit measure μ will be invariant: for any continuous function f on M we have

$$\int_M f(Tx)\,d\mu = \lim_{s\to\infty} \int_M f(Tx)\,d\mu^{(n_s)}$$

$$= \lim_{s\to\infty} \frac{1}{n_s} \sum_{k=0}^{n_s-1} \int_M f(Tx)\,d\mu_k = \lim_{s\to\infty} \frac{1}{n_s} \sum_{k=0}^{n_s-1} \int_M f(x)\,d\mu_{k+1}$$

$$= \lim_{s\to\infty} \frac{1}{n_s} \sum_{k=1}^{n_s} \int_M f(x)\,d\mu_k = \lim_{s\to\infty} \frac{1}{n_s} \sum_{k=0}^{n_s-1} \int_M f(x)\,d\mu_k$$

$$= \int_M f(x)\,d\mu,$$

which obviously signifies the invariance of μ.

If $\{T^t\}$ is a continuous one-parameter group of homeomorphisms of a compact metric space, one can use the same argument to prove that $\{T^t\}$ has at least one invariant measure.

Definition 1.8. A homeomorphism T of a compact metric space M is said to be *uniquely ergodic* if it has precisely one normalized Borel invariant measure. A homeomorphism T is said to be minimal if the trajectory $\{T^n x: -\infty < n < \infty\}$ of any point $x \in M$ is dense in M. A homeomorphism T is said to be topologically transitive if the trajectory of some point $x \in M$ is dense in M.

The notions just introduced characterize in different senses the property of "topological nondecomposability" of T. These notions are sometimes being applied not only to homeomorphisms but also to more general Borel transformations of topological spaces. In Section 3 for the dynamical systems on general measure spaces we shall introduce the notion of ergodicity playing the central role in ergodic theory and characterizing the "nondecomposability" in metric sense.

The properties of unique ergodicity and minimality are close to each other in the sense that there is a great number of natural examples for which both of them are satisfied or not satisfied simultaneously. However, in the general case neither of them implies the other. The minimality of T means that T has no nontrivial invariant closed set. The following theorem makes clear the meaning of the notion of unique ergodicity.

Theorem 1.2 (H. Furstenberg [31]). *Suppose T is a homeomorphism of the compact metric space M and μ is a normalized Borel measure invariant under T. The following statements are equivalent*:
1) *T is uniquely ergodic*;
2) *for any continuous function f on M and any $x \in M$ one has*

$$\lim_{n \to \infty} \frac{1}{n} \sum_{k=0}^{n-1} f(T^k x) = \int_M f \, d\mu;$$

3) *for any continuous function f on M the convergence $\frac{1}{n} \sum_{k=0}^{n-1} f(T^k x) \to \int_M f \, d\mu$ is uniform on M.*

In situations where the Krylov-Bogolyubov theory is applicable it may occur that many invariant measures exist. It seems useful to have a criterion for indicating the most important invariant measures. A new approach to this problem appeared recently in connection with progress in the theory of hyperbolic dynamical systems (cf Part II).

Suppose M is a smooth manifold; $T: M \to M$ is a diffeomorphism, ($\{T^t\}$ is a 1-parameter group of translations along the trajectories of some smooth vector field on M). Consider any absolutely continuous measure μ on M and its translations μ_n, $\mu_n(C) = \mu_0(T^{-n}C)$ (in the case of discrete time), μ_t, $\mu_t(C) = \mu_0(T^{-t}C)$ (in the case of continuous time). It may occur that the translated measures $\mu_n(\mu_t)$ converge to some limit measure μ as $n \to \infty$ (respectively, $t \to \infty$) and μ does not depend on the choice of the initial measure μ_0. This limit measure μ will necessarily be invariant, and it may be considered as the most important invariant measure for the dynamical system under consideration.

The dissipative systems may also possess such measures. Suppose $\{T^t\}$ is a 1-parameter group of translations along the solutions of the system of differential equations

$$\frac{dx_i}{dt} = f_i(x_1, \ldots, x_m), \quad 1 \leq i \leq m < \infty. \tag{1.4}$$

The system (1.4) is said to be dissipative if $\operatorname{div} f = \sum_{i=1}^{m} \frac{\partial f_i}{\partial x_i} < 0$. Liouville's theorem mentioned above implies that for any $C_0 \subset \mathbb{R}^m$ the m-dimensional volume of the set $C_t = T^t C_0$ decreases in time and tends to zero. This fact, however, by no means signifies that the initial measures become degenerate in some sense under the action of dynamics. On the contrary, they may converge to some nontrivial measures concentrated on the invariant sets whose m-dimensional volume vanish. The situation may be interpreted by saying that the dynamics itself produces a natural invariant measure as a result of the evolution of smooth measures. Such a situation arises frequently in many problems concerning so called strange attractors (see Part II Chap. 7). The question of the existence of such measures was raised repeatedly in connection with mathematical approaches to the analysis of turbulence.

Remark on Lebesgue spaces. A number of results presented below necessitate the assumption that the phase space of the dynamical system is the so-called *Lebesgue space*. We shall not give the full definition of the Lebesgue space (cf [Ro2]). Notice only that among all nonatomic measure spaces the Lebesgue spaces are precisely the ones which are isomorphic to the closed interval [0,1] with the Lebesgue measure. The assumption that the measure space is Lebesgue is not restrictive: the spaces arising in applications are, as a rule, automatically Lebesgue. In particular, any separable metric space with a measure defined on the Borel σ-algebra is Lebesgue. On the other hand, under this (Lebesgue) assumption one can apply the theory of measurable partitions which does not exist in the general case.

The partition of the space $(M, \mathcal{M}, \mu)$ is a system $\xi = \{C\}$ of nonempty and pairwise nonintersecting measurable sets such that $\bigcup_{C \in \xi} C = M$. The fact that ξ is measurable means that one can define the measures μ_C on its elements $C \in \xi$ in such a way that they play the role of conditional probabilities. The formal definition is as follows:

Definition 1.9. By a *canonical system of conditional measures* belonging to the partition ξ we mean a system of measures $\{\mu_C\}$, $C \in \xi$, possessing the following properties:

1) μ_C is defined on some σ-algebra $\mathcal{M}_C$ of subsets of C;
2) the space $(C, \mathcal{M}_C, \mu_C)$ is Lebesgue;
3) for any $A \in \mathcal{M}$ the set $A \cap C$ belongs to $\mathcal{M}_C$ for almost all $C \in \xi$, the function $f(x) = \mu_{C_\xi(x)}(A \cap C_\xi(x))$, where $C_\xi(x)$ is the element of ξ containing $x \in M$, is measurable and

$$\mu(A) = \int \mu_{C_\xi(x)}(A \cap C_\xi(x)) \, d\mu.$$

A partition ξ is said to be *measurable* if it possesses a canonical system of conditional measures. If $\{\mu_C\}$, $\{\mu'_C\}$ are two canonical systems for a partition ξ, then $\mu_C = \mu'_C$ for almost all $C \in \xi$; in this sense the canonical system for ξ is unique.

§2. First Corollaries of the Existence of Invariant Measures. Ergodic Theorems[1]

The following theorem due to H. Poincaré gives the basic information about the behavior of trajectories of transformations with invariant measures.

Theorem 2.1 (Poincaré Recurrence Theorem) (cf [CFS]). *Suppose $(M, \mathcal{M}, \mu)$ is a measure space, $T: M \to M$ is its endomorphism. Then for any $C \in \mathcal{M}$, $\mu(C) > 0$, almost all points $x \in C$ return to C infinitely many times. In other words, there exists an infinite sequence $\{n_i\}$ of integers, $n_i \to \infty$ as $i \to \infty$, such that $T^{n_i}x \in C$. Therefore, any set $A \in \mathcal{M}$ for which $T^n A \cap A = \emptyset$ for all n sufficiently large, has zero μ-measure.*

The so-called "Zermelo paradox" in statistical mechanics is closely related to the Poincaré recurrence theorem. Consider the closed box containing N pointlike masses (molecules), that move under the action of interaction forces and reflect elastically from the boundaries. The differential equations describing the dynamics of such a system are Hamiltonian, so the one-parameter group of translations along the trajectories of this system preserves the Liouville measure. The manifolds of constant energy for the system under consideration are compact, and the Liouville measure induces the finite invariant measures concentrated on them. Therefore the Poincaré recurrence theorem may be applied. Suppose now that the set $C \subset M$ consists of such points of the phase space for which all molecules are located in one half of the box at the moment $t = 0$. The Poincaré recurrence theorem says that at some moments $t > 0$ all the moving molecules will be again in the same half of the box. At first glance we have a contradiction, since nobody has ever seen a gas not occupying entirely the volume available.

However, this phenomenon may be explained quite simply. The probability of the event C is estimated as $\exp(-\text{const} \cdot N)$, where N is the total number of molecules, and const depends on temperature, density and so on. In real conditions we have $N \sim 10^{23}$ molecules/cm^3, so $\mu(C)$ is extremely small. It will be shown later (see Sect. 4) that the time intervals between two consecutive realizations of the event C are estimated as $[\mu(C)]^{-1}$, so in our case they are extremely large. On the other hand, if N is sufficiently small, for example, if $N \sim 10$, it is quite possible that at some moment $t > 0$ all the molecules will be again in one half of the box. One can realize such an experiment with the help of a computer.

The important consequence of the existence of an invariant measure for a given dynamical system is the possibility of averaging over time (for the functions on the phase space). The following theorem is one of the cornerstones of ergodic theory.

---

[1] This section was written in collaboration with Ya.B. Pesin.

Theorem 2.2 (Birkhoff-Khinchin Ergodic Theorem, (cf [CFS]). *Suppose $(M, \mathcal{M}, \mu)$ is a space with normalized measure and $f \in L^1(M, \mathcal{M}, \mu)$. Then for almost every $x \in M$ the following limits exists:*
1) *in the case of an endomorphism T*
$$\lim_{n \to \infty} \frac{1}{n} \sum_{k=0}^{n-1} f(T^k x) \stackrel{\text{def}}{=} \bar{f}(x);$$

2) *in the case of an automorphism T*
$$\lim_{n \to \infty} \frac{1}{n} \sum_{k=0}^{n-1} f(T^k x) = \lim_{n \to \infty} \frac{1}{n} \sum_{k=0}^{n-1} f(T^{-k} x) \stackrel{\text{def}}{=} \bar{f}(x);$$

3) *in the case of a flow $\{T^t\}$*
$$\lim_{T \to \infty} \frac{1}{T} \int_0^T f(T^t x)\, dt = \lim_{T \to \infty} \frac{1}{T} \int_0^T f(T^{-t} x)\, dt \stackrel{\text{def}}{=} \bar{f}(x);$$

4) *in the case of a semiflow $\{T^t\}$*
$$\lim_{T \to \infty} \frac{1}{T} \int_0^T f(T^t x)\, dt \stackrel{\text{def}}{=} \bar{f}(x).$$

Moreover, $\bar{f} \in L^1(M, \mathcal{M}, \mu)$ and $\int_M \bar{f}\, d\mu = \int_M f\, d\mu$. The function $\bar{f}$ is invariant, i.e. $\bar{f}(T^n x) = \bar{f}(x)$ for $n \geq 0$ in the case of an endomorphism and for $-\infty < n < \infty$ in the case of an automorphism; $\bar{f}(T^t x) = \bar{f}(x)$ for $-\infty < t < \infty$ in the case of a flow and for $t \geq 0$ in the case of a semiflow.

Consider now the case when $f = \chi_C$, i.e. f is the indicator of some set $C \in \mathcal{M}$. The time mean $\frac{1}{n} \sum_{k=0}^{n-1} f(T^k x)$ is the relative frequency of visits of the points $T^k x$, $0 \leq k < n$, to the measurable set C. By the Birkhoff-Khinchin ergodic theorem, the limit value of such frequencies as $n \to \infty$ exists. This theorem is therefore analogous to the strong law of large numbers of probability theory.

The Birkhoff-Khinchin ergodic theorem was preceded by another important result due to von Neumann and also related to the convergence of the means $\frac{1}{n} \sum_{k=0}^{n-1} f(T^k x), \frac{1}{T} \int_0^T f(T^t x)\, dt$ (for $f \in L^2(M, \mathcal{M}, \mu)$), but instead of almost everywhere convergence, the convergence in the metric of L^2 was studied. The general form of this result deals with the isometric linear operators and groups (semigroups) of such operators in Hilbert space. We shall formulate it only in the case of a single operator.

Theorem 2.3 (The von Neumann Ergodic Theorem cf [CFS]). *Suppose U is an isometric operator in complex Hilbert space H; H_U is the subspace of vectors $f \in H$ invariant with respect to U, i.e. $H_U = \{f \in H : Uf = f\}$; P_U is the (operator of the) orthogonal projection to H_U. Then*

$$\lim_{n\to\infty}\left\|\frac{1}{n}\sum_{k=0}^{n-1}U^kf-P_Uf\right\|_H=0\quad\text{for any }f\in H.$$

To derive from this theorem the convergence in L^2 of the means $\frac{1}{n}\sum_{k=0}^{n-1}f(T^kx)$ for an endomorphism $T:(M,\mathcal{M},\mu)\to(M,\mathcal{M},\mu)$ it suffices to apply it to the operator U given by $U(f(x))=f(Tx)$, $f\in L^2(M,\mathcal{M},\mu)$. The invariance of the measure μ under T implies that U is an isometric operator. There are various generalizations of the von Neumann and Birkhoff-Khinchin Ergodic theorems. They are related to the measurable transformations without invariant measure or with the infinite invariant measure, to general groups of transformations, to the functions taking values in Banach spaces and so on. We will not discuss these results here (cf [Kr] and Sect. 3 of Chap. 4) Note only that in the simplest case (for a single transformation with finite invariant measure, flow or semiflow and a function $f\in L^2$) these general statements are, as a rule, equivalent to the Birkhoff-Khinchin or von Neumann theorems.

There is, however, a recent deep result due to J.F.C. Kingman, giving additional information even in this case.

Suppose T is an automorphism or an endomorphism of the measure space $(M,\mathcal{M},\mu)$ and $f\in L^1(M,\mathcal{M},\mu)$. The Birkhoff-Khinchin ergodic theorem deals with the almost everywhere convergence of the sequences of functions

$$\frac{1}{n}g_n(x)=\frac{1}{n}g_n(x;f)\stackrel{\text{def}}{=}\frac{1}{n}\sum_{k=0}^{n-1}f(T^kx).$$

All sequences $\{g_n(x)\}$ obviously satisfy the relation

$$g_{m+n}(x)=g_m(x)+g_n(T^mx);\quad x\in M;\quad m,n\in\mathbb{Z}_+^1, \tag{1.5}$$

and it is easily seen that (1.5) can be considered as an intrinsic characterization of such sequences. Indeed, if (1.5) holds for some $\{g_n(x)\}$, we have $g_n(x)=g_n(x;f)$ for $f=g_1(x)$.

Now consider the sequence $\{g_n(x)\}$ of the real-valued measurable functions on M satisfying, for almost all $x\in M$, the inequality

$$g_{m+n}(x)\leqslant g_m(x)+g_n(T^mx);\quad m,n\in\mathbb{Z}_+^1. \tag{1.6}$$

instead of the equality (1.5). It turns out that even under this weaker condition $\lim_{n\to\infty}\frac{1}{n}g_n(x)$ exists.

Theorem 2.4 (Subadditive ergodic theorem, J.F.C. Kingman [Kin]). *Suppose T is an endomorphism of the measure space $(M,\mathcal{M},\mu)$, and $\{g_n(x)\}$, $n>0$, is a sequence of measurable functions, $g_n:M\to\mathbb{R}^1\cup\{-\infty\}$, $g_1^+\in L^1(M,\mathcal{M},\mu)$[2] satisfying the condition (1.6). Then there exists a function $\bar{g}:M\to\mathbb{R}^1\cup\{-\infty\}$, $g\in L^1$,*

---

[2] For any function $f(x)$ we set $f^+(x)=\max(0,f(x))$.

invariant under T and such that

$$\lim_{n \to \infty} \frac{1}{n} g_n(x) = \bar{g}(x) \quad \text{for almost every } x \in M,$$

$$\lim_{n \to \infty} \frac{1}{n} \int_M g_n(x) \, d\mu = \inf_n \frac{1}{n} \int_M g_n(x) \, d\mu = \int_M \bar{g}(x) \, d\mu.$$

The following important result is an immediate consequence of Theorem 2.4.

Theorem 2.5 (The Furstenberg-Kesten theorem on the products of random matrices, H. Furstenberg, H. Kesten [FK]). *Suppose we are given an endomorphism T of a measure space $(M, \mathcal{M}, \mu)$ and a measurable function $G(x)$ on M taking values in the space of $m \times m$ real matrices ($m \geq 1$). Set $G_x^{(n)} = G(x) \cdot G(Tx) \cdot \ldots \cdot G(T^{n-1}x)$. If $\log^+ \|G(\cdot)\| \in L^1(M, \mathcal{M}, \mu)$, then the limit*

$$\lambda(x) = \lim_{n \to \infty} \frac{1}{n} \log \|G_x^{(n)}\|$$

exists for almost all $x \in M$, and $\lambda(x)$ is invariant under T. Moreover, $\lambda \in L^1(M, \mathcal{M}, \mu)$ and

$$\int_M \lambda \, d\mu = \lim_{n \to \infty} \frac{1}{n} \int_M \log \|G_x^{(n)}\| \, d\mu = \inf_n \frac{1}{n} \int_M \log \|G_x^{(n)}\| \, d\mu.$$

The next statement considerably strengthens the Furstenberg-Kesten theorem and is, in turn, a special case of the multiplicative ergodic theorem that will be formulated below.

Theorem 2.6 (V.I. Oseledets [Os2]). *Suppose T is an endomorphism of a measure space $(M, \mathcal{M}, \mu)$ and $G(x)$, $x \in M$, is a measurable function on M taking values in the space of $m \times m$ real matrices ($m \geq 1$), $G_x^{(n)} \stackrel{\text{def}}{=} G(x) \cdot G(Tx) \cdot \ldots \cdot G(T^{n-1}x)$. Suppose, further, that $\log^+ \|G(\cdot)\| \in L^1(M, \mathcal{M}, \mu)$. Then*
1) *there exists an invariant set $\Gamma \in \mathcal{M}$, $\mu(\Gamma) = 1$, such that the limit*

$$\Lambda_x = \lim_{n \to \infty} [G_x^{(n)*} G_x^{(n)}]^{1/2n}$$

exists for all $x \in \Gamma$ and Λ_x is a symmetric non-negative definite $m \times m$ matrix (we take the non-negative definite root);
2) *if $\exp \lambda_x^{(1)} < \exp \lambda_x^{(2)} < \cdots < \exp \lambda_x^{(s)}$ $(x \in \Gamma)$ is the ordered set of all different eigenvalues of Λ_x (we have $s = s(x) \leq m$; the case $\lambda_x^{(1)} = -\infty$ is not excluded); $E_x^{(1)}, E_x^{(2)}, \ldots, E_x^{(s)}$ is the corresponding set of eigenspaces, $\dim E_x^{(r)} \stackrel{\text{def}}{=} m_x^{(r)}, 1 \leq r \leq s$, then the functions $x \mapsto s(x)$, $x \mapsto \lambda_x^{(r)}$, $x \mapsto m_x^{(r)} (1 \leq r \leq s)$ are measurable and invariant with respect to T;*
3) *for any $x \in \Gamma$ and any $u \in \mathbb{R}^m$, $u \neq 0$, the limit*

$$\lim_{n \to \infty} \frac{1}{n} \log \|G_x^{(n)} u\|,$$

exists and this limit equals $\lambda_x^{(r)}$, where r, $1 \leq r \leq s$, is uniquely determined by the relations

$$u \in E_x^{(1)} \oplus E_x^{(2)} \oplus \cdots \oplus E_x^{(r)},$$

$$u \notin E_x^{(1)} \oplus E_x^{(2)} \oplus \cdots \oplus E_x^{(r-1)}.$$

Some important notions will be necessary for the formulation of the multiplicative ergodic theorem, and we are going to introduce them.

Definition 2.1. By *a linear measurable bundle* we mean the triple (N, M, π), where N, M are measurable spaces, $\pi \colon N \to M$ is a measurable map, and there exists an isomorphism $\psi \colon N \to M \times \mathbb{R}^m$ such that

1) the partition of N, whose elements are the sets $\pi^{-1}(x)$, $x \in M$, goes under ψ to the partition of $M \times \mathbb{R}^m$ with the elements of the form $\{x\} \times \mathbb{R}^m$, $x \in M$.

2) the map $\pi \circ \psi \circ \pi^{-1}$ is the identity map of M onto itself.

In other words, a linear measurable bundle is the image of the direct product $M \times \mathbb{R}^m$ under some measurable map. The following terminology will be used: N is the space of the bundle, M is the base, π is the projection, $\pi^{-1}(x)$ is the fiber over x. The map ψ induces a structure of normed vector space in each fiber $\pi^{-1}(x)$.

Any continuous subbundle of the tangent bundle of a smooth manifold (in particular, the tangent bundle itself) gives us an example of a linear measurable bundle.

Definition 2.2. *A characteristic exponent* is a measurable function $\chi \colon N \to \mathbb{R}^1$ such that for almost every $x \in M$ and any $v, v_1, v_2 \in \pi^{-1}(x)$ one has

1) $-\infty < \chi(x, v) < \infty$ if $v \neq 0$; $\chi(x, 0) = -\infty$;
2) $\chi(x, \alpha v) = \chi(x, v)$, $\alpha \in \mathbb{R}^1$, $\alpha \neq 0$;
3) $\chi(x, v_1 + v_2) \leq \max\{\chi(x, v_1), \chi(x, v_2)\}$.

It may be shown that for any $x \in M$ the restriction of χ to $\pi^{-1}(x)$ takes at most m values different from $-\infty$. Denote these values by $\chi_i(x)$, $1 \leq i \leq s(x) \leq m$, and assume that

$$\chi_1(x) < \chi_2(x) < \cdots < \chi_{s(x)}(x). \tag{1.7}$$

Let $L_i(x)$ be the subspace $\{v \in \pi^{-1}(x) \colon \chi(x, v) \leq \chi_i(x)\}$. The subspaces $L_i(x)$, $1 \leq i \leq s(x)$ determine the filtering of $\pi^{-1}(x)$, i.e.

$$\{0\} = L_0(x) \subset L_1(x) \subset \cdots \subset L_{s(x)}(x) = \pi^{-1}(x). \tag{1.8}$$

Let $k_i(x) = \dim L_i(x)$, $k_0(x) = 0$. The integer-valued functions $s(x)$, $k_1(x), \ldots, k_{s(x)}(x)$, as well as the families of subspaces $L_i(x)$, $1 \leq i \leq s(x)$ depend measurably upon x. Conversely, suppose we are given the integer-valued measurable function $s(x) \leq m$, the measurable functions $\chi_1(x), \ldots, \chi_{s(x)}(x)$ satisfying (1.7), and the filtering (1.8) depending measurably upon x with $\dim L_i(x) = k_i(x)$. Then the function $\chi(x, v)$ given by $\chi(x, v) = \chi_i(x)$ for $x \in M$, $v \in L_i(x) \setminus L_{i-1}(x)$, is measurable and defines the characteristic exponent on N.

Let T be an endomorphism of M preserving a normalized measure μ.

Definition 2.3. *A measurable multiplicative cocycle* with respect to T is a measurable function $a(n, x)$, $x \in M$, taking values in the space of $m \times m$ matrices and satisfying the relation $a(n + k, x) = a(n, T^k x) \cdot a(k, x)$ (a more general definition of a cocycle will be given in Section 4).

The function $a(n, x) = G(x) \cdot \ldots \cdot G(T^{n-1} x)$, where $G(x)$ is a measurable function taking values in the space of $m \times m$ matrices, is an example of a measurable multiplicative cocycle.

Let $a(n, x)$ be a measurable multiplicative cocycle with respect to T. Consider the function

$$\chi^+(x, v) = \varlimsup_{n \to \infty} \frac{1}{n} \log \|a(n, x)v\|, \quad x \in M, \quad v \in \pi^{-1}(x). \tag{1.9}$$

It may be shown that χ^+ is measurable and satisfies the conditions of Definition 2.2. Thus it defines some characteristic exponent which will be called the Lyapunov characteristic exponent corresponding to T and to the cocycle $a(n, x)$. It is easily seen that the functions $s(x)$, $\chi_i(x)$, $k_i(x)$ and the subspaces $L_i(x)$ corresponding to χ^+ are invariant with respect to T.

Fix $x \in M$ and consider the filtering (1.8) at the point x. A normalized basis $\bar{e}(x) = \{e_i(x)\}$ of the space $\pi^{-1}(x)$ is said to be regular if the vectors $e_i(x)$, $1 \leq i \leq k_1(x)$ belong to $L_1(x)$, the vectors $e_i(x)$, $k_1(x) + 1 \leq i \leq k_2(x)$ belong to $L_2(x) \setminus L_1(x)$ and so on. If $\{e_i(x)\}$ is a regular basis of $\pi^{-1}(x)$, then $\{e_i^{(n)}(x)\}$, where $e_i^{(n)}(x) = a(n, x) e_i(x) / \|a(n, x) e_i(x)\|$ is a regular basis of $\pi^{-1}(T^n x)$.

Definition 2.4. A point $x \in M$ is said to be *forward regular* if for some regular basis $\bar{e}(x) = \{e_i(x)\}$ we have

$$\sum_{i=1}^{m} \chi^+(x, e_i(x)) = \lim_{n \to \infty} \log |\det a(n, x)|.$$

In a somewhat different but equivalent form the definition of forward regularity may be found in [18], [82].

If in the right hand side of (1.9) we take the upper limit as $n \to -\infty$ rather than as $n \to +\infty$, we obtain the definition of the characteristic exponent χ^- on N. A point $x \in M$ is called backward regular if it is forward regular with respect to χ^-.

The notions of forward and backward regularity are classical ones and go back to A.M. Lyapunov and O. Perron who studied the stability properties of the solutions of linear ordinary differential equations with nonconstant coefficients (for the most part, the one-sided solutions—for $t > 0$ and for $t < 0$—were studied). In order to investigate the stability properties of the solutions of the Jacobi equations along the two-sided trajectories of the invertible dynamical systems, it is necessary to consider the points which are not only forward and backward regular but also have the characteristic exponents χ^+ and χ^-, as well as corresponding filterings, compatible with each other. This signifies that there exist subspaces $E_i(x)$, $i = 1, \ldots, s(x)$, such that

a) $L_i^+(x) = \bigoplus_{j=1}^{k_i(x)} E_j(x)$, $L_i^-(x) = \bigoplus_{j=k_i(x)+1}^{s(x)} E_j(x)$, where $\{L_i^+(x)\}$, $\{L_i^-(x)\}$ are the filterings related to χ^+, χ^- respectively;

b) $\lim_{n \to \pm\infty} \frac{1}{|n|} \log \|a(n,x)v\| = \pm \chi_j(x)$ uniformly over $v \in E_j(x)$.

c) $\chi^-(\Gamma_j(x)) = \chi^+(\Gamma_j(x)) = (k_j(x) - k_{j-1}(x))\chi_j(x)$, where $\Gamma_j(x)$ is the volume of the parallelepiped in the space $E_j(x)$ and

$$\chi^\pm(\Gamma_j(x)) \stackrel{\text{def}}{=} \lim_{n \to \pm\infty} \frac{1}{|n|} \log |\Gamma_j(T^n x)|.$$

The points which are both forward and backward regular and satisfy the conditions a), b), c) are called *Lyapunov regular*, or *biregular* (see [Mi1]).

It may be shown that if x is Lyapunov regular, then so are all points of the form $T^n x$, $n \in \mathbb{Z}$ (so it is convenient to speak of Lyapunov regular trajectories), and the subspaces $E_i(T^n x)$ at the point $T^n x$ satisfy the relations $E_i(T^n x) = a(n, x) E_i(x)$.

Denote by M^+, M^-, M_0 the sets of forward, backward and Lyapunov regular points in M respectively. These sets are invariant with respect to T and, by Theorem 2.6, $\mu(M^+) = \mu(M^-) = 1$. We also have $M_0 \subset M^+ \cap M^-$, where the inclusion may in general be strict. Nevertheless, it may be shown that the set M_0 is of full measure. More precisely, the following is true.

Theorem 2.7 (The multiplicative ergodic theorem, V.I. Oseledets [Os2]; in a somewhat different form this theorem was proved by V.M. Millionshchikov [Mi2]). Let $a(n, x)$ be the multiplicative cocycle on the linear measurable bundle (N, M, π). Assume that $\int_M \log \|a(1, x)\| \, d\mu < \infty$ (such a cocycle is called Lyapunov). Then μ-almost every point $x \in M$ is Lyapunov regular.

Note that, unlike the theorem on forward and backward regularity, which is an immediate consequence of the subadditive ergodic theorem, the above statement requires some additional ideas for its proof.

§3. Ergodicity. Decomposition into Ergodic Components. Various Mixing Conditions

The Birkhoff-Khinchin ergodic theorem shows that the only fact of existence of an invariant measure for a given dynamical system guarantees the possibility of averaging along its trajectories almost everywhere. For $f \in L^1(M, \mathcal{M}, \mu)$ denote by $\bar{f}$ the time mean appearing in the Birkhoff-Khinchin theorem. Let T be an endomorphism of the space $(M, \mathcal{M}, \mu)$.

Definition 3.1. A set $A \in \mathcal{M}$ is said to be *invariant mod 0* with respect to T if $\mu(A \triangle T^{-1}A) = 0$.

The invariance mod 0 of A with respect to T implies its invariance mod 0 with respect to all T^n.

Let $\{T^t\}$ be a flow or a semiflow.

Definition 3.2. A set $A \in \mathcal{M}$ is said to be *invariant mod 0 with respect to* $\{T^t\}$ if $\mu(A \triangle T^t A) = 0$ for all t.

All invariant mod 0 sets form a σ-algebra which will be denoted by $\mathcal{M}^{\text{inv}}$. The function $\bar{f}$ in the Birkhoff-Khinchin ergodic theorem is the conditional expectation of f with respect to $\mathcal{M}^{\text{inv}}$.

The next definition plays a fundamental role in ergodic theory.

Definition 3.3. A dynamical system is *ergodic* (with respect to an invariant measure μ) if $\mathcal{M}^{\text{inv}} = \mathcal{N}$ where $\mathcal{N}$ is the trivial σ-algebra consisting of the sets of measure 0 or 1.

In various fields of mathematics it is often useful to single out a class of elementary, in some sense indecomposable objects and then represent the objects of general form as some combinations of these elementary ones (prime numbers in number theory, irreducible representations, extreme points of convex sets etc.). The ergodic systems may be thought of as such elementary objects in ergodic theory. If T is ergodic, for any $f \in L^1(M, \mathcal{M}, \mu)$ we have $\bar{f} = \int_M f \, d\mu$, where $\bar{f}$ is the time mean for f, so the statement of the ergodic theorem coincides in this case with the statement of the strong law of large numbers.

Let T be an ergodic automorphism and $f = \chi_C$ be the indicator of a set $C \in \mathcal{M}$. Then $\frac{1}{n}\sum_{k=0}^{n-1} f(T^k x)$ is the relative frequency of visits to C by the trajectory of the point x in the time interval $[0, n-1]$, and $\lim_{n\to\infty} \frac{1}{n}\sum_{k=0}^{n-1} f(T^k x) = \mu(C)$. In this case therefore, the Birkhoff-Khinchin ergodic theorem may be formulated as follows: the time mean equals the space mean almost everywhere.

If T is a uniquely ergodic homeomorphism of a compact metric space M preserving a normalized Borel measure μ, then T, if considered as an automorphism of the space (M, μ), is ergodic. The convergence in the Birkhoff-Khinchin theorem holds in this case at every point $x \in M$, not only almost everywhere.

The following result concerning the uniquely ergodic realizations of automorphisms was first obtained by R. Jewett under some restrictions and then by W. Krieger in the general case.

Theorem 3.1 (W. Krieger [Kri2], R. Jewett [$\mathcal{J}$]). *For any ergodic automorphic T of the Lebesgue space M there exists a uniquely ergodic homeomorphism T_1 of some compact metric space M_1 such that T_1 as an automorphism of M_1 with its invariant measure is metrically isomorphic to T.*

For an arbitrary, not necessarily ergodic, dynamical system $\{T^t\}$ on the Lebesgue space $(M, \mathcal{M}, \mu)$ introduce the measurable hull ξ of the partition ζ of M into separate trajectories of $\{T^t\}$, i.e. the most refined of those partitions of M whose elements consist of entire trajectories of $\{T^t\}$. Let $\{\mu_C : C \in \xi\}$ be the canonical system of conditional measures for ξ.

Theorem 3.2 (on the decomposition into ergodic components, von Neumann [N], V.A. Rokhlin [Ro2]). *For almost every $C \in \xi$ the dynamical system $\{T^t\}$ induces the dynamical system $\{T_C^t\}$ on (C, μ_C) which is ergodic with respect to μ_C.*

The elements of the partition ξ are sometimes called *ergodic components*.

Examples. 1. Let M be the m-dimensional torus with the Haar measure, T be a group translation on M. Then T is of the form $Tx = (x_1 + \alpha_1, \ldots, x_m + \alpha_m)$ for $x = (x_1, \ldots, x_m)$ (we write the group operation additively). T is ergodic if the numbers $1, \alpha_1, \ldots, \alpha_m$ are linearly independent over the field of rational numbers (rationally independent).

2. More generally, suppose M is a commutative compact group, μ is the Haar measure on M, T is a group translation, i.e. $Tx = x + g$ $(x, g \in M)$. Denote by χ_n the characters of the group $M, (n = 0, 1, \ldots), \chi_0 \equiv 1$. Then T is ergodic if $\chi_n(g) \neq 1$ for any $n \neq 0$. This condition guarantees also the unique ergodicity and minimality of T.

3. As in Example 1, M is the m-dimensional torus with the Haar measure, $\{T^t\}$ is a group translation on M, i.e. $T^t x = (x_1 + \alpha_1 t, \ldots, x_m + \alpha_m t)$. This flow is sometimes referred to as a conditionally periodic flow, or a conditionally periodic winding of the torus. The flow $\{T^t\}$ is ergodic if $\alpha_1, \ldots, \alpha_m$ are rationally independent.

4. Integrable systems of classical mechanics. A Hamiltonian system in $2m$-dimensional phase space is said to be integrable (cf [Ar1]) if it has m prime integrals in involution. The Liouville theorem says that if all trajectories of the system are concentrated in a bounded part of the space, there exist locally m prime integrals $I_1, \ldots, I_m$, such that the m-dimensional manifolds $I_1 = \text{const}, \ldots, I_m = \text{const}$ are m-dimensional tori, and the motions on them are conditionally periodic. The Liouville theorem shows, therefore, that the system considered is not ergodic and its ergodic components are m-dimensional tori. In particular, geodesic flows on the surfaces of rotation having the additional prime integral which is known as the Clairaut integral, are not ergodic, and their ergodic components are two-dimensional tori. It follows from Jacobi's results that the same statement holds for geodesic flows on ellipsoids.

5. Group automorphisms. Suppose M is a commutative compact group, μ is the normalized Haar measure, T is a group automorphism of M. Denote by T^* the adjoint automorphism for T acting in the group M^* of characters of M by $(T^*\chi)(x) = \chi(Tx)$. Then T is ergodic if and only if the equality $(T^*)^n \chi = \chi, n \neq 0$, is impossible unless $\chi \equiv 1$. In particular, if M is the m-dimensional torus and T is given by an integer matrix $\|a_{ij}\|$, the above condition means that there are no roots of unity among the eigenvalues of $\|a_{ij}\|$.

6. Suppose M is the space of sequences $x = (\ldots, x_{-1}, x_0, x_1, \ldots), x_i \in X$, T is the shift in M, μ is an invariant measure. Denote by $\mathscr{A}^{(k)}$, $-\infty < k < \infty$, the σ-

algebra generated by the random variables x_i, $-\infty < i \leq k$. It is easily seen that $\mathcal{M}^{\text{inv}} \subseteq \bigcap_k \mathcal{A}^{(k)}$. A random process $(M, \mathcal{M}, \mu)$ is said to satisfy the zero-one law due to Kolmogorov if $\bigcap_k \mathcal{A}^{(k)} = \mathcal{N}$. The shift T corresponding to such a process is necessarily ergodic. In particular, each Bernoulli shift is ergodic.

If T is an ergodic automorphism, for any pair of functions $f, g \in L^2(M, \mathcal{M}, \mu)$ we have, by the Birkhoff-Khinchin theorem,

$$\lim_{n \to \infty} \frac{1}{n} \sum_{k=0}^{n-1} \int_M f(T^k x) g(x) \, d\mu = \int_M f \, d\mu \cdot \int_M g \, d\mu$$

almost everywhere.

The statistical properties of ergodic systems in the general case are fairly poor. We shall introduce now some notions characterizing the systems having more or less developed statistical properties. For the sake of simplicity we restrict ourselves to the case of automorphisms.

Definition 3.4. An automorphism T is *weak mixing* if for any $f, g \in L^2(M, \mathcal{M}, \mu)$ we have

$$\lim_{n \to \infty} \frac{1}{n} \sum_{k=0}^{n-1} \left[\int_M f(T^k x) g(x) \, d\mu - \int_M f \, d\mu \cdot \int_M g \, d\mu \right]^2 = 0.$$

Definition 3.5. An automorphism T is *mixing* if for any $f, g \in L^2(M, \mathcal{M}, \mu)$

$$\lim_{n \to \infty} \int_M f(T^n x) g(x) \, d\mu = \int_M f \, d\mu \cdot \int_M g \, d\mu.$$

Definition 3.6. An automorphism T is *r-fold mixing* ($r \geq 1$) if for any $f_1, \ldots, f_r$, $g \in L^{r+1}(M, \mathcal{M}, \mu)$

$$\lim_{n_1 \to \infty, \ldots, n_r \to \infty} \int_M f_1(T^{n_1} x) f_2(T^{n_1 + n_2} x) \cdots f_r(T^{n_1 + \cdots + n_r} x) g(x) \, d\mu$$

$$= \prod_{i=1}^{r} \int_M f_i \, d\mu \cdot \int_M g \, d\mu.$$

Definition 3.7. An automorphism T is said to be a *K-automorphism* (*Kolmogorov automorphism*) if for any $A_0, A_1, \ldots, A_r \in \mathcal{M}$, $1 \leq r < \infty$,

$$\lim_{n \to \infty} \sup |\mu(A_0 \cap B^{(n)}) - \mu(A_0) \mu(B^{(n)})| = 0,$$

where the supremum is taken over all sets $B^{(n)}$ in the σ-algebra generated by the sets of the form $T^k A_i$, $k \geq n$, $1 \leq i \leq r$.

A flow $\{T^t\}$ is said to be a *K-flow* if there exists a t_0 such that T^{t_0} is K-automorphism.

The mixing property implies weak mixing, and weak mixing implies ergodicity.

Suppose μ_0 is a measure on $\mathcal{M}$ absolutely continuous with respect to μ and $d\mu_0/d\mu = f_0(x)$. From the physical point of view it is natural to call μ_0 a nonequilibrium state. Denote by μ_n the measure defined by $\mu_n(C) = \mu_0(T^{-n}C)$, $C \in \mathcal{M}$. Then

$$\mu_n(C) = \int_{T^{-n}C} f_0(x)\,d\mu = \int_C f_0(T^n x)\,d\mu.$$

In other words, μ_n is absolutely continuous with respect to μ and $d\mu_n/d\mu = f_0(T^n x)$. In the case of mixing the measure μ_n converge to μ in the following sense: for any $g \in L^2(M, \mathcal{M}, \mu)$ we have $\int g\,d\mu_n \to \int g\,d\mu$. So, under the action of dynamics, any non-equilibrium state converges to some limit, which may be naturally called the equilibrium state.

K-systems are r-fold mixing for any $r \geq 1$. They play an important role in the entropy theory of dynamical systems (cf Chap. 3). The definition of K-automorphism may be also formulated in a somewhat different form, as follows: suppose that $\mathcal{A}_0$ is the σ-algebra generated by the sets A_i, $0 \leq i \leq r$, and $\mathcal{A}_k = T^k \mathcal{A}_0$; let, further, $\mathcal{A}_l^\infty$ be the minimal σ-algebra containing all $\mathcal{A}_k$, $k \geq l$. An automorphism T is K-automorphism if and only if $\bigcap_l \mathcal{A}_l^\infty = \mathcal{N}$ for any initial σ-algebra $\mathcal{A}_0$.

If $\{T^t\}$ is a flow and T^{t_0} is a K-automorphism for some $t_0 \in \mathbb{R}^1$, then so are T^t for all $t \in \mathbb{R}^1$, $t \neq 0$, (cf Chap. 3).

It is not known whether or not the mixing (1-fold mixing) property implies r-fold mixing for $r > 1$. We will now state a deep result, due to H. Furstenberg, on the relationship between weak mixing and a kind of r-fold weak mixing (the formal definition of r-fold weak mixing can be obtained from Definition 3.6 by means of some obvious changements).

Theorem 3.3 (H. Furstenberg [FKO]). *Let T be a weak mixing automorphism of a Lebesgue space $(M, \mathcal{M}, \mu)$ and $r \geq 1$ is an integer. For any $f_0, f_1, \ldots, f_r \in L^\infty(M, \mathcal{M}, \mu)$ one has*

$$\lim_{n \to \infty} \frac{1}{n} \sum_{k=1}^{n} \left[\int_M \prod_{l=0}^{r} T^{kl} f_l\,d\mu - \prod_{l=0}^{r} \int_M f_l\,d\mu \right]^2 = 0.$$

There is a very interesting application of the above theorem to some number-theoretical problems. We begin by the formulation of a general statement based on Theorem 3.2.

Theorem 3.4 (H. Furstenberg [FKO]). *Let T be an automorphism of the Lebesgue space $(M, \mathcal{M}, \mu)$. For any set $A \in \mathcal{M}$, $\mu(A) > 0$, and any natural number r there exists a natural number k such that*

$$\mu\left(\bigcap_{l=0}^{r} T^{-lk} A \right) > 0.$$

This statement is known as the Furstenberg ergodic theorem. It can be easily verified that in the case of weak mixing theorem, 3.4 is an immediate consequence of Theorem 3.3. On the other hand, there is yet another special case, namely that of ergodic group translations of the commutative compact groups, for which the proof is not difficult and can be obtained by standard tools. Turning to the general case, we may assume the automorphism T to be ergodic: otherwise, the decomposition into ergodic components can be used to complete the proof, and for T ergodic the statement can be obtained from two special cases mentioned above.

We are going to formulate now a remarkable result giving an answer to a well-known number-theoretical question which was open for a long time. It will be seen a little later how ergodic theory, namely the Furstenberg ergodic theorem, might work to obtain an independent and very elegant proof of it.

Theorem 3.5 (E. Szemeredi, cf [FKO]). *Suppose $\Lambda \subset \mathbb{Z}^1$ is a set of integers with positive upper density*[3]. *Then for any natural number r it contains some arithmetic progression of length r.*

To derive the Szemeredi theorem from Theorem 3.3 it suffices to apply it to the automorphism T and the set A, where T is the restriction of the shift in the space X of all sequences $x = \{x_i\}$, $x_i = 0$ or 1, to the closure M of the trajectory of the point $x^{(\Lambda)}$ with $x_i^{(\Lambda)} = 1$ if $i \in \Lambda$, $x_i^{(\Lambda)} = 0$ if $i \notin \Lambda$[4], and $A = \{x \in M: x_0 = 1\}$.

The positiveness of the upper density of Λ implies that there exists an invariant measure μ for T such that $\mu(A) > 0$.

§4. General Constructions

In the section the descriptions of the most important general constructions of ergodic theory are collected together. Using these constructions one can obtain new examples of dynamical systems as some combinations of the known ones.

4.1. Direct Products of Dynamical Systems. For the sake of simplicity we restrict ourselves to the case of automorphisms. Suppose we are given the automorphisms T_1, T_2 with phase spaces $(M_1, \mathcal{M}_1, \mu_1)$, $(M_2, \mathcal{M}_2, \mu_2)$ respectively. The automorphism T of the product-space $M = M_1 \times M_2$ given by $Tx = (T_1 x_1, T_2 x_2)$ for $x = (x_1, x_2)$ is called the *direct product* $T_1 \times T_2$ of the automorphisms T_1, T_2. The direct product of several automorphisms is defined in a similar way.

Theorem 4.1. 1) *If T_1 is ergodic and T_2 is weak mixing, then $T_1 \times T_2$ is ergodic.*
2) *If T_1, T_2 are weak mixing, so is $T_1 \times T_2$.*
3) *If T_1, T_2 are mixing, so is $T_1 \times T_2$.*
4) *If T_1, T_2 are K-automorphisms, so is $T_1 \times T_2$.*

---

[3] By the upper density of Λ we mean the number $\bar{\rho}(\Lambda) = \overline{\lim}_{n-m \to \infty} \operatorname{card}(\Lambda \cap [m, n])$.
[4] We consider X as the direct product, $X = \prod_{-\infty}^{\infty} \{0, 1\}$ with the Tikhonov topology.

The direct product of two ergodic automorphisms may be non-ergodic. Example: $M_1 = M_2$ is the unit circle S^1 with the Haar measure and $T_1 = T_2$ is an irrational rotation. Then $T_1 \times T_2$ is not ergodic.

The direct products of endomorphisms and flows can also be defined in an obvious way. The statements of Theorem 4.1 remain valid in the case of flows.

4.2. Skew Products of Dynamical Systems. Suppose $(M, \mathcal{M}, \mu)$ is the direct product of the measure spaces $(M_1, \mathcal{M}_1, \mu_1)$, $(M_2, \mathcal{M}_2, \mu_2)$. Consider an automorphism T_1 of $(M_1, \mathcal{M}_1, \mu_1)$ and a family $\{T_2(x_1)\}$ of automorphisms of M_2 depending measurably on $x_1 \in M_1$. The measurability means here that for any measurable function $f(x_1, x_2)$ on $M_1 \times M_2$ the functions $f_n(x_1, x_2) = f(T_1^n x_1, T_2^n(x_1) x_2)$ are also measurable for all n. It may be easily checked that T preserves the measure μ. The automorphism T is called the *skew product* of the automorphism T_1 and the family $\{T_2(x_1)\}$.

Examples. 1. Suppose M_2 is a commutative compact group, μ_2 is the Haar measure on M_2 and the family $\{T_2(x_1)\}$ consists of group translations, i.e. $T_2(x_1)x_2 = x_2 + \varphi(x_1)$, where φ is a measurable map from M_1 into M_2. Sometimes $T = T_1 \times \{T_2(x_1)\}$ is called a group extension of T_1. It is easy to formulate in this case the criterion of ergodicity for T: T is ergodic if and only if:
1) T_1 is ergodic;
2) for any nontrivial character χ of the group M_2 the equation $c(x_1) = c(T_1 x_1) \cdot \chi(\varphi(x_1))$ has only the trivial solution $c = 0$.

If, in particular, $M_1 = M_2 = S^1$, and T_1 is a translation: $T_1 x_1 = x_1 + \alpha$, $\alpha \in S^1$, then the skew product $T(x_1, x_2) = (x_1 + \alpha, x_2 + \varphi(x_1))$ is known as a skew translation of the torus.

By iterating this construction we get the so called compound skew translations of the m-dimensional torus ($m \geq 2$):

$$T(x_1, \ldots, x_m) = (x_1 + \alpha, x_2 + \varphi_1(x_1), x_3 + \varphi_2(x_1, x_2), \ldots, x_m + \varphi_{m-1}(x_1, \ldots, x_{m-1})).$$

It was proved by H. Furstenberg that the ergodicity of a compound skew translation implies its unique ergodicity as a homeomorphism of the torus (cf [F]).

2. T_1 is a Bernoulli automorphism in the space M_1 of sequences $\{x_n^{(1)}\}_{-\infty}^{\infty}$ of 0's and 1's, $p(0) = p(1) = 1/2$, $M_2 = S^1$ with the Haar measure. For an irrational number $\alpha \in S^1$ consider the family $\{T_2(x_1)\}$ given by

$$T_2(x_1)x_2 = \begin{cases} x_2, & \text{if } x_0^{(1)} = 0 \\ x_2 + \alpha, & x_0^{(1)} = 1, \end{cases}$$

where $x_1 = \{x_n^{(1)}\}_{-\infty}^{\infty}$, $x_2 \in M_2$.

The corresponding skew product is a K-automorphism.

One of the most important notions in the theory of skew products is that of a cocycle. Suppose T_1 is an automorphism of the space $(M_1, \mathcal{M}_1, \mu_1)$ and G is a measurable group, i.e. the set endowed with the structures of both a group and a measurable space compatible with each other. The measurable map $\phi: M_1 \times \mathbb{Z}^1 \to G$ such that $\phi(x_1, m+n) = \phi(x, m) \cdot \phi(T_1^m x_1, n)$, $x_1 \in M_1$, $m, n \in \mathbb{Z}^1$, is called a cocycle for T_1 with values in G. The cocycles ϕ_1 and ϕ_2 are *cohomological* if there exists a measurable map $\psi: M_1 \to G$ such that $\phi_1(x_1, n) = [\psi(T_1^n x_1)]^{-1} \cdot \phi_2(x_1, n) \cdot \psi(x_1)$. To each skew product $T = T_1 \times \{T_2(x_1)\}$ one can associate the cocycle ϕ taking values in the group $\mathfrak{M}$ of automorphisms of the space M_2:

$$\phi(x_1, n) = T_2(x_1) \cdot T_2(T_1 x_1) \cdot \ldots \cdot T_2(T_1^{n-1} x_1).$$

If two such cocycles are cohomological, the associated skew products are metrically isomorphic.

4.3. Factor-Systems.
As before we shall deal with automorphisms only, the other cases being similar. Suppose the automorphisms T, T_1 of the measure spaces $(M, \mathcal{M}, \mu)$, $(M_1, \mathcal{M}_1, \mu_1)$ respectively are given. If there exists a homomorphism $\varphi: M \to M_1$ such that $\varphi(Tx) = T_1 \varphi(x)$ for all $x \in M$, then T_1 is called a *factor-automorphism* of T. Example: if $T = T_1 \times T_2$ is the direct product of two automorphisms, then both T_1 and T_2 are factor-automorphisms of T. Let T_1 be a factor-automorphism of T and $\varphi: M \to M_1$ be the corresponding homomorphism. This homomorphism induces naturally the partition ξ of M into the preimages of the points $x_1 \in M_1$ under φ. If the spaces M, M_1 are Lebesgue, then ξ is measurable, and we may speak of the conditional measures on the elements of ξ. Suppose now that for almost every point $x_1 \in M_1$ the corresponding element of ξ is finite and consists of, say, N points, where N does not depend on x_1. If T is ergodic, the conditional measure on such an element equals $1/N$ at any of its points. In this case the automorphism T can be represented as a skew product over T_1, such that M_2 consists of N points and $\{T_2(x_1)\}$ is a family of permutations of M_2.

Now consider the general case, when the elements of ξ are not necessarily finite. As above, we have the canonical systems of the conditional measures $\{\mu_C: C \in \xi\}$. Since T is ergodic, for almost all $C \in \xi$ the Lebesgue spaces (C, μ_C) are metrically isomorphic. They are either the spaces with non-atomic measure, or the finite measure spaces consisting of $N < \infty$ points. The automorphism T induces the automorphisms of the spaces (C, μ_C). The family of such automorphisms may be identified in a natural way with the measurable family $\{T_2(x_1)\}$, so T is represented as a skew product $T_1 \times \{T_2(x_1)\}$.

4.4. Integral and Induced Automorphisms.
Let T be an automorphism of the measure space $(M, \mathcal{M}, \mu)$ and $E \in \mathcal{M}$, $\mu(E) > 0$. Introduce the measure μ_E on E by $\mu_E(A) = \mu(A)[\mu(E)]^{-1}$, $A \in \mathcal{M}$, $A \subset E$. Then E may be viewed as a space with normalized measure. Consider the integer valued function k_E on E such that $k_E(x) = \min\{n \geq 1: T^n x \in E\}$. It follows from the Poincaré recurrence theorem that k_E is well defined for almost all $x \in E$. The function k_E is known as the *return time function* into E. For T ergodic the following Kac formula is true:

$\int_E k_E(x)\,d\mu_E(x) = [\mu(E)]^{-1}$. This formula signifies that the mean return time into E is equal to $[\mu(E)]^{-1}$. For almost every $x \in E$ define the transformation T_E by $T_E x = T^{k_E(x)} x$, $x \in E$. It may be easily verfied that T_E is an automorphism of the measure space $(E, \mathcal{M}_E, \mu_E)$, where $\mathcal{M}_E$ is the σ-algebra of the sets A, $A \in \mathcal{M}$, $A \subset E$.

Definition 4.1. The automorphism T_E is called the *induced automorphism* constructed from the automorphism T and the set E.

Now consider the "dual" construction. Suppose T_1 is an automorphism of the space $(M_1, \mathcal{M}_1, \mu_1)$ and $f \in L^1(M_1, \mathcal{M}_1, \mu_1)$ is a positive integer valued function. Introduce the measure space M whose points are of the form (x_1, i), where $x_1 \in M_1$, $0 \leqslant i < f(x_1)$, i is an integer. Define the measure μ on M by

$$\mu(A_i) = \mu(A) \cdot \left(\int_{M_1} f\,d\mu_1 \right)^{-1},$$

where $A_i \subset M$ is the set of the form (A, i), $A \in \mathcal{M}_1$. The transformation T^f of M given by

$$T^f(x_1, i) = \begin{cases} (x_1, i+1) & \text{if } i+1 < f(x_1), \\ (T_1 x_1, 0) & \text{if } i+1 = f(x_1) \end{cases}$$

is an automorphism of M.

Definition 4.2. T^f is called the *integral automorphism* corresponding to T_1 and f.

The space M_1 may be naturally identified with the subset of M consisting of the points $(x_1, 0)$. Under this identification one may consider T_1 as the induced automorphism of T corresponding to the set M_1. If an automorphism T is ergodic, so are its integral and induced automorphisms. The situation is much more complicated with mixing properties. For example, any ergodic automorphism has mixing induced automorphisms.

Example. Let T be a Markov automorphism in the space of 2-sided infinite sequences $x = \{x_n\}_{-\infty}^{\infty}$, $x_n \in Y$, where Y is a finite set. Suppose T is ergodic and define $E_y = \{x : x_0 = y\}$, $y \in Y$. Then the induced automorphism of T corresponding to E_y is Bernoulli. The well known Doeblin method in the theory of Markov processes is based on the transition from T to T_{E_y}.

4.5. Special Flows and Special Representations of Flows. Let T_1 be an automorphism of the measure space $(M_1, \mathcal{M}_1, \mu_1)$ and $f \in L^1(M_1, \mathcal{M}_1, \mu_1)$, $f > 0$. Consider the space M whose points are of the form (x_1, s), $x_1 \in M_1$, $0 \leqslant s < f(x_1)$. The measure μ on M is defined as a restriction to M of the direct product $\mu_1 \times \lambda$, where λ is the Lebesgue measure on $\mathbb{R}^1$. We may suppose μ to be normalized. Introduce the flow $\{T^t\}$ on M under which any point $(x_1, s) \in M$ moves vertically upward with unit speed until its intersection with the graph of f, then jumps instantly to the point $(T_1 x_1, 0)$ and continues its vertical motion. Formally $\{T^t\}$

may be defined for $t > 0$ by

$$T^t(x_1, s) = \left(T_1^n x_1, s + t - \sum_{k=0}^{n-1} f(T_1^k x_1) \right),$$

where n is uniquely determined from the unequality

$$\sum_{k=0}^{n-1} f(T_1^k x_1) \leqslant s + t < \sum_{k=0}^{n} f(T_1^k x_1).$$

For $t < 0$ the flow $\{T^t\}$ may be defined analogously, or else by the relation $T^{-t} = (T^t)^{-1}$.

Definition 4.3. The flow $\{T^t\}$ is called the *special flow* corresponding to the automorphism T_1 and the function f.

Theorem 4.2 (W. Ambrose, S. Kakutani, cf [AK]). *Any flow $\{T^t\}$ on a Lebesgue space, for which the set of fixed points is of zero measure, is metrically isomorphic to some special flow.*

The metric isomorphism in Theorem 4.2 is sometimes referred to as a *special representation* of the flow $\{T^t\}$. The idea of special representations goes back to Poincaré who used the so called "first return map" on the transversals to vector fields in his study of the topological behavior of the solutions of ordinary differential equations.

Examples of special representations. 1. Suppose we are given the system of differential equations

$$\frac{du}{dt} = A(u, v), \qquad \frac{dv}{dt} = B(u, v) \qquad (1.10)$$

on the 2-dimensional torus with cyclic coordinates (u, v), and the functions A, $B \in C^r, r \geqslant 2$, $A^2 + B^2 > 0$. Suppose further that the 1-parameter group $\{T^t\}$ of translations along the solutions of this system has an absolutely continuous invariant measure μ, $d\mu = P(u, v) \, du \, dv$, $P(u, v) > 0$. It is known that there exists a smooth non-self-intersecting curve Γ on the torus transversal to the vector field (1.10) and having the property that for any point $p \in \Gamma$ the trajectory starting at p will intersect Γ again at some moment $t = f(p)$ (the so-called Siegel curve). Denote by $q = q(p) \in \Gamma$ the point of this intersection. One may choose the parameter on Γ in such a way that the transformation $p \mapsto q$ becomes a rotation of the circle by a certain angle. This implies that the flow $\{T^t\}$ has a representation as a special flow corresponding to a rotation of the circle and the function f.

2. Suppose $Q \subset \mathbb{R}^d$, $d \geqslant 2$, is a compact domain with piecewise smooth boundary, M is the unit tangent bundle over Q, $\{T^t\}$ is the billiards in Q (cf Sect. 1). Denote by M_1 the set of the unit tangent vectors having supports in ∂Q and directed into Q. Introduce the following transformation $T_1: M_1 \to M_1$ sending any point $x \in M_1$ to some point $y = T_1 x$: x moves along its billiards trajectory until the intersection with the boundary ∂Q and then reflects from the

boundary according to billiards law; the point y is just the result of this reflection. T_1 has an invariant measure μ_1 of the form $d\mu_1 = d\sigma(q)\,d\omega_q \cdot |(n(q), x)|$, where $d\sigma$ is the element of the volume of ∂Q, $d\omega$ is the element of the volume of S^{d-1}, $n(q)$ is the unit normal vector to ∂Q at a point $q \in \partial Q$. Denoting by $f(x)$ the time interval between two consecutive reflections (at x and at y), we obtain the representation of $\{T^t\}$ as a special flow corresponding to T_1 and f.

The following result strengthens considerably Theorem 4.2.

Theorem 4.3 (D. Rudolph [Ru1]). *Suppose $\{T^t\}$ is an ergodic flow on the Lebesgue space $(M, \mathcal{M}, \mu)$, and we are given the positive numbers p, q, ρ; p/q is irrational. There exists a special representation of $\{T^t\}$ such that the function f appearing in this representation takes only two values, p and q, and*

$$\mu_1(\{x_1 \in M_1: f(x_1) = p\}) = \rho \cdot \mu_1(\{x_1 \in M_1: f(x_1) = q\}).$$

The explicit construction of such a representation for a given flow may often be very non-trivial.

4.6. Natural Extensions of Endomorphisms. Suppose T_0 is an endomorphism of the space $(M_0, \mathcal{M}_0, \mu_0)$. Define a new space M whose points are the infinite sequences of the form $x = (x_1^{(0)}, x_2^{(0)}, x_3^{(0)}, \ldots)$, where $x_i^{(0)} \in M_0$ and $T_0 x_{i+1}^{(0)} = x_i^{(0)}$ for any $i > 0$. Denote by $\mathcal{M}$ the σ-algebra generated by the subsets $A_{i,C}$ of M of the form $A_{i,C} = \{x \in M: x_i^{(0)} \in C\}$, where $i > 0$, $C \in \mathcal{M}_0$. Consider the measure μ on $\mathcal{M}$ given by $\mu(A_{i,C}) = \mu_0(C)$, and define the transformation T of $(M, \mathcal{M}, \mu)$ by $T(x_1^{(0)}, x_2^{(0)}, x_3^{(0)}, \ldots) = (T_0 x_1^{(0)}, T_0 x_2^{(0)}, T_0 x_3^{(0)}, \ldots)$. This transformation is invertible and its inverse T^{-1} is given by $T^{-1}(x_1^{(0)}, x_2^{(0)}, x_3^{(0)}, \ldots) = (x_2^{(0)}, x_3^{(0)}, \ldots)$. The measure μ is invariant under T.

Definition 4.4. The automorphism T is the *natural extension* of the endomorphism T_0.

Theorem 4.4. *T is ergodic (mixing, weak mixing) if and only if T_0 is also ergodic.*

Given an endomorphism T_0, one may construct the decreasing sequence of sub-σ-algebras $\mathcal{M}_k$, where $\mathcal{M}_k$ consists of the sets of the form $T_0^{-k}C$, $C \in \mathcal{M}_0$.

Definition 4.5. An endomorphism T_0 is said to be *exact* if $\bigcap_{k \geq 0} \mathcal{M}_k = \mathcal{N}$.

If an endomorphism T_0 is exact, its natural extension T is a K-automorphism. The notion of exact endomorphism plays an important role in the theory of one-dimensional mappings (cf Part II, Chap. 9).

Example. Suppose M_0 is the space of 1-sided sequences $(x_1^{(0)}, x_2^{(0)}, \ldots)$ of 0's and 1's; T_0 is the shift in M_0, μ_0 is an invariant measure for T_0. The phase space M of the natural extension of T_0 is the space of 2-sided sequences of 0's and 1's; the extension itself is the shift in M, and for any cylinder $C \subset M$ the invariant measure (for T) $\mu(C)$ equals $\mu_0(C_0)$, where C_0 is the corresponding cylinder in M_0.

Chapter 2
Spectral Theory of Dynamical Systems

I.P. Cornfeld, Ya.G. Sinai

§1. Groups of Unitary Operators and Semigroups of Isometric Operators Adjoint to Dynamical Systems

We will introduce now the notions of correlation functions and their Fourier transforms which play an important role in various applications of the theory of dynamical systems. Suppose $\{T^t\}$ is a flow on a measure space $(M, \mathcal{M}, \mu)$ and $f \in L^2(M, \mathcal{M}, \mu)$. By the correlation function corresponding to f we mean the function $b_f(t) = \int_M f(T^t x) f(x) \, d\mu$. A number of statistical properties of the dynamical system may be characterized by the limit behavior of the differences $[b_f(t) - (\int_M f \, d\mu)^2]$. In the case of mixing, these expressions tend to zero as $t \to \infty$. If the convergence is fast enough one may write the above expressions in the form

$$b_f(t) - \left(\int_M f \, d\mu\right)^2 = \int_{-\infty}^{\infty} \exp(i\lambda t) \cdot \rho_f(\lambda) \, d\lambda.$$

The function $\rho_f(\lambda)$ in this representation is called the spectral density of f. The set of those λ for which $\rho_f(\lambda)$ is essentially non-zero for typical f, characterizes in some sense the frequencies playing the crucial role in the dynamics of the system under consideration. It is sometimes said that the system "produces a noise" on this set. The investigation of the behavior of the functions $b_f(t)$, as well as of their analogs $b_f(n)$, $n \in \mathbb{Z}$, in the case of discrete time, is very important both for theory and for applications.

In this chapter the basic information concerning the properties of the functions b_f will be given.

Let $\{T^t\}$ (respectively, $\{T^n\}$) be a 1-parameter (respectively, cyclic) group of automorphisms or a semigroup of endomorphisms. It induces the adjoint group or semigroup of operators in $L^2(M, \mathcal{M}, \mu)$ which acts according to the formula $U^t f(x) = f(T^t x)$ (respectively, $U^n f(x) = f(T^n x)$), $f \in L^2$. In the case of automorphisms (endomorphisms) these operators are unitary (isometric). If $\{T^t\}$ is a flow and $L^2(M, \mathcal{M}, \mu)$ is separable, the group $\{U^t\}$ is continuous. The function $b_f(t)$ ($b_f(n)$) may be expressed in terms of U^t (U^n):

$$b_f(t) = (U^t f, f) \quad (t \in \mathbb{R}^1 \text{ or } \mathbb{R}^1_+)$$
$$b_f(n) = (U^n f, f) \quad (n \in \mathbb{Z}^1 \text{ or } \mathbb{Z}^1_+).$$

Definition 1.1. Suppose we are given two dynamical systems in the spaces $(M_1, \mathcal{M}_1, \mu_1)$, $(M_2, \mathcal{M}_2, \mu_2)$ respectively. These systems are said to be *spectrally* equivalent if there exists an isomorphism of the Hilbert spaces $L^2(M_1, \mathcal{M}_1, \mu_1)$, $L^2(M_2, \mathcal{M}_2, \mu_2)$ intertwining the actions of groups (semigroups) $\{U_1^t\}$, $\{U_2^t\}$ (or else $\{U_1^n\}$, $\{U_n^2\}$).

The metric isomorphism of two dynamical systems implies their spectral equivalence. The converse is, generally, false.

The properties of a dynamical system which can be expressed in terms of the spectral properties of the operators $U^t(U^n)$ are said to be its spectral properties. In any case, there exists the 1-dimensional subspace of L^2 consisting of the eigenfunctions of $U^t(U^n)$ which are constant almost everywhere (the corresponding eigenvalue is 1).

Theorem 1.1. 1) *A dynamical system is ergodic if and only if the space of eigenfunctions with eigenvalue 1 is one-dimensional;*

2) *A dynamical system is weak mixing if and only if any eigenfunction of the adjoint group of unitary operators (or the semigroup of isometric operators) is a constant.*

3) *A dynamical system is mixing if and only if for any $f \in L^2(M, \mathcal{M}, \mu)$ one has*

$$\lim_{t \to \infty} b_f(t) = \left(\int f \, d\mu \right)^2$$

in the case of continuous time,

$$\lim_{n \to \infty} b_f(n) = \left(\int f \, d\mu \right)^2$$

in the case of discrete time.

Therefore, ergodicity, mixing and weak mixing are spectral properties.

In what follows we shall consider the invertible case only, i.e. the automorphisms (the cyclic groups of automorphisms) and the flows on a Lebesgue space $(M, \mathcal{M}, \mu)$, as well as corresponding adjoint groups of unitary operators. We begin by recalling the main results of the theory of spectral equivalence of unitary operators.

1. *The case of automorphisms.* There exists a finite Borel measure σ on the circle S^1 such that for any $f \in L^2(M, \mathcal{M}, \mu)$ the function $b_f(n)$ may be expressed in the form

$$b_f(n) = \int_2^1 \exp(2\pi i \lambda n) \cdot p_f(\lambda) \, d\sigma(\lambda), \tag{2.1}$$

and for some $f_0 \in L^2(M, \mathcal{M}, \mu)$ we have $p_{f_0}(\lambda) \equiv 1$. The measure σ is called the measure of maximal spectral type. There exists a partition of the circle S^1 into countably many measurable subsets $A_1, A_2, \ldots, A_k, \ldots, A_\infty$, and a decomposition of $L^2(M, \mathcal{M}, \mu)$:

$$L^2(M, \mathcal{M}, \mu) = \bigoplus_{k=1}^{\infty} H_k \oplus H_\infty,$$

such that the subspaces H_k are pairwise orthogonal, $\dim H_k = k$, and in any H_k one can choose a basis $\{f_{k,i}\}$, $i = 1,\ldots,k$, for which $p_{f_{k,i}}(\lambda) = 1$ for $\lambda \in A_k$, $p_{f_{k,i}}(\lambda) = 0$ for $\lambda \notin A_k$.

The function $m(\lambda)$ such that $m(\lambda) = k$, $\lambda \in A_k$, is called the spectral multiplicity function. If $\sigma(A_k) > 0$ for infinitely many k, the operator U is said to have the spectrum of unbounded multiplicity. Otherwise the spectrum of U is of bounded multiplicity. If there is only one k with $\sigma(A_k) > 0$, the operator U has the homogeneous spectrum. In particular, if $k = \infty$ and σ is the Lebesgue measure, the operator U has the countable Lebesgue spectrum.

2. *The case of flows.* Instead of (2.1), we have the representation

$$b_f(t) = \int_{-\infty}^{\infty} \exp(2\pi i \lambda t) p_f(\lambda)\, d\sigma(\lambda),$$

where σ is a finite Borel measure on $\mathbb{R}^1$. The sets A_k in this case become the subsets of $\mathbb{R}^1$, and in the definition of the Lebesgue spectrum the Lebesgue measure on S^1 should be replaced by the Lebesgue measure on $\mathbb{R}^1$.

Suppose T is an ergodic automorphism, $\{U^n\}$ is the adjoint group of operators and $\Lambda_d(T)$ is the set of eigenvalues of the operator $U_T = U^1$. U_T being unitary, we have $\Lambda_d(T) \subset S^1$. Similarly, for an ergodic flow $\{T^t\}$, denote by $\Lambda_d(\{T^t\}) \subset \mathbb{R}^1$ the set of eigenvalues of the adjoint group of operators $\{U^t\}$.

Theorem 1.2 (cf [CFS]). *$\Lambda_d(T)$ is a subgroup of S^1, any eigenvalue $\lambda \in \Lambda_d(T)$ is of multiplicity one and the absolute value of any eigenfunction is constant almost everywhere. For a flow $\{T^t\}$ the similar assertions are true, but $\Lambda_d(\{T^t\})$ in this case is a subgroup of $\mathbb{R}^1$.*

Theorem 1.3 (cf [CFS]). *Suppose T is a K-automorphism or $\{T^t\}$ is a K-flow. Then the adjoint group of operators has the countable Lebesgue spectrum on the invariant subspace of functions with zero mean.*

The proofs of these spectral theorems are based on the following important property of the operators adjoint to automorphisms: if f, g and $f \cdot g \in L^2(M, \mathcal{M}, \mu)$, then $Uf \cdot Ug = U(f \cdot g)$, i.e. these operators preserve the additional structure in L^2 related to the existence of the partial multiplication in L^2—the structure of the so-called unitary ring.

§2. The Structure of the Dynamical Systems with Pure Point and Quasidiscrete Spectra

Definition 2.1. An ergodic automorphism T (or a flow $\{T^t\}$) is said to be an *automorphism* (*a flow*) *with pure point spectrum* if the cyclic (1-parameter) group $\{U^n\}$ ($\{U^t\}$) has a system of the orthogonal eigenfunctions that is complete in $L^2(M, \mathcal{M}, \mu)$.

Unlike the general case, the unitary equivalence of the groups of operators adjoint to the dynamical systems with pure point spectrum implies the metric isomorphism of the systems themselves. This fact enables us to obtain the complete metric classification of such systems.

Theorem 2.1 (J. von Neumann [N]). *Suppose T_1, T_2 are the ergodic automorphisms with the pure point spectrum of the Lebesgue spaces $(M_1, \mathcal{M}_1, \mu_1)$, $(M_2, \mathcal{M}_2, \mu_2)$. They are metrically isomorphic if and only if $\Lambda_d(T_1) = \Lambda_d(T_2)$, where $\Lambda_d(T)$ is the countable subgroup of the circle S^1 consisting of the eigenvalues of the operator U_T adjoint to T.*

Using the Pontryagin duality theory one can easily construct for arbitrary countable subgroup $\Lambda \subset S^1$ the automorphism T with pure point spectrum such that $\Lambda_d(T) = \Lambda$. This automorphism T is a group translation on the character group M of the group Λ with the Haar measure μ. The group M is compact in the case considered and T is given by $Tg = g \cdot g_0$, $(g, g_0 \in M)$ where $g_0(\lambda) = \lambda$, $\lambda \in \Lambda$. Combining this assertion and Theorem 2.1, we get the following fact: any ergodic automorphism with pure point spectrum is metrically isomorphic to a certain group translation on the character group of its spectrum. In the case of continuous time the similar statement holds.

Theorem 2.2. *For two ergodic flow $\{T_1^t\}$, $\{T_2^t\}$ on the Lebesgue spaces $(M_1, \mathcal{M}_1, \mu_1)$, $(M_2, \mathcal{M}_2, \mu_2)$ with pure point spectrum to be metrically isomorphic, it is necessary and sufficient that $\Lambda_d(\{T_1^t\}) = \Lambda_d(\{T_2^t\})$, where $\Lambda_d(\{T^t\})$ is a countable subgroup of $\mathbb{R}^1$ consisting of the eigenvalues of the 1-parameter group $\{U^t\}$.*

Just as in the case of automorphisms, for any countable subgroup Λ of $\mathbb{R}^1$, one can construct the flow $\{T^t\}$ with pure point spectrum $\Lambda_d(\{T^t\}) = \Lambda$ such that any T^t is a group translation on the character group of Λ. Hence, any ergodic flow with pure point spectrum is metrically isomorphic to a flow generated by the group translations along a certain 1-parameter subgroup of the character group of the spectrum.

Examples. 1. T is an ergodic group translation on the m-dimensional torus:
$Tx = ((x_1 + \alpha_1) \bmod 1, \ldots, (x_m + \alpha_m) \bmod 1)$ for $x = (x_1, \ldots, x_m)$. The functions $\chi_{n_1, \ldots, n_m}(x) = \exp 2\pi i \sum_{k=1}^m n_k x_k$ form a complete system of characters of the torus. The automorphism T has the pure point spectrum consisting of numbers $\exp 2\pi i \sum_{k=1}^m n_k \alpha_k$; $(n_1, \ldots, n_m \in \mathbb{Z}^1)$.

2. Suppose M is the additive group of integer 2-adic numbers with the normalized Haar measure μ. Associating to each point $x = \sum_{k=0}^\infty a_k 2^k \in M$ ($a_k = 0$ or 1) the infinite sequence $(a_0, a_1, \ldots)$ of 0's and 1's, one may identify M with the space of all such sequences. Fix $x_0 = 1 \cdot 2^0 + 0 \cdot 2^1 + 0 \cdot 2^2 + \cdots \in M$. The transformation $T: M \to M$ given by $Tx = x \oplus x_0$, ($\oplus$ stands for the addition of 2-adic numbers) is a group translation and, on the other hand, is an automorphism of the space (M, μ). Evaluate the spectrum of T.

For any n-tuple $i^{(n)} = (i_0, i_1, \ldots, i_{n-1})$ of 0's and 1's define the cylinder set $C^{(n)}_{i^{(n)}} = \{x = \sum_{k=0}^{\infty} a_k 2^k \in M : a_k = i_k \text{ for } 0 \leq k \leq n-1\}$. For each n there are 2^n sets $C^{(n)}_{i^{(n)}}$, and the transformation T permutes them cyclically. We can enumerate these sets $C^{(n)}_0, C^{(n)}_1, \ldots, C^{(n)}_{2^n-1}$, so that we have $TC^{(n)}_p = C^{(n)}_{p+1}$, $0 \leq p < 2^n$. The functions $f_{r/2^n}(x) = \exp\dfrac{2\pi i r p}{2^n}$, $x \in C^{(n)}_p$ ($n = 0, 1, \ldots; 0 \leq r < 2^n$) form a complete system of characters of M, and $U_T \chi_{r/2^n} = \exp\dfrac{2\pi i r}{2^n} \cdot \chi_{r/2^n}$. Hence, T is an automorphism with pure point spectrum consisting of numbers of the form $\exp\dfrac{2\pi i r}{2^n}$.

There is a generalization of the theory of pure point spectrum to a wider class of dynamical systems.

Suppose T is an automorphism of a Lebesgue space $(M, \mathcal{M}, \mu)$; $\Lambda_0 \overset{\text{def}}{=} \Lambda_d(T)$ is the group of eigenvalues of the unitary operator U_T; Φ_0 is the (multiplicative) group of normalized in $L^2(M)$ eigenfunctions of U_T. The groups Λ_0 and Φ_0 may be considered as the subsets of $L^2(M)$, and the inclusion $\Lambda_0 \subset \Phi_0$ (in $L^2(M)$) is obviously true.

For any $n \geq 1$ set $\Phi_n = \{f \in L^2(M) : \|f\| = 1, U_T f = \lambda f \text{ for some } \lambda \in \Phi_{n-1}\}$, $\Lambda_n = \{\lambda \in L^2(M) : \|\lambda\| = 1, U_T f = \lambda f \text{ for some } f \in \Phi_n\}$.

By induction over n, the groups Φ_n, Λ_n are defined for all $n \geq 0$. Their elements are called quasi-eigenfunctions and quasi-eigenvalues of rank n respectively. Since $\Phi_n \subset \Phi_{n+1}$, $\Lambda_n \subset \Lambda_{n+1}$, one may define the groups $\Phi = \bigcup_{n=0}^{\infty} \Phi_n$, $\Lambda = \bigcup_{n=0}^{\infty} \Lambda_n$. Any eigenvalue of rank $n \geq 1$ may be considered as an eigenfunction of rank $n - 1$, so the corresponding eigenvalue of rank $n - 1$ exists. This argument shows that a certain homomorphism $\theta : \Lambda \to \Lambda$ arises in a natural way. It is given by $\theta f = \lambda$ for $f \in \Lambda$, if $U_T f = \lambda f$, $\lambda \in \Lambda$. Restricting θ to Λ_n, one obtains the homeomorphisms $\theta_n : \Lambda_n \to \Lambda_n$.

Definition 2.2. An ergodic automorphism T such that Φ is a complete system of functions in $L^2(M)$ is called an *automorphism with quasi-discrete spectrum*.

Any automorphism with pure point spectrum has, obviously, quasi-discrete spectrum. The ergodic skew translation on the 2-dimensional torus: $T(x, y) = (x + \alpha, y + x)$, $x, y \in S^1$, α is irrational, is the example of the automorphism with quasi-discrete but not pure point spectrum.

We restrict ourselves to the case of the so called totally ergodic automorphisms, i.e. such automorphisms T that all T^n, $n \neq 0$, are ergodic.

Theorem 2.3 (L.M. Abramov [A]). *Suppose T_1, T_2 are totally ergodic automorphisms of the Lebesgue spaces $(M_1, \mathcal{M}_1, \mu_1)$, $(M_2, \mathcal{M}_2, \mu_2)$ with quasi-discrete spectrum. They are metrically isomorphic if and only if the corresponding groups $\Lambda^{(1)} = \bigcup_{n=0}^{\infty} \Lambda_n^{(1)}$, $\Lambda^{(2)} = \bigcup_{n=0}^{\infty} \Lambda_n^{(2)}$ of quasi-eigenvalues and the homeomorphisms $\theta^{(1)} : \Lambda^{(1)} \to \Lambda^{(1)}$, $\theta^{(2)} : \Lambda^{(2)} \to \Lambda^{(2)}$ satisfy the conditions:*
1) $\Lambda_0^{(1)} = \Lambda_0^{(2)} \overset{\text{def}}{=} \Lambda_0$;
2) *there exists an isomorphism V between the groups $\Lambda^{(1)}$, $\Lambda^{(2)}$ for which*

a) $V\lambda = \lambda$ for $\lambda \in \Lambda_0$;
b) $V\Lambda_n^{(1)} = \Lambda_n^{(2)}$, $n = 0, 1, \ldots$;
c) $\theta^{(2)} = V\theta^{(1)}V^{-1}$.

There is a representation theorem for the automorphisms with quasi-discrete spectrum similar to that for pure point spectrum. It says that for any sequence of countable commutative groups $\Lambda_0 \subseteq \Lambda_1 \subseteq \ldots$, $\bigcup_{n=0}^{\infty} \Lambda_n = \Lambda$, where Λ_0 is a subgroup of S^1 which does not contain roots of unity (except 1) there exists a totally ergodic automorphism T with quasi-discrete spectrum acting on the group M of characters of Λ, such that for any n the group of its quasi-eigenvalues is isomorphic to Λ_n.

§ 3. Examples of Spectral Analysis of Dynamical Systems

There are many examples of dynamical systems for which the spectra of the adjoint groups of operators may be completely evaluated.

Theorem 1.3 says that K-systems have the countable Lebesgue spectrum. Beyond the class of K-systems, the dynamical systems having in $L_2^{(0)}$ the countable Lebesgue spectrum also exist, for example, the horocycle flows on compact surfaces of constant negative curvature.

For a wide class of ergodic compound skew products on tori, i.e. the transformations of the form

$$T(x_1, \ldots, x_m) = (x_1 + \alpha, x_2 + f(x_1), \ldots, x_m + f_{m-1}(x_1, \ldots, x_{m-1})),$$

where $x_1, \ldots, x_m \in S^1$, α is irrational, the spectrum may be calculated. If the functions $f_k(x_1, \ldots, x_k)$ are smooth and their derivatives satisfy certain inequalities, the spectrum of U_T in the invariant subspace $H_1 \subset L^2$ consisting of function depending on x_1 only, is pure point, while in the orthogonal complement $H_1^\perp$ it is countable Lebesgue (A.G. Kushnirenko, cf [CFS]).

For a long time the study of spectra of various classes of dynamical systems gave ground to hope that the theory of pure point spectrum might be extended in some sense to the dynamical systems of general form. However, it is now clear that the situation is much more complicated. The examples of systems with very unexpected spectral properties have been constructed, most of them with the help of theory of periodic approximation. Some of these examples will be described now (cf [CFS]).

1. Let M be the measure space, $M = S^1 \times \mathbb{Z}_2$, where S^1 is the unit circle with the Lebesgue measure, $\mathbb{Z}_2 = \{1, -1\}$ with the measure $p(\{1\}) = p(\{-1\}) = 1/2$. Consider the automorphism T of M which is a skew product over a rotation of the circle: $T(x, z) = (x + \alpha, g(x)z)$, $x \in S^1$, $z \in \mathbb{Z}_2$. In the subspace $H \subset L^2(M)$ consisting of functions depending on x only, the unitary operator U_T has a pure point spectrum, while in the orthogonal complement $H^\perp$ the spectrum of T is

continuous. Since any function $f \in H^\perp$ satisfies $f(x, -z) = -f(x, z)$, the product of any two functions from $H^\perp$ lies in H.

2. For any automorphism T with pure point spectrum, the maximal spectral type ρ (i.e. the type of discrete measure concentrated on the group $\Lambda_d(T)$) obviously dominates the convolution $\rho * \rho$. This property is known as the group property of spectrum. It is also satisfied by many dynamical systems with continuous and mixed spectra which arise naturally in applications. Generally, however, this is not true. A counterexample may be constructed, as before, in the phase space $M = S^1 \times \mathbb{Z}_2$. Namely, there are certain irrational numbers α, $\beta \in [0, 1]$ for which the skew product T, $T(x, z) = ((x + \alpha) \mod 1, w(x)z)$, $x \in S^1$, $z \in \mathbb{Z}_2$, where $w(x) = -1$ for $x \in [0, \beta)$, $w(x) = 1$ for $x \in [\beta, 1)$, lacks the group property of spectrum.

3. The spectral multiplicity function has been evaluated for many classes of dynamical systems, and in all "natural" cases this function is either unbounded, or equals 1 on the set of full measure of maximal spectral type. But again this observation cannot be extended to the general case. There exist automorphisms (constructed also as skew products over the rotations of the circle) with non-simple continuous spectrum of finite multiplicity.

The problem of finding the exact conditions to be satisfied by the spectrum of a group of unitary operators for this group to be the adjoint group of operators of some dynamical system, is extremely difficult.

In particular, it is not known whether the dynamical systems with simple Lebesgue spectrum, or even absolutely continuous spectrum of finite multiplicity, exist.

§4. Spectral Analysis of Gauss Dynamical Systems

An important class of dynamical systems for which complete spectral analysis has been carried out, is related to Gauss distributions and stationary Gauss processes of probability theory.

Consider the space M of sequences of real numbers, infinite in both directions, $x(s)$ where s is an integer and $-\infty < s < \infty$. Suppose $\mathcal{M}$ is the σ-algebra generated by all finite dimensional cylinders, i.e. sets of the form $A = \{x(s) \in M : x(s_1) \in C_1, \ldots, x(s_r) \in C_r\}$, where $C_1, \ldots, C_r$ are Borel subsets of $\mathbb{R}^1$. The measure μ on $\mathcal{M}$ is said to be a Gauss measure if the joint distribution of any family of random variables $\{x(s_1), \ldots, x(s_r)\}$ is an r-dimensional Gauss distribution. If the mean value $m = \mathbb{E}x(s)$ does not depend on s and the correlation function $b(s_1, s_2) = \int x(s_1)x(s_2) d\mu(x)$ depends on $s_1 - s_2$ only, the Gauss measure is stationary. Without loss of generality we may assume that $m = 0$.

For a stationary measure μ the sequence $b(s) \stackrel{\text{def}}{=} b(s, 0)$ is positive definite and thus may be represented in the form $b(s) = \int_0^1 \exp(2\pi i s \lambda) d\sigma(\lambda)$, where σ is a finite measure on the circle S^1. The measure σ is known as the spectral measure of the Gauss measure μ.

Definition 4.1. The cyclic group $\{T^n\}$ of shift transformations in the space M provided with a stationary Gauss measure is said to be *a Gauss dynamical system*. The generator T^1 of this group is said to be *a Gauss automorphism*.

One may consider a space M of all real valued functions $x(s)$, $s \in \mathbb{R}^1$, rather than sequences $x(s)$, s is an integer. In this case the σ-algebra $\mathcal{M}$ and Gauss stationary measures on it may be defined in the same way.

Definition 4.2. The one-parameter group $\{T^t\}$ of shift transformations in the space M provided with a stationary Gauss measure is said to be *a Gauss dynamical system with continuous time (Gauss flow)*.

We shall formulate the results for the case of discrete time only. Describe first the unitary operator U_T adjoint to the Gauss automorphism.

The complex Hilbert space $L^2(M, \mathcal{M}, \mu)$ can be decomposed into the countable orthogonal direct sum of subspaces, $L^2(M, \mathcal{M}, \mu) = \bigoplus_{m=0}^{\infty} H_m$, such that all H_m are invariant under U_T. Any H_m is of the form $H_m = H_m^{(r)} + iH_m^{(r)}$, where $H_m^{(r)}$ is a real subspace of the real Hilbert space $L^2_{(r)}(M, \mathcal{M}, \mu)$. $H_0^{(r)}$ (respectively, H_0) is the subspace of real (respectively, complex) constants. $H_1^{(r)}$ is the subspace spanned by the vectors y of the form $y = \sum a_k x(s_k)$, a_k are real. All random variables $y \in H_1^{(r)}$ have Gauss distributions. The space $H_m^{(r)}$, $m > 1$, is spanned by all possible Hermite-Ito polynomials $:y_1 \ldots y_m:$ of Gauss random variables $y_1, \ldots, y_m \in H_1^{(r)}$.[1] Describe now the action of the operator U_T in the subspaces H_m, $m \geq 1$. First we introduce the real Hilbert space $Q_m^{(r)}$ consisting of complex-valued functions $\varphi(\lambda_1, \ldots, \lambda_m)$ defined for $\lambda_1, \ldots, \lambda_m \in S^1$ (i.e. for $\lambda = (\lambda_1, \ldots, \lambda_m)$ belonging to the torus Tor^m), symmetric with respect to their variables, satisfying the relation $\varphi(-\lambda_1, \ldots, -\lambda_m) = \overline{\varphi(\lambda_1, \ldots, \lambda_m)}$ and having finite norms $\|\varphi\|$, where $\|\varphi\| \stackrel{\text{def}}{=} [\int_{Tor^m} |\varphi(\lambda_1, \ldots, \lambda_m)|^2 \, d\sigma(\lambda_1) \ldots d\sigma(\lambda_m)]^{1/2} < \infty$. Next, introduce the complex Hilbert space Q_m by $Q_m = Q_m^{(r)} \oplus iQ_m^{(r)}$.

Theorem 4.1 (cf [CFS]). *For any $m \geq 1$ there exists an isometric map $\theta_m: Q_m \to H_m$ such that*

1) $\theta_m Q_m = H_m$;

2) *under the isomorphism $\theta_m^{-1}: H_m \to Q_m$ the operator U_T is mapped into the operator of multiplication by the function $\exp 2\pi i(\lambda_1 + \cdots + \lambda_m)$.*

Such a complete description of the structure of the operator U_T enables us to connect a number of ergodic and spectral properties of the automorphism T to the properties of the spectral measure σ.

The necessary and sufficient condition for a Gauss automorphism T to be ergodic is that the measure σ be continuous. This condition is also necessary and sufficient for T to be weak mixing. For a Gauss automorphism T to be mixing, as well as mixing of all orders, it is necessary and sufficient that the Fourier

---

[1] The Hermite-Ito polynomial $:y_1 \cdot y_2 \cdot \ldots \cdot y_m:$ ($y_i \in H_1^{(r)}$) is the perpendicular lowered from the extremity of vector $y_1 \cdot y_2 \cdot \ldots \cdot y_m$ to the subspace generated by all possible products $y_1' \cdot y_2' \cdot \ldots \cdot y_p'$, where $p < m$, $y_i' \in H_1^{(r)}$.

coefficients $b_s = \int \exp(2\pi i s\lambda) d\sigma(\lambda)$ of the measure σ tend to zero as $|s| \to \infty$. The maximal spectral type of the operator U_T is the type of the measure $e^\sigma \overset{\text{def}}{=} \sum_{k=0}^{\infty} \frac{\sigma^{(k)}}{k!}$ where, for $k \geq 1$, the measure $\sigma^{(k)}$ is the k-fold convolution of σ with itself, and for $k = 0$, $\sigma^{(0)}$ is the normalized measure supported at the point $\lambda = 1$.

The description of the structure of the operator U_T also enables us to construct the examples of Gauss dynamical systems with non-trivial spectral properties. In particular, there exist Gauss automorphisms with simple continuous spectrum (cf [CFS]).

Chapter 3
Entropy Theory of Dynamical Systems

I.P. Cornfeld, Ya.G. Sinai

The notion of entropy was introduced in the XIX century in the works of founders of statistical mechanics, R. Clausius, J.C. Maxwell, L. Boltzmann and others, in connection with the analysis of irreversibility phenomena. Later, entropy appeared and became the fundamental concept in the information theory created by C. Shannon in the 1940's and was concerned with the problems of the transmission of information in the presence of noise. Though the formal expression for entropy was the same in both cases, there were some differences in its meaning. A.N. Kolmogorov in his work [Kol1] applied the ideas of information theory and the notion of entropy to the analysis of some problems of ergodic theory. This work gave rise to a new branch of ergodic theory with numerous results and applications—the so-called entropy theory of dynamical systems. At the present time one may consider the developing of this theory to be mostly completed. This chapter is devoted to its exposition.

§1. Entropy and Conditional Entropy of a Partition

Entropy theory is based on rather elementary concepts of entropy and conditional entropy of a finite or countable partition of a measure space. Let ξ be such a partition of a Lebesgue space $(M, \mathcal{M}, \mu)$. Denote the elements of ξ by C_i, $i = 1, 2, \ldots$.

Definition 1.1. The *entropy of the partition* ξ is the number

$$H(\xi) = -\sum_i \mu(C_i) \log \mu(C_i).$$

If $\mu(C_i) = 0$, we adopt the convention that $\mu(C_i)\log\mu(C_i) = 0$. Further, all logarithms are to the base e. It is clear that $0 \leqslant H(\xi) \leqslant \infty$. If the partition ξ is uncountable, we set by definition $H(\xi) = \infty$. Thus, $H(\xi)$ is defined for all partitions ξ. The expression for $H(\xi)$ may be rewritten in a somewhat different form. Namely, if $C_\xi(x)$ is the element of ξ containing the point $x \in M$, we have

$$H(\xi) = -\int_M \log\mu(C_\xi(x))\,d\mu(x).$$

The properties of measurable partitions will be used below. For reader's convenience we collect them here.

1. The partition ξ_1 is not finer than ξ_2 (we write $\xi_1 \leqslant \xi_2$) if ξ_2 is a subpartition of ξ_1 (mod 0). Explain the notation (mod 0) that we have just used. It means that one can throw out from the space M a certain set of zero measure in such a way that on the complement to this set each element $C_{\xi_1} \in \xi_1$ is the union of the entire elements $C_{\xi_2} \in \xi_2$. The inequality $\leqslant$ introduces the partial order in the set of all partitions. The maximal element of this set is the partition ε of M into separate points, while the minimal one is the partition v whose unique element is M itself. For any family of measurable partitions $\{\xi_\alpha\}$ the partitions $\sup_\alpha \xi_\alpha$ (notation $\bigvee_\alpha \xi_\alpha$) and $\inf_\alpha \xi_\alpha$ (notation $\bigwedge_\alpha \xi_\alpha$) are well defined and are called the measurable product and measurable intersection of $\{\xi_\alpha\}$ respectively. For any measurable partition ξ denote by $\mathcal{M}(\xi)$ the sub-σ-algebra of the σ-algebra $\mathcal{M}$ whose elements are the subsets of M consisting (mod 0) of the entire elements of ξ. Then $\mathcal{M}(\bigwedge_\alpha \xi_\alpha) = \bigcap_\alpha \mathcal{M}(\xi_\alpha)$, and $\mathcal{M}(\bigvee_\alpha \xi_\alpha)$ is the smallest σ-algebra containing all $\mathcal{M}(\xi_\alpha)$.

Suppose η is a measurable partition. Almost every element D_η of η may itself be considered as a Lebesgue space with the measure $\mu(\cdot|D_\eta)$. Any measurable partition ξ induces the measurable partition on almost each D_η. The entropy of this partition is denoted by $H(\xi|D_\eta)$ and is called the conditional entropy of ξ under the condition D_η.

Definition 1.2. By the *conditional entropy* of the partition ξ with respect to the partition η we mean the number

$$H(\xi|\eta) = \int_{M/\eta} H(\xi|D_\eta)\,d\mu,$$

where M/η is the factor-space of M corresponding to the partition η, and the measure on M/η is induced by μ. If $\mu(C_\xi(x)|\eta)$ is the conditional measure of $C_\xi(x)$ under the condition $D_\eta(x)$, then $H(\xi|\eta) = -\int_M \log\mu(C_\xi(x)|\eta)\,d\mu$. It is also clear that $H(\xi|v) = H(\xi)$.

The partitions ξ, η are said to be independent if for any $A \in \mathcal{M}(\xi)$, $B \in \mathcal{M}(\eta)$ we have $\mu(A \cap B) = \mu(A) \cdot \mu(B)$. If ξ, η are independent, we have $H(\xi|\eta) = H(\xi)$.

We outline now the properties of conditional entropy. It follows from what was said above that the properties of (unconditional) entropy of a partition are the special cases of those for conditional entropy.

1) $H(\xi|\eta) \geq 0$; the equality holds only if $\xi \preceq \eta$;
2) $H(\xi|\eta) \leq H(\xi)$; if $H(\xi) < \infty$, the equality holds only if ξ, η are independent;
3) if $\xi_1 \preceq \xi_2$ then $H(\xi_1|D_\eta) \leq H(\xi_2|D_\eta)$ on almost every D_η; hence, $H(\xi_1|\eta) \leq H(\xi_2|\eta)$; if $\xi_1 \preceq \xi_2$ and $H(\xi_1|\eta) = H(\xi_2|\eta)$, then $\xi_1 = \xi_2 \pmod 0$;
4) $H(\xi_1 \vee \xi_2|\eta) = H(\xi_1|\eta) + H(\xi_2|\xi_1 \vee \eta)$;
5) if $\eta_1 \succeq \eta_2 \pmod 0$, then $H(\xi|\eta_1) \leq H(\xi|\eta_2)$;

It follows immediately from 4) and 5) that

$$H(\xi_1 \wedge \xi_2|\eta) \leq H(\xi_1|\eta) + H(\xi_2|\eta);$$

if $H(\xi_1|\eta), H(\xi_2|\eta) < \infty$, the equality holds if and only if ξ_1 and ξ_2 are independent on almost every D_η;

6) if $\xi_1 \preceq \xi_2 \preceq \ldots, \xi = \bigvee_n \xi_n$, then $H(\xi|\eta) = \lim_{n \to \infty} H(\xi_n|\eta)$;
7) if $\xi_1 \succeq \xi_2 \succeq \ldots, \xi = \bigwedge_n \xi_n$ and $H(\xi_n|\eta) < \infty$ for at least one value of n, then $H(\xi|\eta) = \lim_{n \to \infty} H(\xi_n|\eta)$;
8) if $\eta_1 \preceq \eta_2 \preceq \ldots, \eta = \bigvee_n \eta_n$ and $H(\xi|\eta_n) < \infty$ for at least one value of n, then $H(\xi|\eta) = \lim_{n \to \infty} H(\xi|\eta_n)$;
9) if $\eta_1 \succeq \eta_2 \succeq \ldots$, and $\eta = \bigwedge_n \eta_n$, then $H(\xi|\eta) = \lim_{n \to \infty} H(\xi|\eta_n)$;
10) if T is an endomorphism of a Lebesgue space, then

$$H(T^{-1}\xi|T^{-1}\eta) = H(\xi|\eta).$$

Denote by Z the space of partitions ξ with $H(\xi) < \infty$ and define the metric ρ on Z by $\rho(\xi,\eta) = H(\xi|\eta) + H(\eta|\xi)$. Then (Z, ρ) is a complete metric space.

§2. Entropy of a Dynamical System

In this section we shall formulate the definition of entropy of a dynamical system and calculate the entropy of some dynamical systems. The meaning of the notion of entropy will become more clear step by step in the subsequent sections and in part II. We begin with the case of discrete time.

Suppose T is an endomorphism of the Lebesgue space $(M, \mathcal{M}, \mu)$ and ξ is a measurable partition. Set $\xi_T^- = \bigvee_{n=0}^\infty T^{-n}\xi$. The partition $T^{-1}\xi_T^-$ is sometimes referred to as "the past" of the partition ξ with respect to T.

Definition 2.1. By the *entropy per unit time* of the partition ξ with respect to the endomorphism T we mean the number $h(T,\xi) = H(\xi|T^{-1}\xi_T^-)$.

The properties of $h(T,\xi)$ (cf [Ro3]):
1) $h(T,\xi) \leq h(\xi)$;
2) $h(T, \xi_1 \vee \xi_2) \leq h(T,\xi_1) + h(T,\xi_2)$; if $(\xi_1^-)_T$ and $(\xi_2^-)_T$ are independent, the equality is reached;
3) if $H(\xi) < \infty$, then

$$h(T,\xi) = \lim_{n \to \infty} \frac{1}{n} H(\xi \vee T^{-1}\xi \vee \cdots \vee T^{-n+1}\xi);$$

4) $h(T^n, \bigvee_{k=0}^{n-1} T^{-k}\xi) = n \cdot h(T,\xi)$, $n = 1, 2, \ldots$;
5) if T is an automorphism and $\xi \in Z$, then $h(T,\xi) = h(T^{-1},\xi)$;
6) $h(T,\xi)$ as a function in ξ is continuous on Z;
7) if $\xi_1, \xi_2 \in Z$ and $\xi_1 \leqslant \xi_2$, then $h(T,\xi_1) \leqslant h(T,\xi_2)$;
8) if $H(\xi_1 \vee \xi_2 | T^{-1}\xi_1^-) < \infty$, then

$$h(T,\xi_1) = \lim_{n\to\infty} H(\xi_1 | T^{-1}\xi_1^- \vee T^{-n}\xi_2^-).$$

The next definition plays a central role in the whole entropy theory. Let $\{T^n\}$ be a cyclic semigroup of endomorphisms or a cyclic group of automorphisms of the Lebesgue space with the generator $T = T^1$.

Definition 2.2. The *entropy of T* (of the dynamical system $\{T^n\}$) is the number

$$h(T) = \sup h(T,\xi),$$

where supremum is taken over all measurable partitions ξ of M.

The entropy $h(T)$ is obviously a metric invariant of T, in the sense that $h(T_1) = h(T_2)$ if T_1 and T_2 are metrically isomorphic. $h(T)$ is also referred to as metric entropy, Kolmogorov entropy, Kolmogorov-Sinai entropy.

The properties of $h(T)$ (cf [Ro3]):
1) $h(T) = \sup h(T,\xi)$, where supremum is taken only over finite partitions ξ;
2) if $\xi_1 \leqslant \xi_2 \leqslant \ldots, \xi_n \in Z$ for all $n \geqslant 1$ and $\bigvee_n \xi_n = \varepsilon$, then $\lim_{n\to\infty} h(T,\xi_n)$ exists and equals $h(T)$;
3) $h(T^n) = n \cdot h(T)$ for all $n \geqslant 0$; if T is an automorphism, then $h(T^{-1}) = h(T)$ and $h(T^n) = |n| \cdot h(T)$ for all integers n, $-\infty < n < \infty$;
4) if T_1 is a factor-endomorphism of an endomorphism T, (cf Chap. 1, Sect. 4), then $h(T_1) \leqslant h(T)$;
5) $h(T_1 \times T_2) = h(T_1) + h(T_2)$;
6) if T is an ergodic automorphism, a set $E \in \mathcal{M}$, $\mu(E) > 0$, and T_E is the induced automorphism, then $h(T_E) = [\mu(E)]^{-1} \cdot h(T)$. This relation is known as Abramov's formula;
7) if T^f is the integral automorphism corresponding to an ergodic automorphism T and an integer valued function $f > 0$ (cf Chap. 1, Sect. 4), then $h(T^f) = (\int_M f\, d\mu)^{-1} \cdot h(T)$;
8) the entropy of an endomorphism equals the entropy of its natural extension;
9) if η is the partition of M into ergodic components of T, then $h(T) = \int_{M/\eta} h(T|C_\eta)\, d\mu$.

Define now the entropy of a flow.

Definition 2.3. The *entropy of a flow* $\{T^t\}$ is the number $h(T^1)$.

This definition is motivated by the following theorem.

Theorem 2.1 (L.M. Abramov, cf [CSF]). *If $\{T^t\}$ is a flow, $h(T^t) = |t| \cdot h(T^1)$ for any $t \in \mathbb{R}^1$.*

If $\{T^t\}$ is the special flow corresponding to an ergodic automorphism T_1 and a function f (cf Chap. 1 Sect. 4), then

$$h(\{T^t\}) = \left(\int f\,d\mu_1\right)^{-1} \cdot h(T_1).$$

Definition 2.4. A measurable partition ξ is called a (1-sided) *generating partition* for an endomorphism T if $\xi_T^- = \varepsilon$. A measurable partition ξ is called a 2-sided generating partition or, simply, a generating partition for an automorphism T if $\xi_T \overset{\text{def}}{=} \bigvee_{n=-\infty}^{\infty} T^n \xi = \varepsilon$.

In many cases the evaluation of entropy is based on the following theorem.

Theorem 2.2 (A.N. Kolmogorov [Kol1], Ya.G. Sinai [Si1]). *If $\xi \in Z$ and ξ is a 1-sided generating partition for an endomorphism T or a 2-sided generating partition for an automorphism T, then $h(T) = h(T, \xi)$.*

Examples of entropy computation

1. Suppose T is a periodic automorphism, i.e. there exists a natural number m such that $T^m x = x$ for almost all x. Then $h(T) = 0$. This equality follows from the fact that for any partition ξ of $r < \infty$ elements and any $n > 0$, the partition $\bigvee_{k=0}^{n-1} T^{-k}\xi$ has at most r^n elements. Thus

$$h(T, \xi) = \lim_{n \to \infty} \frac{1}{n} H\left(\bigvee_{k=0}^{n-1} T^{-k}\xi\right) \leq \lim_{n \to \infty} \frac{m}{n} \log r = 0.$$

2. If T is the rotation of the unit circle S^1 by the angle α, then $h(T) = 0$. For α rational this is a special case of Example 1, while for α irrational the partition $\xi = \{[0, \frac{1}{2}), [\frac{1}{2}, 1)\}$ is generating (for T), and $\bigvee_{k=0}^{n-1} T^{-k}\xi$ consists of $2n$ intervals. Thus,

$$h(T) = h(T, \xi) = \lim_{n \to \infty} \frac{1}{n} H\left(\bigvee_{k=0}^{n-1} T^{-k}\xi\right) \leq \lim_{n \to \infty} \frac{1}{n} \log 2n = 0.$$

3. Let T be a Bernoulli automorphism in the space $(M, \mathcal{M}, \mu)$ of sequences $x = (\ldots, x_{-1}, x_0, x_1, \ldots)$, $x_i \in (Y, \mathcal{Y}, \nu)$, $Y = \{a_1, \ldots, a_r\}$, $\nu(\{a_k\}) = p_k$, $1 \leq k \leq r$. The partition $\xi = (C_1, \ldots, C_r)$, where $C_k = \{x \in M: x_0 = a_k\}$, is generating (for T). Therefore, $h(T) = h(T, \xi) = H(\xi) = -\sum_{k=1}^{r} p_k \log p_k$. The immediate consequence of this fact is that for any $h > 0$ there are Bernoulli automorphisms T with $h(T) = h$ and, therefore, there exist continuum of pairwise non-isomorphic Bernoulli automorphisms. All metric invariants of dynamical systems that were known before entropy gave no possibility to distinguish between different Bernoulli automorphisms, and it was not known whether non-isomorphic Bernoulli automorphisms existed.

4. Suppose M is the same space as in Example 3 and T is a Markov automorphism with transition matrix $\prod = \|p_{ij}\|$, $1 \leq i, j \leq r$, and stationary probabilities $(p_1, \ldots, p_r)$. The partition ξ of Example 3 is still generating, and

$$h(T) = h(T, \xi) = -\sum_{i,j=1}^{r} p_i p_{ij} \log p_{ij}.$$

5. Suppose T is an ergodic algebraic automorphism of the m-dimensional torus with the invariant Haar measure. T is given by an integer $m \times m$ matrix A with $\det A = 1$. Let $\lambda_1, \ldots, \lambda_m$ be the eigenvalues of A. Then

$$h(T) = \log \prod_{i:|\lambda_i|>1} |\lambda_i|.$$

6. Suppose M is $[0,1]$ and T is the endomorphism of M given by a function f defined on $[0,1]$, i.e. $Tx = f(x)$. Suppose, further, that f has finitely many discontinuities and its derivative f' exists and satisfies the inequality $|f'| > 1$ on the intervals between any two discontinuities. Let μ be an absolutely continuous Borel measure on $[0,1]$ invariant under T. Then $h(T) = \int_0^1 \log |f'(x)| \rho(x) dx$, where $\rho(x)$ is the density of the measure μ.

For the endomorphism T with $f(x) = \{1/x\}$ (it has a countable number of discontinuities, but the above formula is still valid) we have

$$h(T) = \frac{2}{\log 2} \int_0^1 \frac{|\log x|}{1+x} dx = \frac{\pi^2}{6 \log 2}.$$

This endomorphism is closely related to the decomposition of real numbers into continuous fractions.

7. For an integrable Hamiltonian system $\{T^t\}$ $h(\{T^t\}) = 0$.

8. The entropy of a billards in any polygon or polyhedron is equal to zero.

9. For a geodesic flow $\{T^t\}$ on a surface of negative constant curvature-K we have $h(\{T^t\}) = \sqrt{K}$.

§3. The Structure of Dynamical Systems of Positive Entropy

The following theorem gives important information about the meaning of the concept of entropy.

Theorem 3.1 (The Shannon-McMillan-Breiman theorem, cf [Bi]). *Suppose T is an ergodic automorphism of a Lebesgue space $(M, \mathcal{M}, \mu)$, ξ is a finite partition of M, $C_n(x)$ is the element of the partition $\xi \vee T\xi \vee \cdots \vee T^{n-1}\xi$ containing $x \in M$. Then*

$$\lim_{n \to \infty} \left[-\frac{1}{n} \log \mu(C_n(x)) \right] = h(T, \xi)$$

for almost all $x \in M$.

Another important fact about entropy is the Krieger theorem on generating partitions.

Theorem 3.2 (W. Krieger [Kri1]. *Suppose T is an ergodic automorphism, $h(T) < \infty$. Then for any $\varepsilon > 0$ there exists a finite generating partition for T such that $H(\xi) \leq h(T) + \varepsilon$. If $h(T) < \log k$ for some integer $k > 1$, there exists a generating partition of k elements.*

This theorem shows that any ergodic automorphism of finite entropy can be represented as a stationary random process with discrete time and a finite number of states.

Turning back to arbitrary endomorphisms consider the partitions ξ of the phase space with $h(T,\xi) = 0$. For such partitions we have $\xi_T^- = T^{-1}\xi_T^- = T^{-2}\xi_T^- = \ldots$, i.e. $\xi \preccurlyeq T^{-1}\xi_T^-$. From the probabilistic point of view the equality $h(T,\xi) = 0$ means that we deal with such a random process that the "infinitely remote past" entirely determines its "future".

For an ergodic endomorphism T, M.S. Pinsker (cf [Ro3]) defined the partition $\pi(T)$ which is the least upper bound of all partitions ξ with $h(T,\xi) = 0$. If $\{T^t\}$ is a flow, the partition $\pi(T^t)$ does not depend on t. It is denoted by $\pi(\{T^t\})$. The endomorphisms T with $\pi(T) = v$ are called the automorphisms of completely positive entropy.

Theorem 3.3 (cf [Ro3]). *Suppose T is an ergodic automorphism, $h(T) > 0$. Denote by $\mathcal{H}^+$ the orthogonal complement to the subspace of $L^2(M,\mathcal{M},\mu)$ consisting of functions constant (mod 0) on the elements of $\pi(T)$. Then $U_T \mathcal{H}^+ = \mathcal{H}^+$, and the operator U_T has the countable Lebesgue spectrum on $\mathcal{H}^+$.*

This theorem implies, in particular, that any automorphism with pure point spectrum, singular spectrum, as well as spectrum of finite multiplicity, is of zero entropy.

Theorem 3.4 (cf [Ro3]). *If T is an endomorphism of completely positive entropy, then T is exact. If T is an automorphism of completely positive entropy, then T is a K-automorphism.*

Definition 3.1. A partition ζ is said to be *exhaustive* with respect to an automorphism T, if $T\zeta \succcurlyeq \zeta$, $\bigvee_{k=0}^{\infty} T^k \zeta = \varepsilon$.

From the probabilistic point of view the σ-algebra $\mathcal{M}(\zeta)$ is the analog of the σ-algebra corresponding to "the past" of a random process.

Theorem 3.5 (cf [Ro3]). *If ζ is an exhaustive partition for T, then $\bigwedge_{n=0}^{\infty} T^{-n} \zeta \succcurlyeq \pi(T)$.*

It follows from this theorem that $\pi(T) = v$ for any K-automorphism T. Therefore, the equality $\pi(T) = v$ is equivalent to the fact that T is a K-automorphism.

This implies, in particular, that any factor-automorphism and any power of a K-automorphism also is a K-automorphism.

Definition 3.2. An exhaustive partition ζ is said to be *extremal* if $\bigwedge_{n=0}^{\infty} T^{-n}\zeta = \pi(T)$. An extremal partition is said to be *perfect* if $h(T,\zeta) = h(T)$.

Theorem 3.6 (V.A. Rokhlin, Ya.G. Sinai (cf [Ro3])). *Any automorphism possesses perfect partitions.*

The above definitions can be easily extended to the case of continuous time.

Definition 3.3. A measurable partition ζ is said to be *extremal* with respect to a flow $\{T^t\}$ if 1) $T^t\zeta \geqslant \zeta$ for $t > 0$; 2) $\bigvee_{t>0} T^t\zeta = \varepsilon$; 3) $\bigwedge_t T^t\zeta = \pi(\{T^t\})$. An extremal partition is said to be *perfect* if $H(T^s\zeta|\zeta) = s \cdot h(\{T^t\})$ for any $s > 0$.

Theorem 3.7 (P. Blanchard, B.M. Gurevich, D. Rudolph, [Bl], [Gu], [Ru1]). *Any ergodic flow possesses perfect partitions.*

This theorem implies that if there exists $t_0 \in \mathbb{R}^1$ such that T^{t_0} is a K-automorphism, then $\{T^t\}$ is a K-flow, and, therefore, all T^t, $t \neq 0$, are K-automorphisms.

For the automorphisms of zero entropy we always have $\pi(T) = \varepsilon$; in this sense the properties of such automorphisms are opposite to the properties of K-automorphisms, because $\pi(T) = v$ for them.

Theorem 3.8 (cf [Ro3]. *For an automorphism T to be of zero entropy it is necessary and sufficient that any of the following properties be satisfied:*
i) *the only exhaustive partition for T is ε;*
ii) *any 2-sided generating partition for T is also a 1-sided generating partition;*
iii) *there exists a 1-sided generating partition ξ for T with $H(\xi) < \infty$.*

§4. The Isomorphy Problem for Bernoulli Automorphisms and K-Systems

The problem that will be discussed in this section is the so-called isomorphy problem, i.e. the problem of classifying the dynamical systems up to their metric isomorphism. As for the general setting of this problem, i.e. the one related to the class of all dynamical systems, it is, unfortunately, clear now that there is no complete solution to it in reasonable terms: there are too many examples showing that the metric properties of dynamical systems can be very complicated and, sometimes, very unexpected. When the notion of entropy was introduced and the existence of continuum pairwise non-isomorphic Bernoulli automorphisms had been proved with its help, attention was attracted to a certain special case of the isomorphy problem, namely, to the problem of finding the exact conditions under which two Bernoulli automorphisms and, more generally, two K-automorphisms are metrically isomorphic. The question may be formulated as a specific coding problem.

Suppose T_1, T_2 are two Bernoulli automorphisms with distinct state spaces (distinct alphabets). In order to prove that they are metrically isomorphic one must find a coding procedure translating the sequences written in one alphabet into the sequences written in the other one and having the property that the shifted sequences are coded by the shifted ones.

Suppose the state spaces of T_1, T_2 consist of m_1, m_2 elements ($m_1, m_2 < \infty$) and the measures in them are given by the vectors $(p_1, \ldots, p_{m_1})$, $(q_1, \ldots, q_{m_2})$ respectively. Then T_1, T_2 are obviously non-isomorphic if $h(T_1) \neq h(T_2)$, i.e. $-\sum p_i \log p_i \neq -\sum q_i \log q_i$. The following example due to L.D. Meshalkin shows that the

Bernoulli automorphisms T_1, T_2 with $h(T_1) = h(T_2)$ may be metrically isomorphic even if their state spaces are non-isomorphic.

Suppose $Y_1 = \{a_i\}_{i=1}^4$, $Y_2 = \{b_j\}_{j=0}^4$, and the measures $\sigma^{(1)}$, $\sigma^{(2)}$ on Y_1, Y_2 are given by the vectors $(\frac{1}{4}, \frac{1}{4}, \frac{1}{4}, \frac{1}{4})$ and $(\frac{1}{2}, \frac{1}{8}, \frac{1}{8}, \frac{1}{8}, \frac{1}{8})$ respectively. Let T_l ($l = 1, 2$) be the Bernoulli automorphism acting in $M_l = \prod_{n=-\infty}^{\infty} Y_n^{(l)}$, $Y_n^{(l)} \equiv Y_l$, with the invariant measure $\mu_l = \prod_{n=-\infty}^{\infty} \sigma_n^{(l)}$, $\sigma_n^{(l)} \equiv \sigma^{(l)}$. Consider the map $\phi: M_1 \to M_2$, which sends $x^{(1)} = (\ldots, a_{i_{-1}}, a_{i_0}, a_{i_1}, \ldots)$ to $x^{(2)} = (\ldots, b_{j_{-1}}, b_{j_0}, b_{j_1}, \ldots)$ according to the following rule: if $i_n = 1$ or 2, we set $j_n = 0$; if $i_n = 3$ (respectively 4), we find first the maximal $n_1 < n$ for which $\text{card}(\{k: n_1 \leq k \leq n, i_k = 1 \text{ or } 2\}) = \text{card}(\{k: n_1 \leq k \leq n, i_k = 3 \text{ or } 4\})$. Such an n_1 exists with probability 1. Moreover, it is clear that $i_{n_1} = 1$ or 2, and we set in this case $j_n = 1$ (respectively 3), if $i_{n_1} = 1$; $j_n = 2$ (respectively 4) if $i_{n_1} = 2$. It may be easily verified that ϕ defines the metric isomorphism between T_1 and T_2. There are some generalizations of the above construction, but even in its generalized form the method can be applied only to some special sub-classes of Bernoulli automorphisms. The first general result concerning the isomorphism problem of Bernoulli automorphisms was stated in terms of weak isomorphism.

Definition 4.1. Two dynamical systems are said to be *weakly isomorphic* if each of them is metrically isomorphic to a certain factor-system of the other one.

Theorem 4.1 (Ya.G. Sinai [Si3]). *Any two Bernoulli automorphisms with the same entropy are weakly isomorphic.*

This theorem is an immediate consequence of the following more general result.

Theorem 4.2 (Ya.G. Sinai [Si3]). *If T_1 is an ergodic automorphism of a Lebesgue space, T_2 is a Bernoulli automorphism of finite entropy and $h(T_2) \leq h(T_1)$, then T_2 is metrically isomorphic to some factor automorphism of T_1.*

The weak isomorphism of two dynamical systems does not generally imply its metric isomorphism, but corresponding examples are rather complicated. (cf D. Rudolph [Ru2]).

The final solution to the isomorphism problem of the Bernoulli automorphisms was given by the following theorem due to D. Ornstein.

Theorem 4.3 (D. Ornstein, cf [Or]). *Any two Bernoulli automorphisms with the same entropy are metrically isomorphic.*

The state spaces of T_1, T_2 in this theorem are not assumed to be finite or countable, the case $h(T_1) = h(T_2) = \infty$ is not excluded. Therefore, entropy is a complete metric invariant among the class of Bernoulli automorphisms.

We will outline the ideas involved in the proof of this very important theorem and will then present a series of further results concerning the isomorphy problem for K-automorphisms and K-flows.

Chapter 3. Entropy Theory of Dynamical Systems   45

Definition 4.2. A measurable partition ξ of a Lebesgue space $(M, \mathcal{M}, \mu)$ is said to be *Bernoulli with respect to an automorphism* $T: M \to M$ if all its shifts, i.e. the partitions $T^n\xi$, $-\infty < n < \infty$, are independent. An automorphism T is called *B-automorphism* if there are Bernoulli generating partitions for T.

The class of B-automorphisms obviously coincides with the class of automorphisms which are metrically isomorphic to some Bernoulli automorphisms. The above definition is simply the "coordinates-free" version of the standard definition of Bernoulli automorphisms which is convenient in studying the questions related to the isomorphy problem.

If an automorphism T has a finite Bernoulli generating partition, then the state space of the corresponding Bernoulli automorphism is finite.

We shall often deal in this section with pairs (T, ξ), where T is an ergodic automorphism of $(M, \mathcal{M}, \mu)$, and ξ is a partition of M. It will be assumed without any special mentioning that $\xi = (C_1, \ldots, C_r)$ is a finite measurable partition with the fixed order of its elements, i.e. ξ is an ordered partition, while M is a Lebesgue space with continuous measure. To each such pair (T, ξ) one can naturally associate a random process with r states, or, in other words, a shift-invariant measure in the space $M^{(r)}$ of 2-sided infinite sequences of r symbols. Two pairs (T_1, ξ_1), (T_2, ξ_2) will be called equivalent (notation: $(T_1, \xi_1) \sim (T_2, \xi_2)$) if they correspond to the same measure on $M^{(r)}$. A pair (T, ξ) is sometimes identified with the corresponding random process and, so, (T, ξ) itself is called a process.

The Ornstein distance between the pairs $(T_1, \xi_1), (T_2, \xi_2)$.

We start with the definition of a certain distance ρ between two ordered partitions $\xi_1 = (C_1, \ldots, C_r)$, $\xi_2 = (D_1, \ldots, D_r)$ of the space $(M, \mathcal{M}, \mu)$ by the formula

$$\rho(\xi_1, \xi_2) = \frac{1}{2} \sum_{\substack{1 \leq i,j \leq r \\ i \neq j}} \mu(C_i \cap D_j).$$

Next suppose $\{\xi_1^{(k)}\}_{k=0}^{n-1}$ (respectively $\{\xi_2^{(k)}\}_{k=0}^{n-1}$) is a sequence of $n \geq 1$ partitions of the space $(M_1, \mathcal{M}_1, \mu_1)$ (respectively $(M_2, \mathcal{M}_2, \mu_2)$), each partition being of r elements, $r < \infty$. Consider another measure space $(M, \mathcal{M}, \mu)$ and the maps ϕ_l: $M_l \to M$ ($l = 1, 2$) defining the isomorphisms between M_l and M. The map ϕ_l sends the partition $\xi_l^{(k)}$ to a certain partition, $\phi_l(\xi_l^{(k)})$, of $M (l = 1, 2; k = 0, 1, \ldots, n-1)$. Set

$$\bar{d}_{\phi_1, \phi_2}(\{\xi_1^{(k)}\}_0^{n-1}, \{\xi_2^{(k)}\}_0^{n-1}) = \frac{1}{n} \sum_{k=0}^{n-1} \rho(\phi_1(\xi_1^{(k)}), \phi_2(\xi_2^{(k)})),$$

$$\bar{d}(\{\xi_1^{(k)}\}_0^{n-1}, \{\xi_2^{(k)}\}_0^{n-1}) = \inf_{\phi_1, \phi_2} \bar{d}_{\phi_1, \phi_2}(\{\xi_1^{(k)}\}_0^{n-1}, \{\xi_2^{(k)}\}_0^{n-1}),$$

where inf is taken over all possible maps ϕ_1, ϕ_2 defining the isomorphisms of M_1, M_2 and M. Clearly, $\bar{d}$ does not depend on the choice of M and the inequality $0 \leq \bar{d} \leq 1$ is always true. If T_l is an automorphism of M_l and ξ_l is a partition of

$M_l (l = 1, 2)$, then for any $n \geq 1$ we set

$$\bar{d}_n((T_1, \xi_1), (T_2, \xi_2)) = \bar{d}(\{T_1^k \xi_1\}_0^{n-1}, \{T_2^k \xi_2\}_0^{n-1}).$$

Definition 4.3. The *Ornstein distance* between the pairs (T_1, ξ_1), (T_2, ξ_2) is the number

$$\bar{d}((T_1, \xi_1), (T_2, \xi_2)) = \overline{\lim_{n \to \infty}} \, \bar{d}_n((T_1, \xi_1), (T_2, \xi_2)). \tag{3.1}$$

The right-hand side of the latter equality may be rewritten with $\lim_n$ instead of upper limit, since such limit always exists. This remark shows that the triangle inequality hold for $\bar{d}$. To gain familiarity with the Ornstein metric it is useful to associate with the above sequences $\{\xi_1^{(k)}\}_0^{n-1}$, $\{\xi_2^{(k)}\}_0^{n-1}$ the measures $v_n^{(1)}$, $v_n^{(2)}$ on the space $M_n^{(r)}$ of all sequences $\mathscr{I}$ of the form $\mathscr{I} = (i_0, \ldots, i_{n-1})$, ($i_k$ are integers), given by

$$v_n^{(l)}(\mathscr{I}) = \mu_l(C_{i_0, 0}^{(l)} \cap C_{i_1, 1}^{(l)} \cap \cdots \cap C_{i_{n-1}, n-1}^{(l)}),$$

where $C_{1,k}^{(l)}, \ldots, C_{r,k}^{(l)}$ are the elements of the partition $\xi_l^{(k)}$, $k = 0, 1, \ldots, n - 1$; $l = 1, 2$. If $\{\xi_1^{(k)}\}$, $\{\xi_2^{(k)}\}$ are of the form $\{T_1^k \xi_1\}_0^{n-1}$, $\{T_2^k \xi_2\}_0^{n-1}$, then the measures $v_n^{(1)}$, $v_n^{(2)}$ are precisely the finite dimensional distributions of the random processes corresponding to the pairs (T_1, ξ_1), (T_2, ξ_2). Hence, $\bar{d}_n((T_1, \xi_1), (T_2, \xi_2))$ may actually be regarded as a distance between $v_n^{(1)}$ and $v_n^{(2)}$. The equality $\bar{d}((T_1, \xi_1), (T_2, \xi_2)) = 0$ means that $(T_1, \xi_1) \sim (T_2, \xi_2)$, that is, the factor-automorphisms $T_1|(\xi_1)_{T_1}$, $T_2|(\xi_2)_{T_2}$ are metrically isomorphic (ξ_T stands for $\bigvee_{-\infty}^{\infty} T^n \xi$). Therefore, $\bar{d}$ is the metric on the space of equivalence classes of the pairs (T, ξ). The metric $\bar{d}(\{\xi_1^{(k)}\}_0^{n-1}, \{\xi_2^{(k)}\}_0^{n-1})$ may be obtained from the following general construction due to L.V. Kantorovich and G.S. Rubinstein ([Or], Russian edition). Suppose X is a metrisable topological space, $\text{Mes}(X)$ is the space of all normalized Borel measures on X. Any metric ρ on X may be associated with a metric d_ρ on $\text{Mes}(X)$:

$$d_\rho(v_1, v_2) = \inf_\lambda \int_{X \times X} \rho(x_1, x_2) \, d\lambda(x_1, x_2),$$

where inf is taken over all normalized Borel measures λ on $X \times X$ such that their marginal measures (projections to the factors of $X \times X$) equal v_1, v_2 respectively (i.e. $\lambda(A \times X) = v_1(A)$, $\lambda(X \times B) = v_2(B)$ for all measurable $A, B \subset X$). Now take the space $M_n^{(r)}$ as X, and the Hamming metric χ on X as ρ, i.e. $\chi((i_0^{(1)}, \ldots, i_{n-1}^{(1)}), (i_0^{(2)}, \ldots, i_{n-1}^{(2)}))$ is equal to the relative fraction of those k for which $i_k^{(1)} \neq i_k^{(2)}$. Then the Kantorovich-Rubinstein construction leads to a certain metric d_ρ which is equivalent to $\bar{d}$ (i.e. the corresponding topologies are the same).

Finitely determined partitions. Suppose T is an automorphism of a space $(M, \mathscr{M}, \mu)$, $\xi = (C_1, \ldots, C_r)$ is a partition of M.

Definition 4.4. ξ is said to be *finitely determined* with respect to T if for any $\varepsilon > 0$ there exist $\delta > 0$ and an integer $n \geq 1$ such that for any pair $(\bar{T}, \bar{\xi})$, where $\bar{T}$ is an automorphism of a Lebesgue space $(\bar{M}, \bar{\mathscr{M}}, \bar{\mu})$, $\bar{\xi} = (\bar{C}_1, \ldots, \bar{C}_r)$ is a partition of $\bar{M}$, satisfying the conditions:

1) $|h(\bar{T},\bar{\xi}) - h(T,\xi)| < \delta$,
2) $\sum_{1 \leq i_0, i_1, \ldots, i_{n-1} \leq r} |\bar{\mu}(\bigcap_{k=0}^{n-1} \bar{T}^k \bar{C}_{i_k}) - \mu(\bigcap_{k=0}^{n-1} T^k C_{i_k})| < \delta$, one has $\bar{d}((T,\xi), (\bar{T},\bar{\xi})) < \varepsilon$.

Sometimes the pair (T, ξ) itself is called finitely determined if ξ is finitely determined with respect to T.

The fact that (T, ξ) is finitely determined means (in terms of the corresponding random process) that if any pair (process) $(\bar{T}, \bar{\xi})$ has a sufficiently large collection of finite dimensional distributions which are close to the corresponding distributions of (T, ξ) and, moreover, $h(\bar{T}, \bar{\xi})$ and $h(T, \xi)$ are close to each other, then (T, ξ) and $(\bar{T}, \bar{\xi})$ are $\bar{d}$-close.

Theorem 4.4 (cf [Or]). *If ξ is a finite Bernoulli partition for T, then ξ is finitely determined with respect to T.*

We will outline now the main ideas involved in the proof of Theorem 4.5. Only the case of Bernoulli automorphism with finite state space will be considered. In this case Theorem 4.2 is the consequence of the following general statement.

Theorem 4.5. *If T_1 and T_2 have finitely determined generating partitions and $h(T_1) = h(T_2)$, then T_1 and T_2 are metrically isomorphic.*

To derive Theorem 4.2 from this statement, it suffices to observe that, according to Theorem 4.4, any Bernoulli generating partition is necessarily finitely determined. The proof of Theorem 4.5 splits up into 2 steps.

A. Suppose the pairs $(T_1, \xi_1), (T_2, \xi_2)$ are such that:
1) $h(T_2, \xi_2) \leq h(T_1, \xi_1)$;
2) ξ_2 is finitely determined with respect to T_2. Then there exists a partition $\tilde{\xi}_1$ of the space M_1 such that the factor-automorphisms $T_1|(\tilde{\xi}_1)_{T_1}$, $T_2|(\xi_2)_{T_2}$ are metrically isomorphic, and $\rho(\xi_1, \tilde{\xi}_1) \leq \text{const} \cdot \mathcal{D}^{1/2}$, where $\mathcal{D} = \bar{d}((T_1, \xi_1), (T_2, \xi_2))$.

The assertion A in the case when T_2 is a Bernoulli automorphism and ξ_2 is its Bernoulli generating partition, is a strengthened form of Theorem 4.2. The partition $\tilde{\xi}_1$ appearing in A is obtained as the limit of a certain sequence $\{\xi_1^{(n)}\}_{n=0}^{\infty}$, $\xi_1^{(0)} = \xi_1$, such that $\bar{d}((T_1, \xi_1^{(n)}), (T_2, \xi_2)) \to 0$. For the limit partition $\tilde{\xi}_1$ we have $\bar{d}((T_1, \tilde{\xi}_1), (T_2, \xi_2)) = 0$, that is $T_1|(\tilde{\xi}_1)_{T_1}$ and $T_2|(\xi_2)_{T_2}$ are isomorphic, The inductive choice of $\xi_1^{(n)}$ is based on the following lemma playing the central role in the proof of A.

Lemma. *Suppose $(T_1, \xi_1), (T_2, \xi_2)$ and $\mathcal{D}$ are as in the statement of A. Then for any $\varepsilon > 0$ there exists a partition $\xi^{(\delta)}$ of M_1 such that $\bar{d}((T_1, \xi^{(\delta)}), (T_2, \xi_2)) < \delta$ and $\rho(\xi_1, \xi^{(\delta)}) \leq \text{const} \cdot \mathcal{D}^{1/2}$.*

Applying the statement A to the pairs $(T_1, \xi_1), (T_2, \xi_2)$ appearing in Theorem 4.5 we obtain two partitions ξ_1 and $\tilde{\xi}_1$ of the space M_1 and to complete the proof it suffices to replace $\tilde{\xi}_1$ by a generating partition η (of the space M_1) having the same properties as $\tilde{\xi}_1$. The possibility of this replacement is assured by the following statement.

B. Suppose T is an automorphism of $(M, \mathcal{M}, \mu)$, and the partitions ξ, $\tilde{\xi}$ are finitely determined with respect to T. Suppose, further, that ξ is a generating partition. Then for any $\varepsilon > 0$ there is a partition $\eta^{(\varepsilon)}$ such that the factor-automorphisms $T|\tilde{\xi}_T$, $T|\eta_T^{(\varepsilon)}$ are metrically isomorphic and $\rho(\tilde{\xi}, \eta^{(\varepsilon)}) < \varepsilon$, $\eta_T^{(\varepsilon)} \overset{\varepsilon}{\succ} \xi$.[1]

Using the assertion B with T_1 in the role of T, one can construct the partition η as the limit of a sequence $\{\eta_n\}_{n=0}^{\infty}$, $\eta_0 = \tilde{\xi}$, where $\eta_{n+1} = \eta_n^{(\varepsilon_n)}$, $\varepsilon_n \to 0$. As n goes to infinity, the partitions η_n become "more and more generating", and the limit partition η gives us the needed metric isomorphism.

The structure of B-automorphisms. It follows from Theorems 4.4, 4.5 that any automorphism having a finitely determined generating partition is a B-automorphism. The next result shows that the property of being finitely determined, if held by at least one generating partition, is necessarily inherited by all finite partitions of the phase space.

Theorem 4.6. *If T is a B-automorphism of $(M, \mathcal{M}, \mu)$, $h(T) < \infty$, then any finite partition ξ of M is finitely determined with respect to T.*

Corollary. *Any factor-automorphism T_1 of a B-automorphism T is also a B-automorphism.*

To prove this statement, it suffices to apply Theorem 4.6 to a finite generating partition ξ for T_1 which we consider as a partition of the phase space of T.

The assertion of the corollary is also valid in the case $h(T) = \infty$.

It may be easily deduced from the Ornstein isomorphism theorem that any B-automorphism has the roots of all degrees (the root of n-th degree from T is an automorphism Q_n such that $(Q_n)^n$ is metrically isomorphic to T). One may take as Q_n any Bernoulli automorphism of entropy $\frac{1}{n}h(T)$. Any other root of n-th degree from T is metrically isomorphic to this one.

In order to prove that a given automorphism is B, one must, according to the Ornstein theorem, find a finitely determined generating partition for it. In many cases, however, it is useful to replace the "finitely determined" property by an equivalent one which is often easier to be verified. We shall give now the corresponding definition.

Given an ordered partition $\xi = (C_1, \ldots, C_r)$ of a space $(M, \mathcal{M}, \mu)$ and a set $D \in \mathcal{M}$, $\mu(D) > 0$, denote by $\xi|D$ the ordered partition $(C_1 \cap D, \ldots, C_r \cap D)$ of the space $(D, \mathcal{M}|D, [\mu(D)]^{-1} \cdot \mu)$, where $\mathcal{M}|D = \{A \in \mathcal{M}: A \subseteq D\}$.

Definition 4.5. A partition ξ of a space $(M, \mathcal{M}, \mu)$ is said to be *very weak Bernoulli* (vwB) with respect to an automorphism $T: M \to M$, if for any $\varepsilon > 0$ there is an integer $N = N(\varepsilon) > 0$ such that for any $m \geq 0$, $n \geq N$ there exists a set $A \in \mathcal{M}$ consisting of the entire elements of the partition $\eta_m = \bigvee_{k=-m}^{0} T^k \xi$ and satisfying the conditions:

---

[1] We write $\xi \overset{\varepsilon}{\succ} \eta$ if there exists a partition η' for which $\rho(\eta, \eta') < \varepsilon$ and $\xi \succ \eta'$.

1) $\mu(A) > 1 - \varepsilon$;
2) $\bar{d}(\{T^k\xi|D_1\}_0^{n-1}, \{T^k\xi|D_2\}_0^{n-1}) < \varepsilon$ for any elements D_1, D_2 of η_m such that $D_1, D_2 \subseteq A$.

The condition 2) may be obviously replaced by
2') $\bar{d}(\{T^k\xi\}_0^{n-1}, \{T^k\xi|D\}_0^{n-1}) < \varepsilon$
for any $D \in \eta_m$, $D \subseteq A$.

Theorem 4.7. *For any automorphism T the class of vwB partitions coincides with that of finitely determined partitions.*

Corollary. *If ξ is a vwB partition for T, then the factor-automorphism $T|\xi_T$ is a B-automorphism.*

We shall describe now the so-called weak Bernoulli property of a partition which is stronger than vwB.

Definition 4.6. A partition ξ is said to be *weak Bernoulli* with respect to T, if for any $\varepsilon > 0$ there exists an integer $n > 0$ such that the partitions $\bigvee_{k=-m}^{0} T^k \xi$ are ε-independent from $\bigvee_{k=n}^{n+m} T^k \xi$ for all $m \geqslant 0$.[2]

Any weak Bernoulli partition is vwB. Though the converse is generally false (cf [Sm]), in many important cases the vwB property can be deduced from this, simpler condition.

Let T be a mixing Markov automorphism acting in the space M of 2-sided sequences $x = (\ldots, y_{-1}, y_0, y_1, \ldots)$, $y_i \in Y$, where $Y = \{a_1, \ldots, a_r\}$ is a finite set. The generating partition $\xi = (C_1, \ldots, C_r)$, $C_i = \{x \in M: y_0 = a_i\}$, $1 \leqslant i \leqslant r$, is weak Bernoulli. This implies that any mixing Markov automorphism is B (B-automorphism).

There are examples of finite partitions for B-automorphisms which are not weak Bernoulli (cf Smorodinsky [Sm]). On the other hand, Theorems 4.6 and 4.7 show that such partitions are necessarily vwB.

The next theorem enables us to establish the B-property of a given automorphism without finding a vwB generating partition for it.

Theorem 4.8. *Suppose $\xi_1 \leqslant \xi_2 \leqslant \ldots, \xi_n \to \varepsilon$, is an increasing sequence of finite partitions of a space $(M, \mathcal{M}, \mu)$ which are vwB with respect to an automorphism T acting on M. Then T is a B-automorphism.*

B-flows

Definition 4.7. A flow $\{T^t\}$ on a Lebesgue space $(M, \mathcal{M}, \mu)$ is said to be a *B-flow* if there exists $t_0 \in \mathbf{R}^1$ such that T^{t_0} is a B-automorphism.

By now examples of B-flows of various origins have been constructed (cf part II). It should be noted, however, that even the fact of existence of at least one B-flow is far from being trivial. Only the Ornstein theory gave tools for imbedding Bernoulli automorphisms of finite entropy into flows.

---

[2] The partitions $\xi = \{C_i\}$ and $\eta = \{D_j\}$ are ε-independent if $\sum_{i,j} |\mu(C_i \cap D_j) - \mu(C_i)\mu(D_j)| < \varepsilon$.

Example. Suppose M_1 is the space of 2-sided sequences $x = (\ldots, x_{-1}, x_0, x_1, \ldots)$ of 0's and 1's, and the Bernoulli measure μ_1 on M_1 is given by the vector $(\frac{1}{2}, \frac{1}{2})$. Consider the special flow $\{T^t\}$ corresponding to the shift automorphism T_1 and the function $f: M_1 \to \mathbb{R}^1$ such that $f(x) = \alpha$ if $x_0 = 0$, $f(x) = \beta$ if $x_0 = 1$, and α/β is irrational. The phase space of $\{T^t\}$ is $M = \{(x, t): x \in M_1, 0 \leq t < f(x)\}$. The partition $\xi = (C_1, C_2)$ of M with $C_1 = \{(x, t) \in M: x_0 = 0\}$, $C_2 = \{(x, t) \in M: x_0 = 1\}$ is generating for T^{t_0} if $|t_0|$ is small enough. It may be shown that ξ is vwB, so $\{T^t\}$ is a B-flow.

The Ornstein isomorphism theorem was extended to the case of B-flows.

Theorem 4.9 (D. Ornstein [Or]). *If $\{T_1^t\}$, $\{T_2^t\}$ are B-flows and $h(\{T_1^t\}) = h(\{T_2^t\})$, then $\{T_1^t\}$, $\{T_2^t\}$ are metrically isomorphic.*

The proof of this statement involves the same ideas that the proof of the corresponding theorem about automorphisms. The important tool used in the proof is the continuous-time version of the $\bar{d}$-metric (i.e. the metric in the space of pairs $((\{T_1^t\}, \xi_1), (\{T_2^t\}, \xi_2))$).

The isomorphy problem for K-systems. Unlike the case of Bernoulli automorphisms where a complete metric classification was given by the Ornstein theorem, the situation in the class of K-systems is much more complicated. What we have in this case is a collection of counterexamples. D. Ornstein [Or] produced an example of a K-automorphism which is not isomorphic to any Bernoulli automorphism. Using some modification of the construction of this example, he also constructed continuum of pairwise non-isomorphic K-automorphisms with the same entropy.

M.S. Pinsker conjectured that any ergodic automorphism of positive entropy is metrically isomorphic to a direct product $T_1 \times T_2$, where T is a K-automorphism, while $h(T_2) = 0$. If this conjecture were true, it would signify that the general isomorphism problem might be reduced to the special cases concerned with automorphisms of zero entropy and K-automorphisms. However, the counterexample to the Pinsker hypothesis also due to D. Ornstein [Or] showed that such a reduction is impossible.

Finitary isomorphism. Suppose M_l, $l = 1, 2$, is the space of all sequences $x^{(l)} = (\ldots y_{-1}^{(l)}, y_0^{(l)}, y_1^{(l)}, \ldots)$, $y_i^{(l)} \in Y_l = \{a_1, \ldots, a_{r_l}\}$, $r_1, r_2 < \infty$; $T_l: M_l \to M_l$ is the shift automorphism, i.e. $T_l x^{(l)} = \tilde{x}^{(l)} = (\ldots, \tilde{y}_{-1}^{(l)}, \tilde{y}_0^{(l)}, \tilde{y}_1^{(l)}, \ldots)$, $\tilde{y}_i^{(l)} = y_{i+1}^{(l)}$, with an invariant measure μ_l defined on the Borel σ-algebra $\mathcal{M}_l$ of subsets of M_l.[3] Let $\pi_0: M_l \to Y$ be the projection: $\pi_0 x^{(l)} = y_0^{(l)}$ if $x^{(l)} = (\ldots y_{-1}^{(l)}, y_0^{(l)}, y_1^{(l)} \ldots) \in M_l$. Given $x^{(l)} \in M_l$ and a natural number n, denote by $C_n(x^{(l)})$ the cylinder subset of M_l of the form $\{\bar{x}^{(l)} \in M_l: \pi_0(T^k \bar{x}^{(l)}) = \pi_0(T^k x^{(l)})$ if $|k| \leq n\}$.

Definition 4.8. The shifts T_1, T_2 are said to be *finitarily isomorphic* if they are metrically isomorphic and the maps $\phi_1: M_1 \to M_2$, $\phi_2: M_2 \to M_1$, $\phi_2 = \phi_1^{-1}$, intertwining T_1 and T_2, can be chosen in such a way that there are subsets $A_l \subset M_l$

---

[3] M_l are provided with the direct product topology.

($l = 1; 2$), $\mu_l(A_l) = 1$, such that for any point $x^{(l)} \in A_l$ ($l = 1; 2$) one can find a natural $n = n(x^{(l)})$ for which the inclusion $\bar{x}^{(l)} \in C_n(x^{(l)})$ implies $\pi_0(\phi_l \bar{x}^{(l)}) = \pi_0(\phi_l x^{(l)})$.

In other words, the finitarity of an isomorphism means that for almost every point x of the space $M^{(l)}$ any coordinate of its image under the isomorphism map can be uniquely determined by a finite number of coordinates of x. For example, the map ϕ in the Meshalkin's example (see page 44) defines a finitary isomorphism.

The general Ornstein construction in his proof of isomorphism theorem for Bernoulli automorphisms with the same entropy leads to non-finitary isomorphism. Examples of shift automorphisms which are metrically but not finitarily isomorphic are known. Nevertheless, the following assertion, sharpening the Ornstein isomorphism theorem, is true.

Theorem 4.10 (Keane-Smorodinsky theorem on finitary isomorphism [KeS]). *Any two Bernoulli automorphisms T_1, T_2 with finite state spaces and $h(T_1) = h(T_2)$ are finitarily isomorphic.*

This theorem was also extended to the class of mixing Markov automorphisms (M. Keane, M. Smorodinsky).

§5. Equivalence of Dynamical Systems in the Sense of Kakutani

Since the isomorphism problem of dynamical systems is extremely difficult, various attempts were made to weaken the metric isomorphism condition in order to obtain a more compact picture.

One of the most interesting attempts was based on the notion of equivalence of dynamical systems due to S. Kakutani.

Definition 5.1. Ergodic automorphisms T_1 and T_2 of the Lebesgue spaces $(M_1, \mathcal{M}_1, \mu_1)$, $(M_2, \mathcal{M}_2, \mu_2)$ are said to be *equivalent in the sense of Kakutani* or, simply, *equivalent* if one of the following two conditions is satisfied:

1) there exist subsets $E_1 \in \mathcal{M}_1$, $E_2 \in \mathcal{M}_2$, $\mu_1(E_1) > 0$, $\mu_2(E_2) > 0$ such that the induced automorphisms $(T_1)_{E_1}$, $(T_2)_{E_2}$ are metrically isomorphic.
2) the natural valued functions $f_1 \in L^1(M_1, \mathcal{M}_1, \mu_1)$, $f_2 \in L^1(M_2, \mathcal{M}_2, \mu_2)$ exist such that the integral automorphisms $(T_1)^{f_1}$, $(T_2)^{f_2}$ are metrically isomorphic.

In fact, it may be easily proved that the above conditions 1), 2) are equivalent to each other, and it is for reasons of symmetry only that both of them appear in Definition 5.1. The above relation is transitive, so it is really an equivalence relation.

S. Kakutani introduced this relation in connection with his theorem on special representations of flows. The purpose was to describe the class of all possible special representations for a given flow.

Theorem 5.1 (S. Kakutani, cf [ORW]). *The ergodic automorphisms T_1, T_2 can be considered as base automorphisms in two special representations of the same flow if and only if they are equivalent.*

According to the Abramov theorem (the properties 6, 7 of entropy of an automorphism), the properties of an automorphism to have zero, positive or infinite entropy are invariant under Kakutani equivalence. For a long time only these 3 entropy classes of non-equivalent systems were known.

The examples of non-equivalent systems belonging to the same entropy class were given in the 1970's by J. Feldman who used the methods influenced by the Ornstein isomorphism theory. We shall give a brief exposition of these methods and results.

The distance $\bar{f}$ between the pairs (T_1, ξ_1), (T_2, ξ_2).

It was mentioned in Section 4 that $\bar{d}$-distance between two sequences of partitions $\{\xi_1^k\}_0^{n-1}, \{\xi_2^{(k)}\}_0^{n-1}$, where each $\xi_1^{(k)}, \xi_2^{(k)}$ contains $r < \infty$ elements, may be obtained from the Hamming metric χ in the space $M_n^{(r)}$ by using the Kantorovich-Rubinstein construction. Now we shall define another metric, χ' in $M_n^{(r)}$. For $\mathscr{I}^{(1)}, \mathscr{I}^{(2)} \in M_n^{(r)}, \mathscr{I}^{(1)} = (i_0^{(1)}, \ldots, i_{n-1}^{(1)}), \mathscr{I}^{(2)} = (i_0^{(2)}, \ldots, i_{n-1}^{(2)})$, we set $\chi'(\mathscr{I}^{(1)}, \mathscr{I}^{(2)}) = 1 - s/n$, where s is the maximal integer for which one can find two sequences $k_1 < k_2 < \cdots < k_s, m_1 < m_2 < \cdots < m_s$ with $i_{k_p}^{(1)} = i_{m_p}^{(2)}, 1 \leq p \leq s$. It is clear that

$$\chi'(\mathscr{I}^{(1)}, \mathscr{I}^{(2)}) \leq \chi(\mathscr{I}^{(1)}, \mathscr{I}^{(2)}) \tag{3.2}$$

for any $\mathscr{I}^{(1)}, \mathscr{I}^{(2)} \in M_n^{(r)}$.

The Kantorovich-Rubinstein construction being applied to χ' instead of χ leads to a new metric, $\bar{f}$, measuring the distance between the measures v_1, v_2 on $M_n^{(r)}$, or else, between the corresponding sequences of partitions $\{\xi_1^{(k)}\}_0^{n-1}, \{\xi_2^{(k)}\}_0^{n-1}$:

$$\bar{f}(\{\xi_1^{(k)}\}_0^{n-1}, \{\xi_2^{(k)}\}_0^{n-1}) = \inf_\lambda \int_{M_n^{(r)} \times M_n^{(r)}} \chi'(\mathscr{I}^{(1)}, \mathscr{I}^{(2)}) d\lambda,$$

where inf is taken over all normalized measures λ on $M_n^{(r)} \times M_n^{(r)}$ with $\lambda(A \times M_n^{(r)}) = v_1(A), \lambda(M_n^{(r)} \times B) = v_2(B)$ for all $A, B \subset M_n^{(r)}$. It follows from (3.2) that $\bar{f}(\{\xi_1^{(k)}\}, \{\xi_2^{(k)}\}) \leq \bar{d}(\{\xi_1^{(k)}\}, \{\xi_2^{(k)}\})$.

Definition 5.2. $\bar{f}$-distance between the pairs $(T_1, \xi_1), (T_2, \xi_2)$ is given by the formula

$$\bar{f}((T_1, \xi_1), (T_2, \xi_2)) = \overline{\lim_n} \bar{f}_n((T_1, \xi_1), (T_2, \xi_2)),$$

where

$$\bar{f}_n((T_1, \xi_1), (T_2, \xi_2)) = \bar{f}(\{T_1^k \xi_1\}_0^{n-1}, \{T_2^k \xi_2\}_0^{n-1}).$$

It may be thought that the $\bar{f}$-metric is related to the notion of Kakutani equivalence just in the same way as $\bar{d}$-metric is related to the notion of metric isomorphism. A considerable part of the Ornstein isomorphism theory may be translated from "$\bar{d}$-language" into "$\bar{f}$-language", and this leads to a new "equiva-

Chapter 3. Entropy Theory of Dynamical Systems

lence theory" with many striking results. We begin with the translation of the notion of vwB partition.

LB-partitions and LB-automorphisms

Definition 5.3. A finite partition ξ of a Lebesgue space $(M, \mathcal{M}, \mu)$ is said to be *Loosely Bernoulli (LB)* with respect to an ergodic automorphism T of M, if for any $\varepsilon > 0$ there exists an integer $N = N(\varepsilon) > 0$ such that for any $m \geq 0$ and any $n \geq N$ one can find a set A consisting of the entire elements of the partition $\eta_m = \bigvee_{k=-m}^{0} T^k \xi$ and satisfying
1) $\mu(A) > 1 - \varepsilon$;
2) $\bar{f}(\{T^k \xi\}_0^{n-1}, \{T^k \xi | D\}_0^{n-1}) < \varepsilon$ for every element $D \in \eta_m$ such that $D \subseteq A$.

An automorphism T is said to be Loosely Bernoulli (LB) if it possesses a generating LB-partition.

The inequality connecting $\bar{f}$- and $\bar{d}$-metric yields the fact that the class of LB-automorphisms contains all B-automorphisms of finite entropy. Actually, this class is much wider. Many automorphisms of zero entropy, in particular, ergodic translations on commutative compact groups, ergodic interval exchange transformations (cf Sect. 2, Chap. 4) are LB. The LB-property was introduced by J. Feldman who used it in his construction of new examples of non-Bernoulli K-automorphisms.

Theorem 5.2 (J. Feldman, cf [ORW]). *If T is an LB-automorphism, then*
1) *any induced automorphism of T is LB;*
2) *any integral automorphism of T is LB;*
3) *any factor-automorphism of T is LB.*

It follows from 1) and 2) that the LB-property is stable under Kakutani equivalence.

J. Feldman has constructed the example of an ergodic automorphism T_0 of zero entropy which is not LB, i.e. in particular, is non-equivalent to any group translation. Using this example together with Theorem 5.2, one can construct automorphisms of positive entropy and even K-automorphisms without LB-property.

Consider the Bernoulli automorphism T_1 with two states having probabilities $\frac{1}{2}, \frac{1}{2}$ and acting in the space $(M_1, \mathcal{M}_1, \mu_1)$. Let $\xi = (C_1, C_2)$ be the Bernoulli generating partition for T_1, $\mu_1(C_1) = \mu_1(C_2) = \frac{1}{2}$. For an arbitrary ergodic automorphism T_2 of a space $(M_2, \mathcal{M}_2, \mu_2)$ consider the family of automorphisms $\{T_2(x_1)\}$, $x_1 \in M_1$, of the space M_2: $T_2(x_1) = T_2$ if $x_1 \in C_1$; $T_2(x_1) = Id$ if $x_1 \in C_2$. Let T be the corresponding skew product on $M = M_1 \times M_2$: $T(x_1, x_2) = (T_1 x_1, T_2 x_2)$ if $x_1 \in C_1$; $T(x_1, x_2) = (T_1 x_1, x_2)$ if $x_1 \in C_2$. I. Meilijson (cf [ORW]) proved that all such automorphisms are K-automorphisms. The induced automorphism $T_{C_1 \times M_2}$ is obviously isomorphic to the direct product $(T_1)_{C_1} \times T_2$ and thus T_2 is a factor-automorphism of $T_{C_1 \times M_2}$. In view of Theorem 5.2, the necessary condition for T to be LB is that T_2 be LB. If we take the above mentioned

Feldman automorphism T_0 in the role of T_2, the corresponding skew product T will be a K-automorphism which is not only non-isomorphic to any B-automorphism, but even is non-equivalent to it in the sense of Kakutani.

S. Kalikow [K] showed that the same properties are satisfied by the following, much simpler, example. Let T_1 be, as before, the Bernoulli automorphism of the space $(M_1, \mathcal{M}_1, \mu_1)$ with two states and the probability vector $(\frac{1}{2}, \frac{1}{2})$, and $\xi = (C_1, C_2)$ be its Bernoulli generating partition, $\mu_1(C_1) = \mu_1(C_2) = \frac{1}{2}$. Let T_2 be the same automorphism acting on the space $(M_2, \mathcal{M}_2, \mu_2)$. The automorphism T of the space $M_1 \times M_2$ defined by

$$T(x_1, x_2) = \begin{cases} (T_1 x_1, T_2 x_2) & \text{if } x_1 \in C_1, \\ (T_1 x_1, T_2^{-1} x_2) & \text{if } x_1 \in C_2, \end{cases}$$

is also a K-automorphism and it lacks the LB-property.

FF-partitions. A very important tool in Ornstein's proof of the isomorphism theorem is the notion of a finitely-determined partition. The similar role in the "equivalence theory" is played by its translation into "$\bar{f}$-language".

Definition 5.4. A partition $\xi = (C_1, \ldots, C_r)$ of a Lebesgue space $(M, \mathcal{M}, \mu)$ is said to be *finitely-fixed* (FF) with respect to an ergodic automorphism T of M, if for any $\varepsilon > 0$ there exist $\delta > 0$ and $n \geq 1$ such that any pair $(\bar{T}, \bar{\xi})$, where $\bar{T}$ is an ergodic automorphism of a Lebesgue space $(\bar{M}, \bar{\mathcal{M}}, \bar{\mu})$, $\bar{\xi} = (\bar{C}_1, \ldots, \bar{C}_r)$ is a partition of $\bar{M}$ that satisfies
1) $|h(\bar{T}, \bar{\xi}) - h(T, \xi)| < \delta$,
2) $\sum_{0 \leq i_0, i_1, \ldots, i_{n-1} \leq r} |\bar{\mu}(\bigcap_{k=0}^{n-1} \bar{T}^k \bar{C}_{i_k}) - \mu(\bigcap_{k=0}^{n-1} T^k C_{i_k})| < \delta$,
also satisfies $\bar{f}((T, \xi), (\bar{T}, \bar{\xi})) < \varepsilon$.

An automorphism T is said to be finitely fixed if it possesses a finitely fixed generating partition.

Theorem 5.3. *For any ergodic automorphism T the classes of FF- and LB-partition coincide.*

This assertion together with Theorem 5.2 shows that FF-property is stable under Kakutani equivalence.

Theorems 5.4 and 5.5 below are the main positive results in the theory of equivalence in the sense of Kakutani. They may be viewed as "$\bar{f}$-translations" of Theorems 4.2 and 4.5.

Theorem 5.4 (cf [ORW]). *Suppose that T_1 is an ergodic automorphism of a Lebesgue space, T_2 is a FF-automorphism, and $h(T_2) \leq h(T_1)$. Then there exists an automorphism T_1', Kakutani equivalent to T_1 and having a factor-automorphism which is metrically isomorphic to T_2.*

Theorem 5.5 (cf [ORW]). *If T_1, T_2 are FF-automorphisms and both are either of zero entropy, or of positive entropy, then they are Kakutani equivalent.*

The analogy between the "$\bar{f}$-theory" which was sketched above and the Ornstein "$\bar{d}$-theory" turns out to be so complete that, unfortunately, beyond the class of *LB*-automorphisms (just as in the Ornstein theory—beyond the class of *B*-automorphisms) the results are mostly negative.

There exists an example of an automorphism T which is non-equivalent in the sense of Kakutani to T^{-1} (cf [ORW]). For any h, $0 \leq h \leq \infty$, there exists uncountably many pairwise non-equivalent automorphisms of entropy h (cf [ORW]). The examples of *LB*-automorphisms T whose cartesian squares $T \times T$ are not *LB* have also been constructed (cf [ORW]).

§6. Shifts in the Spaces of Sequences and Gibbs Measures

Suppose M is the space of all sequences $x = \{x_n\}_{-\infty}^{\infty}$, where x_n take values in the finite space $C = \{C_1, \ldots, C_r\}$. Consider the shift T in M: $Tx = x'$, where $x'_n = x_{n+1}$. As was mentioned above, this transformation arises naturally in various problems of ergodic theory. We are now going to describe a wide class of invariant measures for T having strong mixing properties. This class includes Bernoulli and many Markov measures. The thermodynamic formalism (cf [Rue], [Si4]) is used in its construction.

Take an arbitrary function $U = U(x_0; x_1, x_{-1}, x_2, x_{-2}, \ldots)$. It will be called below a potential of interaction of the variable x_0 with all other variables x_n, $n \neq 0$ (the terminology in this section was borrowed from the statistical mechanics). It is natural to require that the main contribution to this interaction should be made by the variables x_n with n sufficiently small. Fix a sequence of non-negative numbers $\alpha = \{\alpha_n\}_{n=1}^{\infty}$, $\alpha_n \to 0$ as $n \to \infty$.

Definition 6.1. A potential U belongs to the class $\mathfrak{U}(\alpha)$ if there exist a number $C = C(U) > 0$ and a sequence of functions $U_n = U_n(x_0; x_1, x_{-1}, \ldots, x_n, x_{-n})$, such that

$$\sup_{x_m : |m| \geq n+1} |U(x_0; x_1, x_{-1}, \ldots, x_n, x_{-n}, x_{n+1}, x_{-n-1}, \ldots)$$
$$- U_n(x_0, x_1, x_{-1}, \ldots, x_n, x_{-n})| \leq C\alpha_n.$$

The latter relation means that the value of the function U alters by no more than $2C\alpha_n$ under the arbitrary variations of all x_m, $|m| \geq n+1$. We shall consider the classes $\mathfrak{U}(\alpha)$ with $\sum \alpha_n < \infty$. Let us explain the meaning of this condition. Take a sequence $x = \{x_n\}$ and a segment $[a, b]$. Set $x' = \{x'_n\}$, where $x'_n = x_n$, $n \notin [a, b]$; $x'_n = C = \text{const}$, $n \in [a, b]$. Introduce the sum

$$H(x_k\}_a^b | \{x_k\}, k \notin [a, b]) = \sum_{k=-\infty}^{\infty} [U(x_k; x_{k+1}, x_{k-1}, \ldots)$$
$$- U(x'_k; x'_{k+1}, x'_{k-1}, \ldots)].$$

If $\sum \alpha_n < \infty$, this sum is finite. We shall see later (cf Definition 6.2) that the exact value of $x'_n, n \in [a,b]$ does not matter for our purpose; one may define x' by fixing another sequence in $[a,b]$. The value $H(\{x_k\}_a^b | \{x_k\}, k \notin [a,b])$ will be called the energy of the configuration $\{x_k\}_a^b$ under the boundary condition $\{x_k\}, k \notin [a,b]$.

Definition 6.2. By the *conditional Gibbs state* in the segment $[a,b]$ with the boundary condition $\{x_k\}, k \notin [a,b]$, corresponding to a potential U, we mean the probability distribution on the space of finite sequences $\{x_k\}, a \leq k \leq b$, such that

$$P(\{x_k\}_a^b | \{x_k\}, k \notin [a,b]) = \frac{\exp(-H(\{x_k\}_a^b | \{x_k\}, k \notin [a,b]))}{\Xi(\{x_k\}, k \notin [a,b])}$$

Here $\Xi(\{x_k\}, k \notin [a,b])$ is a normalizing factor which is called a statistical sum,

$$\Xi(\{x_k\}, k \notin [a,b]) = \sum_{\{x_k\}_a^b} \exp(-H(\{x_k\}_a^b | \{x_k\}, k \notin [a,b])).$$

Now let μ be an arbitrary probability measure on M. For any segment $[a,b]$ consider the measurable partition $\xi_{[a,b]}$ whose elements are obtained by fixing all $x_n, n \notin [a,b]$.

Definition 6.3. A measure μ in the space M is called *a Gibbs measure* with respect to a potential U, if for any segment $[a,b]$ its conditional measure on μ-a.e. $C_{\xi_{[a,b]}}$ is a conditional Gibbs state.

For another way to introduce the Gibbs measures see Part III, Chapter 10.

Theorem 6.1. *If $\sum \alpha_n < \infty$, then at least one Gibbs measure with respect to the potential U exists.*

A Gibbs measure which is not necessarily invariant under T may be constructed rather simply. For any increasing sequence of segments $[a_i, b_i]$, $\bigcup_i [a_i, b_i] = \mathbb{Z}^1$, and any self-consistent sequence of boundary conditions $\{x_k\}$, $k \notin [a_i, b_i]$, one may construct, using Definition 6.2, the sequence of conditional Gibbs states corresponding to the configurations in the segments $[a_i, b_i]$. It can be easily verified that any weak limit point of this sequence is a Gibbs measure. The translations of this measure, i.e. its images under T^n, $n \in \mathbb{Z}^1$, and the arithmetical means of these translations will also be Gibbs. In order to obtain a shift invariant Gibbs measure, we need only take a weak limit point of the sequence of these arithmetical means.

There are some examples of the potentials of classes $\mathfrak{U}(\alpha)$, where α is a slowly decreasing sequence, for which a Gibbs measure is not unique. The problem of construction and analysis of such measures is closely related to the theory of phase transitions in statistical physics, and we shall not discuss it. Instead, we shall be interested in the opposite case when α_n decrease rapidly.

Theorem 6.2 (cf [Rue]). *If $\alpha = \{\alpha_n\}$ satisfies $\sum n\alpha_n < \infty$, then the Gibbs measure μ_0, corresponding to a potential $U \in \mathfrak{U}(\alpha)$, is unique. The dynamical system (M, μ_0, T) is Bernoulli.*

Chapter 3. Entropy Theory of Dynamical Systems

Explain the meaning of the condition $\sum n\alpha_n < \infty$. Fix a large segment $[a, b]$ and consider the energy $H(\{x_k\}_a^b | \{x_k\}, k \notin [a, b])$. Then, if $\sum n\alpha_n < \infty$, the difference $H(\{x_k\}_a^b | \{\bar{x}_k\}, k \notin [a, b]) - H(\{x_k\}_a^b | \{\bar{\bar{x}}_k\}, k \notin [a, b])$, corresponding to boundary values $\{\bar{x}_k\}, \{\bar{\bar{x}}_k\}$, is bounded from above by a constant not depending on $[a, b]$, i.e. the conditional Gibbs states under the distinct boundary conditions are equivalent to each other, and the density is uniformly bounded from above and from below. The uniqueness of the Gibbs measure is an easy consequence of this fact.

The variational principle for Gibbs measures. For an arbitrary invariant under T measure μ, consider the expression

$$P(\mu) = -\int U(x) \, d\mu + h_\mu(T),$$

where $h_\mu(T)$ is the entropy of T with respect to μ. The Gibbs measure μ_0 is uniquely determined by the property

$$P(\mu_0) = \max_\mu P(\mu),$$

where max is taken over all measures invariant under T. There is a similar variational principle in the theory of smooth dynamical systems (cf Ch. 7, Section 3 (part II)).

The convenience of dealing with the Gibbs measures is related to the fact that they are determined by a simple functional parameter—the potential U.

Consider a sequence α with $\sum n\alpha_n < \infty$ and $U_1, U_2 \in \mathfrak{U}(\alpha)$. Suppose that the corresponding Gibbs measures are equal to each other. Then U_1, U_2 satisfy the so-called homological equation

$$U_1(x) = U_2(x) + V(Tx) - V(x),$$

where $V \in \mathfrak{U}(\alpha')$, $\alpha'_n = n\alpha_n$ (cf [99]). For the Gibbs measures the rate of decay of correlations has been studied in sufficient detail. The simplest case is the exponential one: $\alpha_n = \rho^n$, $0 < \rho < 1$.

Theorem 6.3 (cf [Rue]). *Suppose $U \in \mathfrak{U}(\{\rho^n\})$, and μ is the corresponding Gibbs measure. If for some function $f(x)$, $x \in M$, there exist a sequence of functions $f_n(x) = f_n(x_{-n}, \ldots, x_n)$ and positive numbers $C < \infty$, $\rho_1 < 1$ satisfying*

$$\sup_x |f(x) - f_n(x)| \leq C\rho_1^n,$$

then

$$\left| \int f(T^n x) f(x) \, d\mu - \left(\int f \, d\mu \right)^2 \right| \leq K\lambda^n$$

for some positive K and $\lambda < 1$.

In applications, the Gibbs measures arise in a more general framework. Let $\Pi = \|\pi_{ij}\|$ be a square $r \times r$-matrix of 0's and 1's. We shall consider the so-called transitive case when for some $m > 0$ all entries of the matrix Π^m are strictly positive. Introduce the space M_Π consisting of such sequences $x = (\ldots x_{-1}, x_0, x_1, \ldots) \in M$ that $\pi_{x_n, x_{n+1}} = 1$, $-\infty < n < \infty$. In other words, the matrix Π dictates which symbols may occur as neighbours in the sequences $x \in M_\Pi$. Our initial situation may now be considered as a special case, when $\pi_{ij} \equiv 1$.

The shift T on M_Π is called a topological Markov chain. By a measure of maximal entropy for the shift T we mean an invariant measure μ_0 such that $h_{\mu_0}(T) = \max_\mu h_\mu(T)$, where max is taken over all invariant measures μ. In the case considered there is an explicit way to construct such a measure μ_0. Suppose $e = \{e_i\}_1^r$ is the eigenvector for Π with positive coordinates corresponding to a positive eigenvalue λ, i.e. $\Pi e = \lambda e$, and $e^* = \{e_i^*\}_1^r$ is a similar vector for Π^*, i.e. $\Pi^* e^* = \lambda e^*$. The existence of e, e^* follows from the well-known Perron-Frobenius theorem. Since all π_{ij} are integers, we have $\lambda > 1$. The measure μ_0 is the Markov measure with the transition probabilities $p_{ij} = \pi_{ij} e_j / \lambda e_i$ and the stationary distribution $\{e_i e_i^*\}_1^r$, where e, e^* are normalized in order to have $\sum_i e_i e_i^* = 1$.

Theorem 6.4. *Let $\mathcal{O}_n (n = 1, 2, \ldots)$ be the set of all points of period n for T (i.e. $T^n x = x$, $x \in \mathcal{O}_n$). Then*

1) $\lim\limits_{n \to \infty} \dfrac{1}{n} \log \operatorname{card}(\mathcal{O}_n) = \log \lambda$,

2) *for any $f \in C(M_\Pi)$ one has*

$$\lim_{n \to \infty} \frac{1}{\operatorname{card}(\mathcal{O}_n)} \sum_{x \in \mathcal{O}_n} f(x) = \int f \, d\mu_0.$$

In other words, the periodic points for T are uniformly distributed in the space M_Π with respect to μ_0.

A considerable generalization of this theorem will be given in Part II, Chapter 7.

The definitions of conditional Gibbs states and Gibbs measures may be carried over to the case of shifts in M_Π without any changement. Theorems 6.1 and 6.2 are still valid in this situation as well as the variational principle. In particular, the measure of maximal entropy may be defined by means of the variational principle with $U \equiv 0$.

Chapter 4
Periodic Approximations and Their Applications. Ergodic Theorems, Spectral and Entropy Theory for the General Group Actions[1]

I.P. Cornfeld, A.M. Vershik

§1. Approximation Theory of Dynamical Systems by Periodic Ones. Flows on the Two-Dimensional Torus

Dynamical systems whose trajectories all have the same period are usually called periodic. It seems natural to consider these systems, i.e. the ones with the simplest possible behavior of the trajectories, as an appropriate tool for approximating the systems of general form. In this regard their role is similar to that of polynomials and rational functions in the constructive theory of functions, where the functions of general form are approximated by them.

The starting point in the study of the periodic approximations of dynamical systems is the following fact which may be considered as "the existence theorem" for such approximations.

Definition 1.1. An automorphism T is said to be *aperiodic* if the set of its periodic points is of zero measure.

Theorem 1.1 (The Rokhlin-Halmos Lemma, cf P. Halmos [H]). *If T is an aperiodic automorphism of a Lebesgue space $(M, \mathcal{M}, \mu)$, then for any $\varepsilon > 0$ and any natural number n there is a set $E \in \mathcal{M}$ such that*
1) $T^i E \cap T^j E = \emptyset, 0 \leq i \neq j \leq n - 1$;
2) $\mu(\bigcup_{k=0}^{n-1} T^i E) > 1 - \varepsilon$.

An immediate consequence of the Rokhlin-Halmos lemma is the following assertion.

Corollary. *The set of periodic automorphisms is dense in the space of all automorphisms of the Lebesgue space $(M, \mathcal{M}, \mu)$ provided with uniform topology i.e. the one defined by the metric $d(T_1, T_2) = \sup_{E \in \mathcal{M}} \mu(T_1 E \triangle T_2 E)$.*

Various properties of the dynamical systems are connected to the rapidity of their approximation by the periodic ones. To obtain concrete results of this kind, it is necessary to specify the notion of speed of approximation.

Definition 1.2. Suppose $f(n) \searrow 0$. An automorphism T of the space $(M, \mathcal{M}, \mu)$ admits an approximation of the first type by periodic transformations (aptI) with

---

[1] Sections 1 and 2 were written in collaboration with E.A. Sataev.

speed $f(n)$, if one can find a sequence of partitions $\{\xi_n\}$, $\xi_n \to \varepsilon$, and a sequence of automorphisms $\{T_n\}$, such that

1) T_n preserves ξ_n, i.e. it sends each element of ξ_n to an element of the same partition;

2) $\sum_{i=1}^{q_n} \mu(TC_i^{(n)} \triangle T_n C_i^{(n)}) < f(q_n)$, $n = 1, 2, \ldots$, where $\{C_i^{(n)}\}$, $1 \leq i \leq q_n$, is the collection of all elements of ξ_n. If T satisfies 1), 2) and

3) T_n cyclically permutes the elements of ξ_n, then T is said to admit a cyclic approximation with speed $f(n)$.

If for the sequences of partitions $\{\xi_n\}$, $\xi_n = \{C_i^{(n)}\}_{i=1}^{q_n}$, and periodic automorphisms T_n we have

1') T_n preserves ξ_n;

2') $\sum_{i=1}^{q_n} \mu(TC_i^{(n)} \triangle T_n C_i^{(n)}) < f(p_n)$, where $p_n = \min\{p \geq 1: T_n^p = Id\}$;

3') $U_{T_n} \to U_T$ in strong operator topology in $L^2(M, \mathcal{M}, \mu)$, where U_{T_n}, U_T are the unitary operators adjoint to T_n, T respectively,

then T is said to admit an approximation of the second type (aptII) with speed $f(n)$.

A similar definition of various types of approximation can be given in the case of continuous time.

Definition 1.3. Suppose $g(u) \searrow 0$. A flow $\{T^t\}$ on a Lebesgue space $(M, \mathcal{M}, \mu)$ is said to admit an approximation of the first type by periodic transformations (aptI), if one can indicate sequences of real numbers t_n, of partitions ξ_n of the space M into q_n sets $C_i^{(n)} \in \mathcal{M}$, and of automorphisms S_n of the space M such that

1) $\xi_n \to \varepsilon$;

2) S_n preserves ξ_n;

3) $\sum_{i=1}^{q_n} \mu(T^{t_n} C_i^{(n)} \triangle S_n C_i^{(n)}) < g(q_n)$;

4) $p_n t_n \to \infty$, where p_n is the order of S_n as a permutation of the sets $C_i^{(n)}$ i.e. $p_n = \min\{p \geq 1: S_n^p C_i^{(n)} = C_i^{(n)}, 1 \leq i \leq q_n\}$.

If we have, in addition,

5) S_n cyclically permutes the sets $C_i^{(n)}$,

then the flow $\{T^t\}$ is said to admit a cyclic approximation with speed $g(u)$.

If the sequences of real numbers t_n, of partitions ξ_n and of periodic automorphisms S_n satisfy:

1') $\xi_n \to \varepsilon$;

2') S_n preserves ξ_n;

3') $p_n t_n \to \infty$, where p_n is the order of S_n as a permutation of the elements $C_i^{(n)}$ of ξ_n;

4') $\sum_{i=1}^{q_n} \mu(T^{t_n} C_i^{(n)} \triangle S_n C_i^{(n)}) < g(p_n)$;

5') for any element $f \in L^2(M, \mathcal{M}, \mu)$ we have

$$\lim_{n \to \infty} \|U_{T^{t_n}} f - U_{S_n} f\| = 0,$$

then the flow $\{T^t\}$ is said to admit approximation of the second type by periodic transformations (aptII) with speed $g(u)$.

Chapter 4. Periodic Approximations and Their Applications

Example. Suppose $T = T_\alpha$ is a rotation of the circle S^1 by an irrational angle α, i.e. $Tx = (x + \alpha) \pmod 1$, $x \in [0, 1)$, and $\alpha_n = p_n/q_n$ is a sequence of irreducible fractions such that $\lim_{n \to \infty} \alpha_n = \alpha$. Suppose, further, that for some function $f(n)$ satisfying $n \cdot f(n) \searrow 0$, we have

$$|\alpha - p_n/q_n| < f(q_n), \qquad n = 1, 2, \ldots . \tag{4.1}$$

Taking $\xi_n = \{C_i^{(n)}\}_1^{q_n}$, $C_i^{(n)} = \left[\dfrac{i-1}{q_n}, \dfrac{i}{q_n}\right)$, and a sequence of the rotations of the circle $T_n = T_{\alpha_n}$ by the angles α_n, we can prove that T_α admits a cyclic approximation with speed $2n \cdot f(n)$. It is known (from the theory of continuous fractions) that for every α there exists a sequence $\{p_n/q_n\}_1^\infty$ such that (4.1) holds with $f(n) = \dfrac{1}{\sqrt{5} \cdot n^2}$. Any rotation T_α therefore admits a cyclic approximation with speed $f(n) = \dfrac{2}{\sqrt{5} \cdot n}$.

It is natural to suggest that the faster a dynamical system can be approximated by periodic transformations, the worse are its statistical properties.

We will formulate a series of rigorous results confirming this suggestion and concerning the relationship of mixing, spectral and entropy properties with the speed of approximation.

Theorem 1.2 (cf [KS]). *If an automorphism T (a flow $\{T^t\}$) admits an aptII with speed $f(n) = \theta/n$, where $\theta < 2$, then $T(\{T^t\})$ is not mixing.*

Theorem 1.3 (cf [KS]). *If an automorphism T (a flow $\{T^t\}$) admits a cyclic approximation with speed $f(n) = \theta/n$, $\theta < \frac{1}{2}$, then the unitary operator U_T (the group of unitary operators $\{U_{T^t}\}$) has a simple spectrum.*

Theorem 1.4 (cf [KS]). *If an automorphism T (a flow $\{T^t\}$) admits an aptII with speed $f(n) = \theta/n$, $\theta < \frac{1}{2}$, then the maximal spectral type of the unitary operator U_T (of the group of unitary operators $\{U_{T^t}\}$) is singular with respect to the Lebesgue measure.*

Theorem 1.5 (cf [KS]). *If T is an ergodic automorphism of entropy $h(T)$, then $h(T) = \frac{1}{2}c(T)$, where $c(T)$ is the infimum of the set of positive numbers θ, for which T admits aptI with speed $f(n) = \theta/\log n$.*

The situation becomes quite different when we are interested in relations between ergodicity and the speed of (cyclic) approximation: a sufficiently fast cyclic approximation guarantees the ergodicity of an automorphism.

Theorem 1.6 (cf [KS]). *If an automorphism T admits cyclic approximation with speed $f(n) = \theta/n$, $\theta < 4$, then T is ergodic.*

There is an estimate from below for the speed of approximation (of aptI) which is valid for all automorphisms. This estimate may be considered as a sharpened version of Rokhlin-Halmos lemma.

Theorem 1.7 (cf [KS]). *Any automorphism T admits aptI with speed $f(n) = a_n/\log n$, where a_n is an arbitrary monotonic sequence of real numbers tending to infinity.*

As for the estimates from above, no such estimate valid for all automorphisms can be obtained. Moreover, there exist automorphisms such that for any $f(n) \searrow 0$, they admit cyclic approximation with speed $f(n)$. This property is satisfied, for example, by the automorphism T with pure point spectrum consisting of the numbers of the form $\exp 2\pi i p/2^q$ (see Chap. 2, Sect. 2, p. 32).

To a great extent the significance of the approximation method in ergodic theory is due to the fact that it enables us to construct the concrete examples of dynamical systems having various non-trivial metric and spectral properties.

With the help of the approximation theory, the automorphism T such that the maximal spectral type σ of the operator U_T does not dominate its convolutional square $\sigma * \sigma$ (i.e. the automorphism with a spectrum lacking the group property) was constructed. We now explain the meaning of this example. For an ergodic automorphism T with pure point spectrum, to say that the maximal spectral type σ of U_T dominates the type $\sigma * \sigma$ is simply to reformulate the fact that the set of eigenvalues of U_T is a subgroup of S^1. The above example showed that in this problem, like in many others, the situation with continuous and mixed spectra is more complicated. Another application of the approximation method is the example of the automorphism with continuous spectrum which has no square roots (cf [KS]).

The complete investigation of spectral properties of smooth dynamical systems on the 2-dimensional torus was also carried out with the help of the approximation theory.

Suppose $M = \mathbb{R}^2/\mathbb{Z}^2$ is the two-dimensional torus with cyclic coordinates u, v and normalized Lebesgue measure $du\,dv$. Consider the system of differential equations

$$\frac{du}{dt} = A(u,v), \qquad \frac{dv}{dt} = B(u,v) \tag{4.2}$$

on it with right-hand sides of class C^r, $r \geq 2$. Define a flow $\{T^t\}$ on M as the one-parameter group of translations along the solutions of (4.2). We will assume that T preserves a measure μ that is absolutely continuous with respect to $du\,dv$ with density $P(u,v)$ of class C^5 and that $A^2 + B^2 > 0$, i.e. that system (4.2) has no fixed points. The number $\lambda = \lambda_1/\lambda_2$, where $\lambda_1 = \iint_M PA\,du\,dv$, $\lambda_2 = \iint_M PB\,du\,dv$, is called a rotation number of the system (4.2). Spectral properties of the flow $\{T^t\}$ are connected with the speed of approximation of λ by rational numbers.

If λ is rational or if at least one of the numbers λ_1, λ_2 is equal to zero, then the flow $\{T^t\}$ cannot be ergodic. If λ is irrational, the study of the properties of the flow $\{T^t\}$ is based on a special representation of this flow.

Theorem 1.8 (A.N. Kolmogorov [Kol2]). *If λ is irrational, then the flow $\{T^t\}$ is metrically isomorphic to the special flow constructed from the automorphism T_1*

of rotation of the circle S^1 by a certain irrational angle α, where α is of the form

$$\alpha = \frac{m\lambda + n}{p\lambda + q}; \qquad m, n, p, q \in \mathbb{Z}^1, \qquad \det \|{}^m_p {}^n_q\| = 1,$$

and a function $F: S^1 \to \mathbb{R}^1$ of class C^5.

The proof of Theorem 1.8 is based on the fact that there is a smooth closed non-self-intersecting curve Γ on the torus, which is transversal to the trajectories of the flow $\{T^t\}$ at all its points, and such that for any trajectory of $\{T^t\}$ there are infinitely many moments $t > 0$ and infinitely many moments $t < 0$, when it intersects the curve Γ. Such a Γ is known as the Siegel curve for $\{T^t\}$. The transformation of Γ which sends any point $x \in \Gamma$ to a point $T^{t_0}x$, where $t_0 = \min\{t > 0: T^t x \in \Gamma\}$, is conjugate to a certain rotation of S^1, which is the base automorphism of the special representation of $\{T^t\}$. If the number λ and, therefore, all numbers α of the form $\alpha = \frac{m\lambda + n}{p\lambda + q}$, are poorly approximable by rational numbers, then the special flow $\{T^t\}$, appearing in the statement of Theorem 1.8, is metrically isomorphic to the special flow constructed from the same base automorphism and a constant function. With such an argument, we can obtain the following result.

Theorem 1.9 (A.N. Kolmogorov [Kol2]). *If λ satisfies*

$$|\lambda - p/q| \geq \mathrm{const} \cdot q^{-4}, \qquad \mathrm{const} > 0,$$

for all integers p, q, $q \neq 0$, then the group of unitary operators adjoint to the flow $\{T^t\}$ has pure point spectrum consisting of numbers of the form $\mathrm{const}(k + l\lambda)$, $-\infty < k, l < \infty$, where k, l are integers.

A sufficiently fast approximation of the number λ by rational numbers guarantees a sufficiently fast cyclic approximation of the special flow from Theorem 1.8.

Theorem 1.10 (A. Katok [Kat1]). *If for the number λ there is a sequence $\{p_n/q_n\}$ of irreducible fractions such that*

$$q_n^4 |\lambda - p_n/q_n| \to 0$$

when $n \to \infty$, then the flow $\{T^t\}$ admits a cyclic approximation with speed $g(u) = o(u^{-2})$.

Theorems 1.9 and 1.10, together with general theorems about approximations (Theorems 1.2, 1.3, 1.4) show that for any irrational λ the flow $\{T^t\}$ is not mixing, the spectrum of the adjoint group $\{U^t\}$ of unitary operators is simple and the maximal spectral type is singular with respect to the Lebesgue measure.

It was indicated by A.N. Kolmogorov [Kol2] that a smooth flow $\{T^t\}$ given by the equations (4.2) and having no fixed points may be weak mixing, i.e. the spectrum of the adjoint group $\{U^t\}$ may be continuous in the orthogonal complement to the subspace of constant functions. The explicit construction of such examples was developed by M.D. Shklover [Sh]. There is a modification of this construction which enables one to obtain more general examples (D.V. Anosov [An]).

Up to this point it was assumed that the flow $\{T^t\}$ has no fixed points. If we omit this assumption, the situation becomes quite different. Such flows may be mixing [Ko2]. If the right-hand sides of (4.2) are continuous but not necessarily smooth, then the spectrum of the adjoint group $\{U^t\}$ may have both continuous and discrete components (even under the "no fixed point" assumption [Kry].

§2. Flows on the Surfaces of Genus $p \geqslant 1$ and Interval Exchange Transformations

The approximation method turned out to be an appropriate tool in the study of ergodic properties of smooth flows not only on the 2-dimensional torus but also on general orientable surfaces of genus $p \geqslant 1$. For such a flow under some weak conditions, a transversal closed curve (similar to the Siegel curve for the flows on torus) can also be constructed, and this curve, in turn, enables one to construct the special representation of the flow. However, the base automorphisms in this case are of more general form than the rotations of the circle—they are the so-called interval exchange transformations. The study of such transformations is also of intrinsic interest.

Suppose the space M is the semi-interval $[0, 1)$, $\xi = (\Delta_1, \ldots, \Delta_r)$ is a partition of M into r, $2 \leqslant r < \infty$, disjoint semi-intervals numbered from left to right, $\pi = (\pi_1, \ldots, \pi_r)$ is a permutation of the integers $1, 2, \ldots, r$.

Definition 2.1. Suppose the transformation $T: M \to M$ is a translation on each of the semi-intervals Δ_i, $1 \leqslant i \leqslant r$, and "exchange" these semi-intervals according to the permutation π, i.e. the semi-intervals $\Delta'_i = T\Delta_i$ adhere to each other in the order $\Delta'_{\pi_1}, \ldots, \Delta'_{\pi_r}$. Then T is said to be the *interval exchange transformation* corresponding to the partition ξ and the permutation π.

Any interval exchange transformation is an invertible transformation of M with finitely many discontinuities, preserving the Lebesgue measure. If T has at least one periodic point, it cannot be ergodic with respect to the Lebesgue measure, since there exists in this case a non-trivial union of finite number of intervals which is invariant with respect to T.

Theorem 2.1 (M. Keane [Ke1]). *The following properties of T are equivalent:*
i) *T is aperiodic (i.e. T has no periodic points)*
ii) *T is topologically transitive (i.e. T has an everywhere dense trajectory);*
iii) *T is minimal (i.e. all its trajectories are dense in M);*
iv) $\max_{1 \leqslant i_0, i_1, \ldots, i_n \leqslant r} \operatorname{diam}(\Delta_{i_0} \cap T\Delta_{i_1} \cap \ldots \cap T^n \Delta_{i_n}) \to 0$ *when* $n \to \infty$;
v) *the trajectories of the discontinuity points of T are infinite and distinct.*

If $r = 2$ (recall that r stands for the number of the interval exchanged), the interval exchange transformation is isomorphic to some rotation of the circle; for $r = 3$ it is isomorphic to the induced transformation constructed from a rotation automorphism and some interval $\varDelta \subset [0, 1)$. This implies that for $r = 2, 3$ any aperiodic (minimal) interval exchange transformation is ergodic with respect to the Lebesgue measure, and, moreover, is uniquely ergodic. In other words, the Lebesgue measure is the unique invariant normalized Borel measure for it. For any $r \geqslant 4$ this assertion is not true (Keane [Ke2]).

Even the ergodicity with respect to the Lebesgue measure does not guarantee the unique ergodicity of the interval exchange transformation [Ke2]. However, the number of pairwise distinct ergodic normalized invariant measures for an arbitrary interval exchange transformation is always finite. The following estimate holds.

Theorem 2.2 (W. Veech [V1]). *There are at most $[r/2]$ ergodic normalized invariant measures for any exchange transformation of r intervals.*

The cases when the interval exchange transformations have more than one invariant normalized measure (i.e. they are not uniquely ergodic) may be considered in some sense as exceptional ones. In order to explain the exact meaning of this assertion, note that any interval exchange transformation is entirely determined by the pair (λ, π), where $\lambda = (\lambda_1, \ldots, \lambda_r)$, $\lambda_i \geqslant 0$, $\sum_{i=1}^{r} \lambda_i = 1$, is the vector of the lengths of the exchanged intervals, and π is the corresponding permutation.

Denote by $T_{\lambda, \pi}$ the transformation corresponding to the pair (λ, π). Then $T_{\lambda, \pi}$ is obviously non-ergodic if the permutation π is reducible, that is $\pi(\{1, 2, \ldots, j\}) = \{1, 2, \ldots, j\}$ for some j, $1 \leqslant j < r$.

Theorem 2.3 (M. Keane, G. Rauzy, V.A. Chulajevsky [Ch]). *Let an integer $r \geqslant 2$ and an irreducible permutation π of the set $\{1, \ldots, r\}$ be fixed. Then the set of all $\lambda \in \mathbb{R}^r$ for which $T_{\lambda, \pi}$ is non-uniquely ergodic is of first category in the sense of Baire.*

M. Keane conjectured that the measure-theoretic version of Theorem 2.3 is also true, i.e. that "almost all" interval exchange transformations are uniquely ergodic. This conjecture was settled by H. Measur and W. Veech independently, and both proofs use very deep methods.

Theorem 2.4 (H. Masur [Mas], W. Veech [V2]). *If $r \geqslant 2$ and π is an irreducible permutation of the set $\{1, 2, \ldots, r\}$, then the set of all $\lambda \in \mathbb{R}^r$ for which $T_{\lambda, \pi}$ is not uniquely ergodic, is of zero Lebesgue measure.*

At the present time several independent proofs of this theorem are known. A rather "elementary" proof was proposed recently by M. Boshernitzan. In this proof a certain "property $\mathscr{P}$" of an interval exchange transformation was defined in such a way that a) $\mathscr{P}$ is satisfied "almost everywhere", b) minimality of an interval exchange transformation together with property $\mathscr{P}$ imply its unique ergodicity. We shall now give the exact definition of $\mathscr{P}$.

A subset A of positive integers will be called essential if for any integer $l \geq 2$ there is a real number $c > 1$ for which the system of inequalities

$$\begin{cases} n_{i+1} > 2n_i, & 1 \leq i \leq l-1, \\ n_l < cn_1 \end{cases}$$

has infinitely many solutions $(n_1, \ldots, n_l)$, all $n_i \in A$.

Let T be an interval exchange transformation. Denote by $\lambda^{(n)} = (\lambda_1^{(n)}, \ldots, \lambda_{r_n}^{(n)})$ the vector of the lengths of intervals exchanged by T^n, $n = 1, 2, \ldots$. Set $m_n(T) = \min_{1 \leq i \leq r_n} \lambda_i^{(n)}$.

Definition 2.2. *An interval exchange transformation has Property $\mathscr{P}$ if for some $\varepsilon > 0$ the set $A = A(T, \varepsilon) = \{n \in N: m_n(T) \geq \varepsilon/n\}$ is essential.*

Theorem 2.5 (M. Boshernitzan). *Let T be a minimal interval exchange transformation which satisfies Property $\mathscr{P}$. Then T is uniquely ergodic. For any $r \geq 2$ and any permutation π of r symbols, the set of those $\lambda \in \mathbb{R}^r$, for which $T_{\lambda, \pi}$ does not satisfy Property $\mathscr{P}$, is of zero Lebesgue measure.*

Further results on the ergodic and spectral properties of "typical" interval exchange transformations were obtained by W. Veech [V3]. He proved that almost all such transformations (in the same sense as in Theorem 2.4) are totally ergodic (i.e. all powers T^n are ergodic with respect to the Lebesgue measure) and have simple spectrum [V3]. On the other hand, there is the following negative result concerning the strong mixing property.

Theorem 2.6 (A.B. Katok, cf [CFS]). *Suppose T is an interval exchange transformation, μ is an arbitrary invariant Borel measure for T. Then T is not mixing with respect to μ.*

Some properties of the flows on surfaces of genus $p \geq 1$ were established by methods similar to those of the theory of interval exchange transformations or else were deduced from the corresponding results about interval exchanges using the above mentioned special representation of such flows.

Suppose $\{T^t\}$ is a topologically transitive flow of class C^1 on the 2-dimensional compact oriented manifold of genus $p \geq 1$ (cf Vol. 1, Part II) with a finite number of fixed points all of them being non-degenerate saddles. Assume that $\{T^t\}$ has no wandering points (i.e. points x such that for some neighborhood $U \ni x$ and some t_0 one has $U \cap T^t U = \varnothing$ for $|t| \geq t_0$). It was proved in [Kat2] that the number of non-trivial normalized ergodic measures for such flows (i.e. measures such that any trajectory of the flow is of zero measure) does not exceed p. This estimate is exact: for any pair of natural numbers $p, k, p \geq k$, there is a topologically transitive flow of class C^∞ on a surface M_p of genus p having k non-trivial ergodic normalized measures and $2p - 2$ fixed points which are non-degenerate saddles (E.A. Sataev [S]). In [Blo] the examples of uniquely ergodic flows have been constructed on all surfaces except for the sphere, the projective plane and the Klein bottle, where the non-existence of such flows was already known. In

[Ko2] the examples of mixing flows of class C^∞ for which the invariant measure has the density of class C^∞, have been constructed on all surfaces, again except for the three mentioned above.

§3. General Group Actions

3.1. Introduction. Ergodic theory of general group actons (theory of dynamical systems with "general time") deals with arbitrary groups of transformations with invariant and quasi-invariant measures. The classical cases are those of groups $\mathbb{Z}^1$ and $\mathbb{R}^1$. The investigation of other group actions was initially motivated mainly by its applications to the study of classical ones. One of the earliest and most interesting examples of this kind is the geodesic flow on a closed surface of constant negative curvature. It was shown by I.M. Gelfand and S.V. Fomin [GF] that this flow can be represented as an action of a certain 1-parameter (hyperbolic) subgroup of the group $SL(2, \mathbb{R})$ on the homogeneous space $M = SL(2, \mathbb{R})/\Gamma$, where Γ is a discrete group. The information about the unitary representations of $SL(2, \mathbb{R})$ in $L^2(M)$ enabled them to prove easily that the spectrum of a geodesic flow is countable Lebesgue, a fact that was first obtained in the 1940's by E. Hopf and G. Hedlund with the use of rather difficult methods (cf [Mau]). The same idea was employed later in some other problems (the horocycle flow on the surface of constant negative curvature), and it stimulated the systematic study of the flows on homogeneous spaces in connection with the actions of Lie groups on them (cf [AGH]).

Another example is related to the approximation theory. It is useful for the study of approximations to consider the actions of locally finite groups, such as $\Sigma \mathbb{Z}_2$, and of the quasicyclic group. The general theory of actions of such groups (the isomorphy problem, spectral theory, the construction of metric invariants) is no simpler that the corresponding theory for $\mathbb{Z}$ and in many points is parallel to it; on the other hand, there are many questions about approximations which are much simpler in these cases than in the case of $\mathbb{Z}$. Therefore, $\Sigma \mathbb{Z}_2$ and the quasicyclic group may be taken as natural model examples in the approximation theory.

In the 1970's the advantages of systematic study of general group actions became quite evident, and the corresponding theory, closely related to the group representations theory, the theory of Lie groups and differential geometry, was intensely developed. The ergodic methods, in turn, were applied to some problems of Lie groups (i.e. the Mostow-Margulis theory of arithmetic subgroups) and group representatons theory. It should be noted that some metric problems concerning the actions of $\mathbb{Z}^n$, $\mathbb{R}^n$ found their applications to mathematical physics. In recent years, much effort has been devoted to the study of infinite dimensional ("large") groups, such as the groups of diffeomorphisms and of currents. There is a very important difference between the properties of actions of locally-compact and of non-locally compact groups: namely, in non-locally compact case it may occur that even a quasi-invariant measure does not exist and, so, we may have no correct definitions of such notions as the decomposition into ergodic components, the trajectory partition. For locally compact groups these questions may be settled by the same methods as for $\mathbb{Z}^1$ and $\mathbb{R}^1$. We will restrict ourselves to the case of locally compact groups and consider some general questions: the definition of group actions, ergodic theorems, the characterization of discrete spectrum.

3.2. General Definition of the Actions of Locally Compact Groups on Lebesgue Spaces.
Let G be an arbitrary locally compact separable group and $(M, \mathcal{M}, \mu)$ be a Lebesgue space. There are two natural ways to define a measure preserving action of G on $(M, \mathcal{M}, \mu)$. These two definitions obviously coincide in the case of discrete groups, while in the case of continuous groups (including the case $G = \mathbb{R}^1$) they differ considerably.

1) *Definition of a measurable action* (cf Chap. 1, Sect. 1). Consider a map T: $G \times M \to M$, $T(g, x) = T_g x$, satisfying the following conditions: a) measurability, i.e. T is measurable as a map from $(G \times M, m \times \mu)$ into (M, μ), where m is a Haar measure on G (we need not specify it, since it is the measurability itself, and not the numerical value of the measure, that counts); b) invariance: $\mu(T_g^{-1} A) = \mu(A)$, $A \in \mathcal{M}$; c) the group property: $T_{g_1 g_2} x = T_{g_1} T_{g_2} x$. $T_e x = x$. In these equalities x runs through some set of full measure which does not depend on g_1, g_2.

2) *Definition of a continuous action.* Consider a map $g \mapsto \tilde{T}_g$, associating to any $g \in G$ some element of the group of coinciding (mod 0) automorphisms of the space $(M, \mathcal{M}, \mu)$ provided with weak topology.

Assume that the following conditions are satisfied: a) the function $g \mapsto \mu(\tilde{T}_g A \cap B)$ is continuous on G for any $A, B \in \mathcal{M}$; b) $\mu(\tilde{T}_g A) = \mu(A)$; c) $\tilde{T}_{g_1 g_2} = \tilde{T}_{g_1} \tilde{T}_{g_2}$, where $\tilde{T}_g$ is a class of coinciding (mod 0) automorphisms.

It follows from general facts that to define a continuous action is just the same as to define the continuous homomorphism $g \mapsto U_g$ of G into the group of real unitary operators of Hilbert space preserving the structure of "partial multiplication" in L^2: $(U_g f)(x) = f(T_g^{-1} x)$. Hence, a continuous action determines in a standard way a certain weakly continuous representation $g \mapsto U_g$ of the group G. It is known as the Koopman representation.

It is easily seen that any measurable action is continuous. To prove this, consider the operators U_g and use the fact that if for all pairs f_1, f_2 of elements of the Hilbert space H the functions $F(g) = (U_g f_1, f_2)$ are measurable on G, then they are necessarily continuous.

The converse of our statement is also true, but it is more difficult. This problem is known as the group lifting problem. One has to prove that if the relation

$$T_{g_1 g_2} x = T_{g_1} T_{g_2} x$$

holds for almost every x for any pair (g_1, g_2), but the set of such x may depend on g_1, g_2, then there exists a measurable group $g \mapsto T'_g$, for which $T_g = T'_g \pmod 0$ for all $g \in G$.

In the case of countable groups the proof of this statement is rather simple (J. von Neumann, P. Halmos, cf [Ro2]), since we have only a countable number of $g \in G$, and for each g we can alter the definition of T_g on a set of zero measure. In the case of continuous groups more delicate methods for constructing the measurable realization are needed, since there may be continuum sets of zero measure, where the group condition is not satisfied, and their union may coincide with the entire space. For $G = \mathbb{R}^1$ and the pure point spectrum the question was settled positively by V.A. Rohlin in [Ro2]. In the general case of locally compact groups, the positive solutions (in somewhat different terms) were given by G. Mackey [M2] and A.M. Vershik [Ve1]; in the case $G = \mathbb{R}^1$—by C. Maruyama [Ma]. The uniqueness was proved in [Ve1]. A simpler proof was given in [Ve6].

The final statement is as follows:

Theorem 3.1. *Any continuous action of a separable locally compact group G with an invariant measure on the Lebesgue space $(M, \mathcal{M}, \mu)$ possesses a unique (mod 0) measurable realization.*

Therefore, one need not distinguish between continuous and measurable actions and may deal with the kind of action which is more convenient in a given situaton. So, in entropy theory it seems natural to deal with continuous actons, while in trajectory theory (cf Chap. 5) the measurable actions are usually considered. The statement of Theorem 3.1 is also valid for the action with quasi-invariant measures. However, it is not true without the assumption of local compactness of G: the existence of Haar measure is essential for its proof.

3.3. Ergodic Theorems. First investigations of the actions of general groups on measure spaces were concerned, in particular, with the generalizations of ergodic theorems (see, for example [C]). From the probabilistic point of view these theorems may be considered as the great numbers laws for stationary in narrow sense random processes, while from the physical point of view they justify the interchangeability of the "time" means (the integrals over G) and the space means.

Suppose that G is a group acting on $(M, \mathcal{M}, \mu)$ in the sense of definitions given in 3.2. Fix a family $\{G_n\}$, $n = 1, 2, \ldots$, of compact subsets of G. We say that the individual ergodic theorem is satisfied for the action $g \mapsto T_g$ with respect to $\{G_n\}$ if for any $f \in L^1(M)$ the limit

$$\lim_{n \to \infty} \frac{1}{\operatorname{card}(G_n)} \sum_{g \in G_n} f(T_g x) = \bar{f}(x)$$

(in the case of discrete groups), or else

$$\lim_{n \to \infty} \frac{1}{m(G_n)} \int_{G_n} f(T_g x)\, dm(g) = \bar{f}(x),$$

where m is the Haar measure (in the case of continuous groups) exists almost everywhere. The function $\bar{f}$ is the projection of f to the subspace of G-invariant functions; if the action is ergodic, the $\bar{f} = \text{const} = \int_M f d\mu$.

Recall that for $\mathbb{Z}^1$ and $\mathbb{R}^1$ the Birkhoff-Khinchin ergodic theorem claims that:

$$\lim_{n \to \infty} \frac{1}{n} \sum_{k=0}^{n-1} f(T^k x) = \bar{f}(x) \quad \text{almost everywhere,}$$

$$\lim_{T \to \infty} \frac{1}{T} \int_0^T f(T^t x) dt = \bar{f}(x) \quad \text{almost everywhere.}$$

A new and very simple proof of this theorem for the group $\mathbb{Z}^1$ was given recently by Y. Katznelson and B. Weiss [KW2]. Their proof uses parts of the Kamae arguments which he used to obtain a proof of Birkhoff-Khinchin theorem based on of the non-standard analysis.

A sequence $\{G_n\}$, $G_n \subset G$, is called a universal averaging sequence for G if for all measure preserving actions of G the individual ergodic theorem is satisfied with respect to $\{G_n\}$. There are universal sequences for the groups $\mathbb{Z}^m$, $\mathbb{R}^m$: one can take as $\{G_n\}$ the sequence of cubes whose sides increase to infinity. The existence of such sequences is also known for solvable groups, but their construction is more difficult. A general method of proving such theorems (based on the properties of the martingales) was proposed by A.M. Vershik.

Unfortunately, at the present time there is no final solution to this problem for arbitrary locally compact groups.

A natural hypothesis that in the case of the amenable groups, the so-called Følner sequences, i.e. the sequences $\{G_n\}$, $G_n \subset G$, such that for any $h_1, \ldots, h_k \in G$

$$\lim_{n \to \infty} \frac{m\left(\bigcap_{i=1}^k h_i G_n\right)}{m(G_n)} = 1,$$

(m is a Haar measure), should be universal, turned out to be false.

A series of works by A.A. Tempelman (cf. [T1], [T2]) contains, in particular, a proof of the individual ergodic theorem for the groups with polynomial growth of number of words; some sufficient conditions for a Følner sequence to be universal have been given in his theorems. These results seem to be the most general known "universal" (i.e. not depending on a specific action) individual ergodic theorems for the locally compact groups. However, more general methods are needed for arbitrary amenable groups.

There are various generalizations of the individual ergodic theorem dealing with the so-called weighted averages. We will formulate a result of this kind.

Theorem 3.2 (V.I. Oseledets [Os1]). *Suppose $\{T_g\}$ is a measure preserving action of a countable group G on a Lebesgue space $(M, \mathcal{M}, \mu)$, and v is a measure on G. Denote by Γ the support of v and assume that*

   1) $\mathcal{U}_\Gamma = \mathcal{U}_{\Gamma \cdot \Gamma^{-1}}$, *where $\mathcal{U}_\Gamma$ (respectively, $\mathcal{U}_{\Gamma \cdot \Gamma^{-1}}$) is the σ-algebra of the subsets $A \in \mathcal{M}$ invariant under all T_g, $g \in \Gamma$ (respectively, $g \in \Gamma \cdot \Gamma^{-1}$);*

2) v is symmetric, i.e. $v(E) = v(E^{-1})$, $E \subset G$. Then for any $f \in L^1(M, \mathcal{M}, \mu)$ the limit

$$\lim_{n \to \infty} \sum_{g \in G} f(T_g x) v^{(n)}(g)$$

exists, where $v^{(n)}$ is the n-fold convolution of v.

Mention also the multiplicative ergodic theorem [Os2] related to the study of random products of the elements of a group.

A strong generalization of the Birkhoff-Khinchin theorem in the case of a single operator is the so-called ratio ergodic theorem.

Theorem 3.3 (D. Ornstein, R. Chacon [CO]). *Let $(M, \mathcal{M}, \mu)$ be a space with σ-finite measure and T be a positive linear operator in $L^1(M, \mathcal{M}, \mu)$ with $\|T\| \leq 1$. For any pair $f, g \in L^1(M, \mathcal{M}, \mu)$, $g \geq 0$, the limit*

$$\lim_{n \to \infty} \frac{\sum_{k=0}^{n-1} T^k f(x)}{\sum_{k=0}^{n-1} T^k g(x)}$$

exists almost everywhere on the set $\{x \in M : \sup_k T^k g(x) > 0\}$.

Numerous papers were devoted to modifications and generalizations of this statement (cf [Kr]). In particular, its continuous-time analog has been proved (M. Akcoglu, J. Console).

General ergodic theorems for the amenable semigroups have been established in [T1].

The von Neumann statistical ergodic theorem on the strong convergence in $L^2(M)$ of the operators $\frac{1}{n} \sum_{k=0}^{n-1} U^k$, where U is the operator adjoint to an automorphism T, was also generalized by many authors. These generalizations can in a natural way be considered as part of operator theory rather than of ergodic theory. They are usually formulated for the operators in Banach spaces. The most general results are due to A.A. Tempelman [T1]. He proved, in particular, that for all connected simple groups with finite center the statistical ergodic theorem is satisfied for any sequence of averaging sets whose measures tend to infinity [T1].

The problem of finding the exact conditions for the sequence of subsets of a group to be an averaging sequence is very delicate. Some interesting counter-examples exist: if W_2 is the free group with two generators, G_n is the set of words of length $\leq n$, then there exists an action for which neither individual nor statistical ergodic theorem is satisfied with respect to $\{G_n\}$.

3.4. Spectral Theory. The spectral theory of measure preserving group actions is part of representation theory. Suppose that we are given an action $\{T_g\}$ of a group G on $(M, \mathcal{M}, \mu)$ with invariant or quasi-invariant measure. Associate to any

$g \in G$ an operator U_g in $L^2(M)$ by the formula

$$(U_g f)(x) = f(T_g^{-1}x)\sqrt{p_g(x)},$$

where $p_g(x) = \dfrac{d\mu_g(x)}{d\mu(x)}$ is the density of μ_g with respect to μ. Note that p_g is a multiplicative 1-cocycle f the group G taking values in L^1, and it is cohomological to zero if and only if there is a finite invariant measure equivalent to μ; if μ itself is invariant, then $(U_g f)(x) = f(T_g^{-1}x)$. The correspondence $g \mapsto U_g$ is a unitary representation of G.

The principal questions are as follows:

a) to describe the decomposition of the above representation into irreducible ones;

b) to describe the representations which may appear in this situation.

In the case $G = \mathbb{Z}^1$ question a) is equivalent to the calculation of the spectral measure and the multiplicity function, while question b) requires the description of all possible spectra of the dynamical systems. Both problems are very delicate, although important information has been obtained in the case $G = \mathbb{Z}^1$ (cf [CFS]). In the general locally compact case the situation is, of course, much more complicated.

One of the first works devoted to the spectral theory of general actions was the paper of G.W. Mackey [M2], where the exact generalization of the von Neumann theorem on the pure point spectrum has been obtained. We will formulate this result.

A unitary representation of a group G has, by definition, a pure point (or a discrete) spectrum, if it is a direct sum of finite-dimensional irreducible representations. For $G = \mathbb{Z}^1$ this is equivalent to the standard definition of the pure point spectrum (we need not mention the "finite-dimensional" property). The von Neumann theorem says that for any ergodic action of $\mathbb{Z}^1$ the set of eigenvalues (the spectrum) is a countable subgroup of S^1; moreover, the action itself can be reconstructed from its spectrum up to metric isomorphism and can be represented as a translation on a commutative compact group, more precisely, on the group of characters of the spectrum. Hence, any countable subgroup of S^1 is the spectrum of some dynamical system (cf Chap. 2, Sect. 2). The Mackey generalization is as follows.

Theorem 3.4. *Suppose G is a separable locally compact group which acts ergodically on a Lebesgue space $(M, \mathcal{M}, \mu)$ with invariant measure, and the spectrum of G is discrete. Then there exist a compact subgroup K, a homomorphism $\varphi: G \to K$ onto a dense subgroup in K, as well as a closed subgroup $H \subset K$ such that the action of G on $(M, \mathcal{M}, \mu)$ is metrically isomorphic to the action of G on the homogeneous space K/H by the translations by the elements $\varphi(g)$, $g \in G$, the invariant measure being the image of the Haar measure on K.*

This theorem gives a complete solution to the problem of description of the dynamical systems with "time" G having a discrete spectrum. Unlike the commutative case, the representation does not uniquely determine the dynamical system because there is an example of a group (even a finite group) with two of its non-conjugate subgroups H_1, H_2 for which the representations of K in $L^2(K/H_1)$ and in $L^2(K/H_2)$ are equivalent. However, the group K can be uniquely reconstructed from the representation: it is the closure of the group $\{U_g: g \in G\}$ in the group of unitary operators in $L^2(M)$ with weak topology.

Such a complete analogy cannot be extended from the case of discrete spectrum to the general case. It is not clear even, what are the representations that can appear in the spectra of actions. The interesting special case is $G = \text{SL}(2, \mathbb{R})$ and, more generally, the groups having additional series (which are not contained in the regular representation), i.e. the non-amenable groups.

The role of the countable Lebesgue spectrum is played by the so-called countable Plancherel spectrum.

Let us give an example. Suppose G is a discrete group and $M = [0,1]^G$ is the space of all sequences $\{x_g\}$, $g \in G$ such that $x_g \in [0,1]$. Take some measure μ_0 on $[0,1]$ and consider the corresponding product-measure μ on M. Then the action of G on itself by the left translations brings about an action of G on M preserving the measure μ. It is called a Bernoulli action, by analogy with the Bernoulli shifts in the case $G = \mathbb{Z}^1$.

Theorem 3.5. *The spectrum of a Bernoulli action is countable Plancherel, i.e. $L^2([0,1]^G, \mu)$ can be decomposed into a direct sum of subspaces in such a way that on each of them, the left regular representation of G acts.*

The spectral theory of Gauss dynamical systems (cf [CFS]) can be easily generalized to general locally compact groups. For the groups of type II[2] the spectral theory is closely related to the theory of factors [Kir].

§4. Entropy Theory for the Actions of General Groups

The definition of entropy-type invariants for the group actions generalizing in a natural way the notion of entropy of an automorphism, was given in [Kir] (cf also [Con], [KW1], [OW]). Like the situation in general ergodic theorems, a rather developed theory exists for the class of amenable groups. We consider only the groups $\mathbb{Z}^m$, $m \geqslant 2$, for which the most complete results have been obtained. For notational reasons we set $m = 2$ and begin with the definition of entropy in this case.

Let (T_1, T_2) be a pair of commuting automorphisms of a Lebesgue space $(M, \mathcal{M}, \mu)$ with continuous measure. They define the action $\{T_g\}$ of the group

---

[2] The groups of type II are the groups having such representations that the algebra spanned by the operators of the representation is of type II, i.e. in its central decomposition the factors of type II appear on a set of positive measure.

$G = \mathbb{Z}^2$ on M. Denote by T_g the automorphism $T_1^{n_1} T_2^{n_2}$ corresponding to the element $g = (n_1, n_2) \in \mathbb{Z}^2$.

The first step is the definition of entropy of a measurable partition ξ with respect to $\{T_g\}$. It is convenient to take as a starting point property 3) of $h(T, \xi)$ (cf page 38) rather than the definition of $h(T, \xi)$ in the classical case of actions of $\mathbb{Z}^1$ (Definition 2.1, Chap. 3). We retain the notations introduced in Section 2, Chapter 3.

For any $E \subset \mathbb{Z}^2$ denote by ξ_E the partition $\bigvee_{g \in E} T_g \xi$. The set $\Pi \subset \mathbb{Z}^2$ will be called a parallelogram if $\Pi = \tilde{\Pi} \cap \mathbb{Z}^2$, where $\tilde{\Pi}$ is a parallelogram on $\mathbb{R}^2 \supset \mathbb{Z}^2$. Let $m(\Pi)$ be the length of the minimal side of $\tilde{\Pi}$ and $|\Pi|$ be the cardinality of Π.

Theorem 4.1 (cf [Con]). *Suppose that $\xi \in Z$ and $\{\Pi_n\}$ is a sequence of parallelograms on $\mathbb{Z}^2$ such that $m(\Pi_n) \to \infty$. There exists the limit*

$$h(G, \xi) = \lim_{n \to \infty} \frac{1}{|\Pi_n|} H(\xi_{\Pi_n}),$$

not depending on the choice of $\{\Pi_n\}$.

Note that we often use in this section the notations like $h(G, \xi)$, where under G we mean the action $\{T_g\}$ of G rather than the group G itself.

Definition 4.1. The number $h(G, \xi)$ is said to be the *entropy of the partition ξ with respect to the action $\{T_g\}$*.

In the case of general discrete amenable group G any Følner sequence $\{\Phi_n\}$ of subsets of G can be taken in the role of $\{\Pi_n\}$. The corresponding limit also exists and does not depend on the choice of $\{\Phi_n\}$.

The properties of $h(G, \xi)$ (cf J. Conze [Con]):

1) $h(G, \xi) = H(\xi | \xi_{T_1}^- \vee (\xi_{T_2})_{T_1}^-)$. This property shows that the partition $\xi_G^- \stackrel{\text{def}}{=} \xi_{T_1}^- \vee (\xi_{T_2})_{T_1}^-$ may be thought of as "the past" of the partition ξ with respect to G. "The past" depends not only on the action of G, but also on the ordered set (T_1, T_2) of generators of G. All natural possibilities for defining "the past" for the actions of countable amenable groups are described in [P];

2) $h(G, \xi) \leq H(\xi)$;

3) $h(G, \xi_1 \vee \xi_2) \leq h(G, \xi_1) + h(G, \xi_2)$;

4) suppose G_n is a subgroup of index n of the group $G (n \geq 1)$, Γ_n is the fundamental domain for G_n containing the unity of G. Then $h(G_n, \xi_{\Gamma_n}) = n \cdot h(G, \xi)$;

5) for any automorphism T_g in the group G

$$h(G, \xi) \leq h(T_g, \xi);$$

6) $h(G, \xi)$ as a function of ξ is continuous on Z;

7) if $\xi_1 \leq \xi_2$, then $h(G, \xi_1) \leq h(G, \xi_2)$;

8) for any $\xi_1, \xi_2 \in Z$

$$h(G, \xi_1) = \lim_{n \to \infty} H(\xi_1 | (\xi_1)_G^- \vee T_1^{-n}(\xi_2)_G^-);$$

9) if ξ is a generating partition for G, i.e. $\xi_G \stackrel{\text{def}}{=} \xi_{\mathbb{Z}^2} = \varepsilon$, then for any $\eta \in \mathbb{Z}$

$$h(G, \eta) \leq h(G, \xi).$$

Definition 4.2. By the *entropy of the action* $\{T_g\}$ we mean the number

$$h(G) = \sup_{\xi \in Z} h(G, \xi).$$

The entropy $h(G)$ is a metric invariant in the sense that if two actions G_1, G_2 of the same group are metrically isomorphic, then $h(G_1) = h(G_2)$.

The properties of entropy $h(G)$:
1) If $G = G_1 \times G_2$, i.e. the action G of a certain group is the direct product of the actions G_1 and G_2 of the same group, (the formal definition of a direct product is a simple generalization of the one for the actions of $\mathbb{Z}^1$), then

$$h(G) = h(G_1) + h(G_2);$$

2) if G_1 is a factor-action of an action G (the definition is still the same as in the case of $\mathbb{Z}^1$), then $h(G_1) \leq h(G)$;
3) for any subgroup G_n of index n

$$h(G_n) = n \cdot h(G);$$

4) for any automorphism T_g in G we have

$$h(T_g) \geq h(G);$$

5) if T_1, T_2 are the generators of an action of $\mathbb{Z}^2$ and either $h(T_1) < \infty$ or $h(T_2) < \infty$, then $h(G) = 0$.

Examples. 1) Suppose M is the space of all sequences $x = \{x_{n_1, n_2}\}$, $(n_1, n_2) \in \mathbb{Z}^2$, where each x_{n_1, n_2} is an element of a finite set X. The measure λ on X is given by a probability vector $(p_1, \ldots, p_m)$, $p_k \geq 0$, $\sum_{k=1}^m p_k = 1$. Introduce the measure μ in M which is the product-measure of λ, and set $T_1 x = x'$, $T_2 x = x''$, where $x'_{n_1, n_2} = x_{n_1+1, n_2}$; $x''_{n_1, n_2} = x_{n_1, n_2+1}$. It is clear that T_1 and T_2 commute and that they define the action G of $\mathbb{Z}^2$ on M. This action is called a Bernoulli action. For its entropy we have the formula $h(G) = -\sum_{k=1}^m p_k \log p_k$. It can be easily checked that in this case $h(T_1) = h(T_2) = \infty$.

2) Suppose M is the space of all sequences $x = \{x_n\}$, $n \in \mathbb{Z}^1$ where each x_n is a point of a Lebesgue space $(X, \mathscr{X}, \lambda)$. Let S be an automorphism of $(X, \mathscr{X}, \lambda)$. The measure μ on M is, as before, the product-measure of measure λ. Define the automorphisms T_1, T_2 of M by the formulae $T_1(\{x_n\}) = \{Sx_n\}$, $T_2(\{x_n\}) = \{x_{n+1}\}$. The entropy of the action of $\mathbb{Z}^2$ generated by T_1 and T_2 is equal to $h(S)$.

Many fundamental facts of entropy theory for the actions of $\mathbb{Z}^1$ can be carried over to the case of general group actions. More precisely, for a free and ergodic action of a countable amenable group G with finite entropy, there exists a finite generating partition (cf [Šu]). (Note that an action $\{T_g\}$ is free if $\mu(\{x \in M : T_g x = x\}) = 0$ for any $g \in G \setminus \{e\}$). There are generalizations of the Ornstein isomorphism theorem for the Bernoulli actions with the same entropy (A.M. Stepin [St]). A group generalization of the Shannon-McMillan-Breiman theorem exists, as far as we know, only in the case of the action of the so-called quasi-cyclic group (the group of all dyadic rationals of the unit circle) (cf B.S. Pickel, A.M. Stepin [PS]). If we are interested in the L^1-convergence in this theorem rather than in almost everywhere convergence, the corresponding result (an analog of the McMillan theorem) has been proved for any discrete amenable group (cf J.C. Kieffer [Ki]).

In [Av] the notion of entropy of a random walk on a group was introduced. For the detailed study of this notion and its applications to the problem of the boundaries of random walks, see [KV].

The theory of dynamical systems with positive entropy has been extended to the actions of groups $\mathbb{Z}^m$, $m \geq 2$. For notational reasons we set again $m = 2$.

Just as in the case of the actions of $\mathbb{Z}^1$, the Pinsker partition π can be defined: $\pi(G) = \sup\{\xi: \xi \in Z, h(G, \xi) = 0\}$. The actions G with $\pi(G) = \nu$ are called the actions of completely positive entropy. The next definition is very important for entropy theory of the actions of $\mathbb{Z}^2$ and has no analogs in $\mathbb{Z}^1$-theory.

Definition 4.3. Suppose (T_1, T_2) is an ordered pair of commuting automorphisms of the Lebesgue space $(M, \mathcal{M}, \mu)$; ζ is a measurable partition. The partition ζ is said to be (T_1, T_2)-*strong invariant* if 1) $T_1 \zeta \leq \zeta$; 2) $\bigwedge_{n=0}^{\infty} T_1^{-n} \zeta = T_2^{-1} \zeta_{T_1}$.

The strong invariance of ζ implies that $T_1^{n_1} T_2^{n_2} \zeta \leq \zeta$ if $(n_1, n_2) < (0, 0)$, the pairs (n_1, n_2) being ordered lexicographically. The partitions ζ with this last property are called (simply) (T_1, T_2)-invariant. Generally, they need not be strong invariant. However, any action G with $h(G) > 0$ has non-trivial strong invariant partitions. In particular, if $\xi \in Z$ is such that all $T_g \xi, g \in \mathbb{Z}^2$, are independent, then the partition $\zeta = \xi_G^-$ is strong invariant.

In the theory of actions of $\mathbb{Z}^2$ with positive entropy the strong invariant partitions play the role similar to that of invariant partitions in the classical case $G = \mathbb{Z}^1$.

Definition 4.4. A partition $\zeta \in Z$ is said to be (T_1, T_2)-*exhaustive* if ζ is (T_1, T_2)-strong invariant and $\bigvee_{n=0}^{\infty} T_2^n \zeta_{T_1} = \varepsilon$.

Theorem 4.2. (B. Kaminski [Ka2]). *If ζ is (T_1, T_2)-exhaustive, then*

$$\bigwedge_{n=0}^{\infty} T_2^{-n} \zeta_{T_1} \geq \pi(G).$$

This assertion becomes false if ζ is assumed only to be (T_1, T_2)-invariant rather than (T_1, T_2)-strong invariant.

Definition 4.5. A (T_1, T_2)-exhaustive partition ζ is said to be (T_1, T_2)-*extremal* if $\bigwedge_{n=0}^{\infty} T_2^{-n} \zeta_{T_1} = \pi(G)$. A (T_1, T_2)-extremal partition ζ is said to be (T_1, T_2)-*perfect* if $h(G) = h(G, \zeta) = H(\zeta | T_1^{-1} \zeta)$.

Theorem 4.3 (B. Kaminski [Ka2]). *For any ordered pair (T_1, T_2) of generators of the group G there are (T_1, T_2)- perfect partitions.*

Definition 4.6. An action G of the group $\mathbb{Z}^2$ is said to be a K-action (G is said to be a K-group) if for any pair (T_1, T_2) of generators of G and for any $A_0, A_1, \ldots, A_r \in \mathcal{M}$ ($1 \leqslant r < \infty$) we have

$$\lim_{n \to \infty} \sup_{B^{(n)} \in \mathcal{B}^n} |\mu(A_0 \cap B^{(n)}) - \mu(A_0) \cdot \mu(B^{(n)})| = 0,$$

where $\mathcal{B}^{(n)}$ is the σ-algebra corresponding to the partition $T_1^{-n}\xi_{T_1}^- \vee T_2^{-n}(\xi_{T_1})_{T_2}^-$, and ξ is the partition generated by $A_1, \ldots, A_r$.

With the help of the theory of strong invariant partitions the following statement was proved.

Theorem 4.4. (B. Kaminski [Ka2]). *G is a K-group if and only if G is of completely positive entropy.*

Chapter 5
Trajectory Theory

A.M. Vershik

§1. Statements of Main Results

The starting point for trajectory theory of the dynamical systems is a natural question which goes back to H. Poincaré and is concerned with the classical (smooth) dynamical systems. Consider two topological dynamical systems, i.e. 1-parameter groups of continuous transformations of a compact space. We call them topologically (orbitally) equivalent if there is a homeomorphism of the phase space intertwining the orbits (the trajectories) of these systems and preserving the orientation (the order of the points) on the orbits. Such a rough notion of equivalence is useful for the study of phase portraits of the dynamical systems, i.e. the structure of the partitions of phase spaces into separate trajectories. Such properties as the existence (or non-existence) of the periodic trajectories, of invariant submanifolds and so on, turn out to be stable under the above equivalence. This notion is also very useful for the investigation of the so-called rough properties of the dynamical systems (cf [Ar2]).

In ergodic theory it is natural to consider the measure-theoretical orbital equivalence rather than the topological one.

The exact definition will be given first in the case of a single automorphism (for the group $\mathbb{Z}^1$), and then will be extended to the general case.

Let $(M, \mathcal{M}, \mu)$ be a Lebesgue space and T be its automorphism preserving the measure μ. Denote by $\tau(T)$ the partition of M into separate trajectories of T. In other words, the element of $\tau(T)$ containing a point $x \in M$ is a finite or countable set $\{y: y = T^n x\}$, $n \in \mathbb{Z}^1$. The partition $\tau(T)$ is well defined in the sense that for any pair of coinciding (mod 0) automorphisms T_1, T_2 the corresponding partitions $\tau(T_1)$, $\tau(T_2)$ also coincide (mod 0). It should be stressed that $\tau(T)$ measurable in general case, and therefore the factor-space $(M, \mathcal{M}, \mu)/\tau(T)$ is not necessarily Lebesgue; moreover, there might be no nontrivial measurable sets at all in this factor-space. By the same reason, there might be no measurable functions which are constant on the elements of $\tau(T)$. However, all these facts do not signify that $\tau(T)$ should be visualized as a kind of pathological object like non-measurable sets. On the contrary, they only signify that some special methods are needed in order to study the trajectory properties of dynamical systems. For example, there are no conditional measures on the elements of $\tau(T)$ (as it would be if $\tau(T)$ were measurable) but there are the so-called ratio set and the cocycle which play a similar role in many questions (cf. below).

Definition 1.1. Two automorphisms T_1, T_2 of the space $(M, \mathcal{M}, \mu)$ are said to be *orbitally equivalent* if their trajectory partitions $\tau(T_1)$, $\tau(T_2)$ are metrically isomorphic, i.e. if $S\tau(T_1) = \tau(T_2)$ for some automorphism S of $(M, \mathcal{M}, \mu)$.

To make this definition more clear we reformulate it in a somewhat different form. Assume first that the trajectory partitions for T_1 and T_2 coincide. This yields that if for $x, y \in M$ we have $y = T_2^n x$ for some n, then there is a number $m = m(n, x)$ for which $y = T_1^m x$. Thus, there are measurable (in x) functions $m_1 = m_1(n, x)$, $m_2 = m_2(n, x)$, $x \in M$, $n \in \mathbb{Z}^1$, taking value in $\mathbb{Z}^1$ and satisfying $T_1^n x = T_2^{m_1(n, x)} x$, $T_2^n x = T_1^{m_2(n, x)} x$; in particular, $T_1 x = T_2^{m_1(x)} x$, $T_2 x = T_1^{m_1(x)} x$, where $m_1(x) = m_1(1, x)$, $m_2(x) = m_2(1, x)$. In other words, T_1 (respectively, T_2) can be obtained from T_2 (respectively, T_1) by means of some measurable change of time. Therefore, two automorphisms are orbitally isomorphic if each of them is metrically isomorphic to some automorphism which can be obtained from the other one by means of a measurable change of time. It can be easily seen that $m_1(n, x)$ is uniquely determined by $m_1(x)$: $m_1(n, x) = \sum_{k=0}^{n-1} m_1(T_2^k x)$, $n > 0$; $m_1(-1, x) = -m_1(T_2^{-1} x)$. However, in order to obtain the generalizations to other groups, it is more convenient to deal with the function $m_1(\cdot, \cdot) \mathbb{Z}^1 \times M \to \mathbb{Z}^1$ (rather than with $m_1(\cdot)$) which will be called the function of the change of time.

Suppose now that we are given a measurable action of some locally compact group G on a Lebesgue space $(M, \mathcal{M}, \mu)$ by the automorphisms preserving the measure μ.

In order to define correctly the trajectory partition in this case, we use the theorem on the uniqueness (mod 0) of the measurable realization (cf [Ve1]). In view of this theorem, the trajectory partitions for two coinciding (mod 0) automorphisms may differ only on a set of zero measure. Denote by $\tau(T)$ the trajectory partition of the action T of group G. Identifying the action T with the group itself, we may write $\tau(G)$ instead of $\tau(T)$.

Definition 1.2. The groups G_1, G_2 of automorphisms are said to be *orbitally isomorphic* if the partitions $\tau(G_1)$, $\tau(G_2)$ are metrically isomorphic.

Note that we do not assume in this definition that G_1 and G_2 are isomorphic as abstract groups.

Just as in the case of the group $\mathbb{Z}^1$, the equality $\tau(G_1) = \tau(G_2)$ implies the existence of such measurable functions $m_1: G_1 \times M \to G_2$, $m_2: G_2 \times M \to G_1$, that for $g_i \in G_i$, $i = 1, 2$, one has $T_1(g_1)x = T_2(m_1(g_1,x))x$, $T_2(g_2)x = T_1(m_2(g_2,x))x$. If only one of these equalities holds (for example, the first one) then we have $\tau(G_1) \succcurlyeq \tau(G_2)$, i.e. the trajectories of the group G_2 are the unions of some trajectories of the group G_1. The functions m_1, m_2 may be considered as the change-of-time functions. In particular, for the flows ($G_1 = G_2 = \mathbb{R}^1$) we have:

$$S_1^t x = S_2^{m_1(t,x)} x, \qquad S_2^t x = S_1^{m_2(t,x)} x;$$

$m_1, m_2: \mathbb{R}^1 \times M \to \mathbb{R}^1$ are measurable functions.

Note, further, that the above definition signifies (just as in the case $G = \mathbb{Z}^1$) that if two groups of automorphisms are orbitally equivalent, then each of them is metrically isomorphic to a group obtained from the other one by means of a change of time.

The orbital isomorphism of general dynamical systems is much weaker as an equivalent relation (i.e. the corresponding equivalence classes are larger) than topological orbital equivalence considered at the beginning of this section: the requirement of continuity of the intertwining map is replaced by that of its measurability and the invariance of the measure. Furthermore, this map need not preserve the order of points on the trajectories, and, finally, the sets of zero measure are ignored.

The central problem of measure-theoretical trajectory theory is the description of the equivalence classes (under the orbital isomorphism) of the measure preserving automorphisms groups G (the most important cases are, as usual, $G = \mathbb{Z}^1$ and $G = \mathbb{R}^1$). The groups of automorphisms which are not measure-preserving but have only a quasi-invariant measure can also be considered; in this case the interwining map S in the definition of the orbital isomorphism is also assumed to have only a quasi-invariant measure.

Since almost all non-trivial classification problems in the theory of dynamical systems are well known to have no solutions giving a sufficiently compact picture, the reader might be prepared to learn the same thing about the problem considered. This point of view was initially shared by a number of specialists. Actually, however, the situation is quite different, and in the cases of $\mathbb{Z}^1$, $\mathbb{R}^1$ and, more generally, of the amenable groups, there is an unexpectedly simple answer to our question.

However, in the general case (for the arbitrary non-amenable groups), there is no complete solution this problem and it is clear now that a simple classification cannot be obtained.

At the present time, trajectory theory is a vast area of the theory of dynamical systems. It is intimately related to the study of the so-called ergodic equivalence relations and measurable grouppoids generalizing the trajectory partitions. Other topics related to trajectory theory are the foliations on the manifolds, the classificational partitions and so on. The most detailed information has been obtained about the trajectory partitions.

It is difficult to say now who was the first to state the problem concerning the orbital classification for the actions of the group $\mathbb{Z}^1$. In any case, this question is contained implicitly in the paper by J. von Neumann and F. Murrey [NM] (1944) in connection with their study of the rings of operators. Other formulations were given by C.W. Mackey [M3] (in context of the theory of virtual subgroups) and by V.A. Rokhlin in the late 1950's.

H. Dye [D] in 1963 using some ideas of J. von Neumann gave the solution to this problem for $G = \mathbb{Z}^1$ in the case of measure preserving transformations. Another, purely measure-theoretical solution was obtained by A.M. Vershik [Ve2] and R.M. Belinskaya [B] in 1966.

Following J. von Neumann, H. Dye connected the problem considered with the theory of II_1-factors, namely, the problem was reduced to the proof of the hyperfiniteness of the factor corresponding to a given automorphism.

We will formulate now the final result and give the sketch of the proof in the case $G = \mathbb{Z}^1$.

Theorem 1.1. *Any two measure preserving ergodic automorphisms of a Lebesgue space are orbitally isomorphic.*

Corollaries. 1) *For two measure preserving automorphisms of a Lebesgue space to be orbitally isomorphic, it is necessary and sufficient that their measurable partitions into ergodic components be isomorphic.*

This assertion can be proved in a standard way: we fix first the isomorphism of the partitions into ergodic components and then apply the above theorem to each component.

Theorem 1.1 shows that the notion of orbital isomorphism is in some sense degenerate in the case $G = \mathbb{Z}^1$: we have no dynamical invariants of an automorphism (such as its spectrum, entropy and so on) which are stable under orbital isomorphism; the type of the partition into ergodic components (this type may be regarded as a geometric invariant of an automorphisms) is the only invariant of the trajectory isomorphism.

2) *For any ergodic automorphism T there exists an automorphism S metrically isomorphic to T which can be constructed from the automorphism Q with pure point spectrum of the form $\{2\pi i p/2^r\}$, p, q are integers, $q \neq 0$, by means of some change of time:*

$$Sx = Q^{m(x)}x, \qquad S = VTV^{-1}.$$

The automorphism Q has been chosen in this corollary as the simplest one. Therefore, by rearrangements of the points within the trajectories of Q one can obtain, for example, a Bernoulli automorphism (in the role of S). The function

$m(\cdot)$ in this case will certainly be very complicated and its distribution function will decrease very slowly.

All this enables us to study the invariants of the change-of-time function $m(\cdot)$ appearing in the construction of a given automorphism from Q, instead of the automorphism itself.

Now consider the groups of more general form. The following theorem which was obtained quite recently by A. Connes, D. Ornstein J. Feldman and B. Weiss [CFW] is a final result of the longterm investigations by these authors, as well as by A.M. Vershik and W. Krieger, devoted to the generalizations of Theorem 1.1.

Theorem 1.2. *Suppose we are given an ergodic, free and measure preserving action of a countable discrete group G on the space $(M, \mathcal{M}, \mu)$ (the property of being free for an action means that $\mu(\{x: gx = x\}) = 0$ for all $g \in G\setminus\{e\}$). For this action to be orbitally isomorphic to some action of the group $\mathbb{Z}^1$ it is necessary and sufficient that the group be amenable (i.e. G admits an invariant mean).*

We deduce immediately from this fact that any two ergodic and free actions of countable amenable groups are orbitally isomorphic; on the other hand, the orbital class of any such action of a non-amenable group is not similar to that for amenable groups.

This remarkable theorem (together with many other facts) shows that the class of the amenable groups should be viewed as a natural class of groups to which the most general theorems on the actions of $\mathbb{Z}^1$ may be extended.

A similar result concerning the actions of continuous groups is also true.

Theorem 1.3. *Suppose G_1 and G_2 are two non-discrete locally compact amenable (i.e. possessing a topological invariant mean) groups having a countable base of open sets. Any two ergodic and free measure preserving actions of these groups are orbitally isomorphic. In particular, any of such actions is orbitally isomorphic to the action of $\mathbb{R}^1$ by group translations on the two-dimensional torus.*

This statement is a rather easy consequence of the deep Theorem 1.2.

Theorems 1.1–1.3 give a solution to the problem of study of orbital isomorphism of the measure preserving actions of locally compact groups. Beyond this class of groups the situation is much more complicated. We shall formulate at a later stage some results in this direction, and we now proceed to the sketch of the proof of Theorem 1.2. Some new important notions will be needed; our exposition follows the papers [Ve2], [B] (cf also [FM], [Ve3], [Kri3], [Mo], [HO], [FV]).

§2. Sketch of the Proof. Tame Partitions

Definition 2.1. A partition τ of a Lebesgue space $(M, \mathcal{M}, \mu)$ is said to be *tame* if it can be represented as a measure-theoretical intersection of some decreasing sequence of measurable partitions.

Recall that a partition ξ is smaller that η (ξ is not finer than η) if each element of ξ is the union of some elements of η[1]. Thus, the sequence of the partitions $\{\xi_n\}$ is decreasing if the sequence of the elements $C_n(x)$ containing a point $x \in M$ is increasing. The measure-theoretical intersection $\bigcap_n \xi_n$ is the partition with the elements $C(x) = \bigcup_n C_n(x)$, where $C_n(x)$ is the element of ξ containing $x \in M$. The partition $\bigcap_n \xi_n$ is not necessarily measurable.

Recall further that the measurable intersection $\bigwedge_n \xi_n$ is the maximal measurable partition, which is smaller than all ξ_n. Such a partition always exists, and in the case considered, it coincides with the measurable hull of the partition $\bigcap_n \xi_n$. A tame partition is said to be ergodic if its measurable hull is trivial, i.e. is equal to the partition ν having only one nonempty element.

Any measurable partition is obviously tame, but the example below playing an important role in the theory shows that the converse is not true.

Example. Suppose $M = [0, 1]$, $\xi_n^{(0)}$ is the partition of M such that the points $x, y \in M$ belong to the same element of $\xi_n^{(0)}$ iff $x - y$ is of the form $k/2^n$, $k = 0, 1, \ldots, 2^n - 1$. It is clear that $\xi_1^{(0)} > \xi_2^{(0)} > \cdots$. The partition $\tau^{(0)} = \bigcap_n \xi_n^{(0)}$ can be defined by postulating that $x, y \in M$ belong to the same element iff $x - y$ is of the form $k/2^n$, k, n are integers. It may be easily seen that $\tau^{(0)}$ is ergodic. The sequence $\{\xi_n^{(0)}\}$ (as well as any sequence isomorphic to it) is called the standard dyadic sequence of partitions.

The canonical system of conditional measures cannot be correctly defined for the tame partitions which are not measurable; its role is played by the so-called projective system. In order to define it, observe first that the conditional measures satisfy the following transitivity property: if $\xi < \zeta$ and $x, y \in C \in \xi$, $x, y \in D \in \zeta$, then $\mu_\xi^C(x)/\mu_\xi^C(y) = \mu_\zeta^D(x)/\mu_\zeta^D(y)$ almost everywhere; here $\mu_\eta^E(x)$ is the conditional measure of the point x in the element E of a partition η. Thus, for $\tau = \bigcap_n \xi_n$ the number $\mu^C(x)/\mu^C(y) \equiv \mu_{\xi_n}^{C_n}(x)/\mu_{\xi_n}^{C_n}(y)$ (n is sufficiently large) is well defined. Moreover, this number does not depend on the specific representation of the partition τ as the intersection of a decreasing sequence of measurable partitions (due to the same transitivity property of the conditional measures). Denote $\beta(x, y) = \mu^C(x)/\mu^C(y)$ for $x, y \in C \in \tau = \bigcap_n \xi_n$. The function β satisfies the following properties:

1) β is defined on the pairs of points belonging to the same elements of the tame partition; $\beta(x, y) \in \mathbb{R}_+$;
2) $\beta(x, x) = 1$, $\beta(x, y) = \beta(y, x)^{-1}$;
3) $\beta(x, y) \cdot \beta(y, z) = \beta(x, z)$.

Any function satisfying 1)–3) will be called a $\mathbb{R}_+$-valued 1-cocycle. The above mentioned cocycle is known as the Radon-Nicodym cocycle of the tame partition. If the sum $\sum_{x \in C(y)} \beta(x, y) = [\mu^C(y)]^{-1}$ were finite, then $\mu^C(\cdot)$ might be considered as a conditional measure. Hence, the fact of its finiteness is equivalent to the measurability of τ. If τ is not measurable, the above sum is infinite, and

---

[1] Another ordering of the partitions is usually employed in combinatorics.

the conditional measure cannot be defined. However, the cocycle β can substitute this measure in various questions. It turns out that the full information concerning the metric type of a tame partition is contained in the Radon-Nicodym cocycle similarly to the fact that the system of conditional measures contains the full information about the metric type of a measurable partition. If $\beta \equiv 1$, the corresponding partition is called homogeneous.

We shall now formulate the results on the relationship between measurable and tame partitions.

Proposition 1. *The trajectory partition for an arbitrary automorphism T of a Lebesgue space with quasi-invariant measure is tame. This tame partition is homogeneous if and only if T is measure-preserving. Ergodicity of T is equivalent to ergodicity of its trajectory partition.*

Theorem 2.1. *There is only one (up to isomorphism) ergodic homogeneous tame partition.*

The last two facts imply Theorem 1.1. There are many ways to prove proposition 1; the most clear one is based on the iterated applications of the periodic approximations (analog of the Rokhlin-Halmos Lemma).

Since the trajectory partitions for periodic automorphisms (or for the periodic parts of non-aperoidic automorphisms) are measurable and, thus, are tame, T may be assumed to be aperiodic. Let μ be a quasi-invariant measure for T. Choose a set A_1, $\mu(A_1) < 1/2$, such that $\mu(\bigcup_k T^k A_1) = 1$, and denote by ξ_1 the partition with the elements of the form $\{x, Tx, \ldots, T^{k(x)}x\}$, where $x \in A_1$, $T^{k(x)+1}x \in A_1$, $T^i x \notin A_1$, $1 \leq i \leq k(x)$. We may assume that $k(x)$ is finite almost everywhere; if not, the set A_1 should be slightly increased to meet this condition.

Now consider the induced automorphism T_{A_1} on A_1, choose a set $A_2 \subset A_1$, $\mu(A_2) < \frac{1}{4}$, and repeat the above procedure with A_2 in the role of A_1 and T_{A_1} in the role of T.

Let ξ_2' be the corresponding partition of A_1 (the analog of ξ_1). The projection $\pi: M \to A_1$ (along ξ_1) enables us to lift this partition to M. Set $\xi_2 = \pi^{-1}\xi_2'$. It is clear that $\xi_1 \succcurlyeq \xi_2$. Continuing this procedure, we get the decreasing sequence $\xi_1 \succcurlyeq \xi_2 \succcurlyeq \cdots$, $\bigcap_n \xi_n = \tau(T)$. This proves that $\tau(T)$ is tame. It may be checked that $\tau(T)$ is homogeneous whenever T is measure preserving.

The proof of Theorem 2.1 is more difficult. Consider first the representation of an arbitrary tame partition τ with countable elements in the form $\tau = \bigcap_n \xi_n$ and modify this representation: $\tau = \bigcap_n \tilde{\xi}_n$, so that almost each element of $\tilde{\xi}_n$ contains 2^n points. The sequence $\tilde{\xi}_1 \succ \tilde{\xi}_2 \succ \cdots$ is dyadic. Theorem 2.1 will then follow from the statement below on dyadic sequences.

Theorem 2.2 ([Ve2]). *Let $\{\xi_n\}_1^\infty$ be a homogeneous ergodic (i.e. $\bigwedge_n \xi_n = \nu$) dyadic sequence of measurable partitions. There is a sequence $\{n_k\}$, $n_k \to \infty$, of natural numbers such that the subsequence $\{\xi_{n_k}\}_{k=1}^\infty$ is metrically isomorphic to $\{\xi_{n_k}^{(0)}\}_{k=1}^\infty$, where $\{\xi_n^{(0)}\}$ is the standard dyadic sequence (cf above).*

Corollaries. 1) (on lacunary isomorphism)

Any two homogeneous ergodic dyadic sequences $\{\xi_n\}$, $\{\xi'_n\}$ are lacunarily isomorphic, i.e. there exists a sequence $\{n_k\}$, $n_k \to \infty$, for which $\{\xi_{n_k}\}$ and $\{\xi'_{n_k}\}$ are metrically isomorphic.

2) *The partitions $\bigcap_n \xi_n$ and $\bigcap_n \xi'_n$ (we retain the notations used in corollary 1) are metrically isomorphic (since $\bigcap_n \xi_n = \bigcap_k \xi_{n_k}$ if $n_k \to \infty$).*

The statement of Theorem 2.1 follows from Corollary 2 and Proposition 1. The proof of Theorem 2.2 is more technical. Other proofs of Theorem 2.1 were given by H. Dye [D], S. Kakutani et al [H1K]. Note that the study of problems of the trajectory theory with the use of decreasing sequences of partitions initiated in [Ve2] turned out to be useful and productive in its own right; some new invariants of automorphisms have been found in this way.

Proposition 1 shows that trajectory theory for the automorphisms with quasi-invariant measure may be reduced to the classification problem of the tame partitions with countable elements. The following results have been obtained in this direction: a class of partitions for which the solution of the classification problem is as simple as in the homogeneous case have been determined; on the other hand, the classification in general case turned out to be equivalent to the metric classification of flows and, thus, it cannot be obtained in reasonable terms.

We proceed now to a more detailed exposition of the results (essentially due to W. Krieger) concerning the above mentioned reduction. These results are important on their own right and the methods employed are very elegant. We start off with an example.

Consider a space $M = \prod_{n=1}^{\infty} X$, $\text{card}(X) = k$, and a probability vector $p = (p_1, \ldots, p_k)$. Let μ be the Bernoulli measure on M which is the product-measure of the measure on X defined by p. Define the so-called tail partition κ of M. Its elements are the equivalence classes under the following relation: $x = \{x_n\}$ and $y = \{y_n\}$ are equivalent if $x_n = y_n$ for n sufficiently large. It may be shown that κ is tame and ergodic with respect to μ. We will call κ the Bernoulli partition. If $p_i = \dfrac{1}{k}$, then κ is homogeneous.

In the statements below, under "isomorphism" we mean an isomorphism with quasi-invariant measure (i.e. sending a measure to an equivalent one).

Theorem 2.3. 1) *An ergodic tame partition has a homogeneous ergodic sub-partition if and only if it is Bernoulli.*

2) *Any Bernoulli partition is uniquely determined (up to isomorphism) by the subgroup F of the multiplicative group $\mathbb{R}_+$, which is the closure of the group generated by the numbers p_i/p_j, $1 \leq i, j \leq k$.*

If $p_i = \dfrac{1}{k}$, then $F = \{1\}$, so we get the statement of Theorem 2.1.

The second part of Theorem 2.3 was obtained independently by W. Krieger and A.M. Vershik.

Since any closed subgroup of $\mathbb{R}_+$ is of one of the forms: $\{1\}$, $\mathbb{R}_+$ and $\{\lambda^n\}_{n \in \mathbb{Z}}$, $0 < \lambda < 1$, we have an explicit classification of tame partitions possessing homogeneous subpartitions. The above mentioned subgroups of $\mathbb{R}_+$ are sometimes called the ratio sets for trajectory partitions.

Note that a statement similar to Theorem 2.1 is true for Lebesgue spaces with σ-finite measure (i.e. the spaces which can be represented as countable unions of spaces with finite measure). In this case there also exists only one (up to isomorphism) ergodic homogeneous tame partition.

Now consider the general case. We begin with the definition of the Poincaré flow introduced by G. Mackey and applied to the problem considered by W. Krieger. Set $\Delta(x, y) = \log \beta(x, y)$, where $\beta(x, y)$ is the Radon-Nicodym cocycle. Δ is sometimes called a modular function. If S is an ergodic automorphism of a measure space (M, μ) with quasi-invariant measure, and $y = Sx$, then $\Delta(x, y) = \log \dfrac{d\mu(Sx)}{d\mu(x)}$. Let $Y' = M \times \mathbb{R}^1$ and $\mu' = \mu \times m$, where m is the Lebesgue measure on $\mathbb{R}^1$. Define a flow $\{T'_t\}$ on Y' by $T'_t(x, a) = (x, a + t)$, and an automorphism $\tilde{S}$ by $\tilde{S}(x, a) = (Sx, a + \Delta(x, Sx))$. It is clear that T'_t and $\tilde{S}$ commute. Let $Y = Y'/\tau(\tilde{S})$ be the factor-space of Y' with respect to the action $\tilde{S}$, i.e. with respect to the measurable hull of the partition into separate trajectories of $\tilde{S}$. Then Y, as a measure space with the factor-measure $\mu'/\tau(\tilde{S})$, is a Lebesgue space, (maybe with σ-finite measure) and the flow $\{T^t\} = \{T'_t/\tau(\tilde{S})\}$ acts on it. This flow is know as the Poincaré flow, or the flow associated with the trajectory partition of the automorphism S.

We will now state the main property of the above construction.

Theorem 2.4. *Two ergodic automorphisms S_1, S_2 with quasi-invariant measures are orbitally isomorphic (in the "quasi-invariant" sense) if and only if their Poincaré flows are metrically isomorphic as flows with quasi-invariant measures.*

This theorem reduces the classification problem of tame partitions (or else of trajectory partitions for the automorphisms with quasi-invariant measures) to the problem of metric isomorphism of flows with quasi-invariant measure. Unfortunately, the last problem has no complete solution in reasonable terms, but the reduction that we have described, enables one to obtain some nontrivial invariants of the trajectory partitions with the aid of the invariants of flows.

In the case of homogeneous tame partition ($\beta \equiv 1$, $\Delta \equiv 0$) the corresponding Poincaré flow is of the form $T^t a = a + t$, $a \in \mathbb{R}^1$; the Bernoulli partition with the group $\{\lambda^n\}$ corresponds to the periodic flow of period $(-\log \lambda)$, the Bernoulli partition with the group $\mathbb{R}^1$—to the trivial flow on the single-point space. Other tame partitions are associated with aperiodic flows (cf [Kri3], [Mo], [HO]).

These results may be considered as the answers to the main questions concerning the general problems of trajectory theory for the group $\mathbb{Z}^1$. It is interesting to note that a number of constructions and even statements appeared first in the theory of operator algebras which is closely related to trajectory theory. Although our exposition is purely measure-theoretic, the operator analogs of such notions as main cocycle, modular function etc, exist, and some of them actually appeared in ergodic theory as measure-theoretical translations from the algebraic language. The paper of H. Dye was already mentioned in this connection; there are also the papers of H. Araki and E. Woods [AW], A. Connes [Co1]. These works seem to be the first examples of applications of operator algebras to ergodic theory; earlier, conversely, the measure-theoretical constructions were being used in the theory of factors to solve some algebraic problems.

§3. Trajectory Theory for Amenable Groups

We shall now consider the groups of more general form, and begin with the continuous case.

Two partitions, τ_1, τ_2, of a Lebesgue space $(M, \mathcal{M}, \mu)$ are said to be *stably equivalent* if the partitions $(\tau_1 \times \varepsilon)$, $(\tau_2 \times \varepsilon)$ of the space $(M \times [0,1], \mu \times m)$, where m is the Lebesgue measure, ε is the partition of $[0,1]$ into separate points, are isomorphic.

Theorem 3.1 (A. Ramsey [R]). *The trajectory partition of an arbitrary locally compact non-discrete group G with a countable base, acting on a Lebesgue space with quasi-invariant measure, is stably equivalent to the trajectory partition of some discrete group.*

The simplest case is $G = \mathbb{R}^1$. The Ambrose-Kakutani theorem on special representation of flows yields in this case that the trajectory partition of a flow is stably equivalent to that of the base automorphism.

On the other hand, it may be shown that the stable equivalence of two tame partitions with countable elements is equivalent to their metric isomorphism (with quasi-invariant measure). Hence, the problem for flows and, more generally, for locally compact groups is reduced to the case of discrete groups. In particular, the above stated (Section 1) Theorem 1.3 follows from Theorems 1.2 and 3.1.

We will give a separate formulation for flows, i.e. one-parameter groups of measure preserving automorphisms.

Theorem 3.2 *Any two ergodic flows with invariant measures (either finite or σ-finite) are orbitally equivalent. Therefore, any ergodic flow with a finite invariant measure is metrically isomorphic to a flow which can be constructed from the conditionally periodic flow on the two-dimensional torus by means of some measurable change of time.*

Now we turn to the following question: what is the class of discrete groups, for which the trajectory partitions of the actions with invariant measure are tame? It is clear that the actions themselves in this case can be obtained from the actions of $\mathbb{Z}^1$ by means of changes of time. It was proved by H. Dye [D] that this class includes the countable groups for which the number of words of length $\leqslant n$ in any finitely generated subgroup grows polynomially in n. It was suggested at the same time that this class includes all amenable groups. The necessity of the amenability condition follows easily from Følner's criterion (cf [D]), but the converse is much more difficult. The converse statement was first proved in some special cases: for the solvable groups—by A. Connes and W. Krieger, for the completely hyperfinite groups—by A.M. Vershik.

A complete solution based on some generalization of Følner's criterion, was obtained recently by J. Feldman, A. Connes, D. Ornstein and B. Weiss (Theorem 1.2). They consider the problem as a special case of a more general question about the hyperfiniteness of a grouppoid. The notions of tame (= hyperfinite) and amenable grouppoids were proved to be equivalent. We shall not discuss here the details of the proof but note the following fact. The theorem shows that the amenable groups are just the class of groups having the property that the fundamental facts of ergodic theory for $\mathbb{Z}^1$ such as Rokhlin-Halmos lemma, Birkhoff-Khinchin theorem, the existence of entropy as a metric invariant, can be generalized to their free actions. It turned out that these metric facts depend only on the algebraic nature of the group but not on its specific action (this action is assumed only to be free and measure-preserving). In particular, all ergodic actions of such groups are orbitally equivalent to each other and orbitally isomorphic to the actions of $\mathbb{Z}^1$. Beyond the class of the amenable groups[2] the situation is much more complicated (cf Section 4).

To end the consideration of tame partitions we will formulate some other results. It is natural to ask whether or not the class of tame partitions coincides with that of the trajectory partitions of discrete and continuous locally compact groups. The answer is no.

Theorem 3.3 (A.M. Vershik [Ve4], V. G. Vinokurov, N. Ganikhodgaev [VG]). *There exists a unique, up to isomorphism, ergodic tame partition which is not stably isomorphic to any tame partition with countable elements.*

We shall call it the special tame partition. It can be constructed in a following way. Let $\{\xi_n\}_1^\infty$ be a decreasing sequence of measurable partitions such that almost all elements of the partitions ξ_n/ξ_{n-1} have no atoms. Then $\bigcap_n \xi_n$ is the required partition. For example, the tail partition of $M = \prod_1^\infty [0, 1]$ with the product measure m^∞ is special. Recall that the elements of the tail partition are the classes of sequences which differ from each other by a finite number of coordinates only.

---

[2] Note that all finite groups, all commutative groups are amenable, that amenability is stable under the transition to subgroups, factor-groups and extensions (hence, the solvable groups are amenable). Free groups with two or more generators as well as the lattices of semisimple groups are non-amenable. The examples of non-amenable group which do not contain the free group with 2 generators are known [O]. There also exists examples of amenable groups of subexponential but not polynomial growth of the number of words, which cannot be obtained by the standard constructions from the commutative and finite groups (R.I. Grigorchuk [G]).

Here is an equivalent realization of this partition. Let $T^\infty = \prod_1^\infty S^1$ be the infinite-dimensional torus and $\sum_1^\infty S^1$ be the direct sum of the groups S^1 which acts on T^∞ in a natural way. The trajectory partition of this action is also the special tame partition. It can be represented as a trajectory partition for an inductive limit of compact groups but not for a locally compact group. At the present time there is no general trajectory therory of actions of non-locally-compact groups. It is unclear even what the fundamental concepts in this case are.

Our last remark is about the semigroups of endomorphisms of a Lebesgue space. Suppose G is a countable semigroup of endomorphisms of a Lebesgue space with quasi-invariant measure. Define the trajectory partition $\tau(G)$ by assuming that the points x, y are in the same element if there are $g_1, g_2 \in G$ such that $g_1 x = g_2 y$. In the case $G = \mathbb{Z}_+^1$ this definition is due to V.A. Rokhlin; if G is a group, this definition is equivalent to the standard one, since $y = g_2^{-1} g_1 x$. There is another useful partition which might be treated as tail partition: two points x, y are in the same element if there exists $g \in G$ for which $gx = gy$. There are no analogs of this notion for groups. The study of the trajectory partitions of semigroups is now in its very beginning; the general results have been obtained only for $\mathbb{Z}_+^1$, i.e. for the case of a single endomorphism.

Theorem 3.4 (R. Bowen [Bo2], A.M. Vershik [Ve5]). *The trajectory partition for any endomorphism is tame.*

§4. Trajectory Theory for Non-Amenable Groups. Rigidity

Except for the important works of R. Zimmer generalizing G.A. Margulis' results on the rigidity of arithmetic subgroups to the actions of semisimple groups, and a series of interesting examples, little is known about the trajectory partitions of non-amenable groups.

It is clear, however, that the situation differs considerably from the amenable case (cf Sects. 1–3). We shall consider mostly the measure-preserving actions. If the measure is assumed only to be quasi-invariant, then for any countable group (not necessarily amenable) one can construct actions for which the trajectory partitions are tame. (Recall that in measure-preserving case this partition is tame only if the group is amenable). To construct such examples, it is enough to observe that the action of any countable group G on itself by (left) group translations is tame. The phase space (G, m) (m stands for the Haar measure) in this example is discrete. In order to construct a non-discrete example, consider the space $M = G \times [0, 1]$ and an arbitrary free action of G on $[0, 1]$. The direct product of these two actions with a finite quasi-invariant measure equivalent to $m \times \rho$ (ρ is the Lebesgue measure on $[0, 1]$) is the required action. What is essential is the fact that the actions of non-amenable groups having tame trajectory partitions often arise in natural situations. Here is a non-trivial example.

Example 1 (R. Bowen [Bo2], A.M. Vershik [Ve5]). The action of $\mathrm{PSL}(2,\mathbb{Z})$ on $P_1\mathbb{R}$ by fractional-linear transformations.

Consider the action of the group of fractional-linear transformations with integer coefficients on $\mathbb{R}^1$: $x \to \dfrac{ax+b}{cx+d}$, $a, b, c, d \in \mathbb{Z}^1$, $ad - bc = 1$. Its trajectory partition is the partition into the equivalence classes in the sense of continuous fraction expansions (cf [Ca]). Although this action is free with respect to the Lebesgue measure and the group $\mathrm{PSL}(2,\mathbb{R})$ is non-amenable, the trajectory partition is tame (also with respect to the Lebesgue measure). This means, in particular, that there is a measurable transformation T such that any two points x, y are equivalent if and only if $T^n x = y$ for some $n \in \mathbb{Z}$.

This example is closely related to Theorem 3.4, since the above partition and the trajectory partition of the endomorphism are essentially the same. This fact was recently obtained by A. Connes, J. Feldman and B. Weiss (with the use of some Zimmer's results) as a corollary from the following general statement: if G is a locally compact group, P is its closed amenable subgroup and Γ is a discrete subgroup, then the partition $\tau(\Gamma)$ into the trajectories of the action of Γ on G/P is tame. In the above example we have $G = \mathrm{SL}(2,\mathbb{R})$, $P = \begin{pmatrix} * & * \\ 0 & * \end{pmatrix}$, $\Gamma = \mathrm{SL}(2,\mathbb{Z})$.

Example 2. A tame action of the free group (the boundary). Let W_k be the free group with k generators and $\widetilde{W}_k$ be the set of all infinite (in one direction) irreducible words, i.e. sequences $\{x_i\}_0^\infty$, where $x_i \in \{g_1, \ldots, g_k, g_1^{-1}, \ldots, g_k^{-1}\}$ and $x_i x_{i+1} \ne e$.

The group W_k acts on $\widetilde{W}_k$ (by concatenation with the subsequent elimination of pairs of the form xx^{-1}). Introduce the product-measure μ on $\widetilde{W}_k$ with uniform factors (x_i takes $2k$ values for $i = 0$ and $(2k-1)$ values for $i > 0$). The partition $\tau(W_k)$ turns out to be a tame partition of $(\widetilde{W}_k, \mu)$. It is also related to the trajectory partition of the shift endomorphism. Note that $\widetilde{W}_k$ is the boundary of the random walk on W_k with a finite measure.

Unlike the amenable case, a non-amenable group may have many non-orbitally isomorphic measure-preserving actions.

Example 3. Suppose W_2 is the free group with two generators. Consider the following two actions of W_2: 1) by the rotations of the 2-dimensional sphere S^2 with the Lebesgue measure ($W_2 \subset \mathrm{SO}(3)$); 2) the Bernoulli action, i.e. action by left shifts in the space $\mathbb{Z}_2(W_2) = \mathbb{Z}_2^{W_2}$ of all functions on W_2 taking their values in $\mathbb{Z}_2$ and provided with the product-measure with factors $(1/2, 1/2)$. Both actions are free (mod 0), but they are not orbitally isomorphic. The last statement is a corollary from the following theorem.

Theorem 4.1. *Consider a measure-preserving action of a countable group G on the space $(M, \mathcal{M}, \mu)$. Let $g \mapsto U_g$ be the unitary representation of G given by $(U_g f)(x) = f(g^{-1} x)$. The following property of the representation $\{U_g\}$ is stable* under the orbital equivalence: for any $\varepsilon > 0$ and any n-tuple $g_1, \ldots, g_n \in G$ there exists $f \in L^2(M)$, $f \perp 1$, such that $\max_i \| U_{g_i} f - f \| < \varepsilon$ (the almost invariant vector).

The non-equivalence of the two actions from Example 3 follows from the fact that the first of them satisfies this last property, while the second one does not satisfy.

The proof is based on the decomposition of the corresponding unitary representations into irreducible ones. In view of Theorem 1.2, both trajectory partitions are not tame.

There is yet another and more geometrical way to describe the above property: it is equivalent to the existence of a non-trivial almost invariant set, i.e. for any $\varepsilon > 0$ and any $g_1, \ldots, g_n \in G$ there must be a set $B \in \mathcal{M}$ for which $|\mu(B) - \frac{1}{2}| < \varepsilon$, $\max_i \mu(g_i B \triangle B) < \varepsilon$.

Some other trajectory properties, for example, the existence of almost invariant finite partitions and so on, may be formulated in similar terms. V. Ya. Golodets and S.I. Bezuglyi [BG] produced the example of countinuum groups and their measure preserving actions which are pairwise non-orbitally isomorphic. This example may be considered as an analog of D. McDuff's result concerning the existence of continuum of countable groups such that the factors generated by their regular representations are pairwise non-isomorphic.

It would be very interesting to find non-isomorphic trajectory partitions among the different actions of the same group. It seems natural that the Bernoulli actions of different entropies form such a family. Another open question is whether or not the Bernoulli actions of the groups W_k, W_s ($k \neq s$) are orbitally isomorphic.

We will now formulate some results due to R. Zimmer on the rigidity of the trajectory partitions. The main property of the actions of "large" groups discovered by R. Zimmer on the base of Margulis' and others' investigations on the arithematical subgroups of semisimple Lie grouups is that the orbital isomorphism and the metric isomorphism for such groups are essentially the same (unlike the amenable case).

Theorem 4.2 (R. Zimmer [Z]). *Suppose G_1, G_2 are two connected centerless semisimple groups of real rank > 1 without compact factor-groups. Suppose, further, that the restrictions of their free and measure preserving actions to any non-trivial normal subgroup are ergodic (this property is sometimes called irreducible ergodicity). These actions are orbitally isomorphic if and only if there exists an isomorphism between the groups G_1 and G_2 under which the actions are metrically isomorphic.*

Corollary. *Suppose we are given two free and ergodic actions T_1, T_2 of groups $SL(m, \mathbb{R})$, $SL(n, \mathbb{R})$, $(m, n \geq 3)$ respectively on a Lebesgue space $(M, \mathcal{M}, \mu)$. Then the orbital isomorphism between T_1 and T_2 yields that $m = n$ and $V T_1 V^{-1} = T_2(r(g))$, where V is an automorphism of $(M, \mathcal{M}, \mu)$ and r is an automorphism of $SL(n, \mathbb{R})$.*

This statement is not true for $n = 2$.

The following fact about the actions of discrete groups can also be obtained from Theorem 4.2.

Example 4. The actions of $SL(n, \mathbb{R})$ on the tori $\mathbb{R}^n/\mathbb{Z}^n$ are orbitally non-isomorphic for distinct n. The actions of $SL(n, \mathbb{R})$ on $P_{n-1}\mathbb{R}$ are also pairwise orbitally non-isomorphic. The corresponding rigidity theorem is as follows.

Theorem 4.3 (R. Zimmer [Z]). *Suppose G, G' are two centerless semisimple groups of real rank > 1 without compact factor groups, and S, S' are their discrete subgroups. Suppose further, that T, T' are irreducibly ergodic actions of S, S' respectively. If T and T' are orbitally isomorphic, then $G = G'$ and there exists an isomorphism between S and S' under which T and T' are metrically isomorphic.*

The above theorems show that the trajectory class of an action "almost uniquely" determines the action itself. This property of the action is called its rigidity. A similar property appeared earlier in Margulis-Mostow results saying that under certain conditions the fundamental group "almost uniquely" determines the Riemannian metric on the manifold.

The essential part of Zimmer's method is his reformulations of certain notions related to Lie groups such as amenability, rigidity, discrete spectrum and so on, in terms of group actions (the amenability of the group action etc.). It is interesting that such reformulations enabled him to simplify the proofs of some purely group results.

§5. Concluding Remarks. Relationship Between Trajectory Theory and Operator Algebras

Although our exposition of trajectory theory is purely geometrical, there also exists its functional-algebraic version, and both methods were elaborated simultaneously. Recall that there is a natural correspondence between the measurable partitions and the rings of functions (from L^∞ or from L^2) which are constant on their elements. Therefore, the theory of measurable partitions or of sub-σ-algebras of the σ-algebra of measurable sets may be treated as a theory of subrings of the commutative ring L^∞. Since in most interesting cases the trajectory partitions are not measurable, they are not associated with subrings of L^∞.[3] The idea is to consider non-commutative rings in this case.

---

[3] However, there is yet another possibility. Tp any measurable partition one can associate the set of L^2-functions for which the integrals over almost all elements of the partition (the conditional expectations) vanish. This set is a closed subspace of L^2 orthogonal to the above subring. It turns out that such a correspondence can be extended to a wider class of partitions. Namely, one can

Historically, this construction first appeared in [NM], where the cross-product corresponding to group actions were introduced, but it took a long period of time to understand that the above-mentioned cross-product depends only on the trajectory partitions of the action but not on the action itself.

This, now well-known construction, can be used for studying C^*- and W^*-algebras, factors, representations and for constructing the algebraic examples with the use of ergodic methods, as well as for studying the dynamical systems and their trajectory partitions, foliations and so on with the use of algebraic methods. Let us now describe this construction.

Suppose G is a discrete group and $T = \{T_g\}$ is its measure preserving action on a Lebesgue space $(M, \mathcal{M}, \mu)$. Consider the space $M \times G$ and the measure $\mu \times m$ on it, where m is a Haar measure on G. Let $W(G, M)$ be the weakly closed operator algebra in $L^2(M \times G)$ generated by the operators M_φ and U_h of the form

$$(M_\varphi f)(x, g) = \varphi(x) f(x, g), \qquad (U_h f)(x, g) = f(T_h x, hg),$$

where $\varphi \in L^\infty(M)$; $h, g \in G$; $f \in L^2(M \times G)$. This algebra is called the cross-product algebra of $L^\infty(M)$ and $l^1(G)$ with respect to the action T.

This construction is just the one contained in [NM], and it is orbitally invariant, i.e. if we consider, instead of the action T of G some orbitally equivalent action T' of G', this does not affect $W(G, M)$. Moreover, this algebra can be defined in the invariant terms, i.e. the specific action $\{T_g\}$ need not be mentioned [FM]. Hence, the algebraic invariants of $W(G, M)$ are the trajectory invariants of $\{T_g\}$. By the present time a widely developed theory generalizing the above construction to quasi-invariant measures, to foliations, to C^*-algebras, making use of cohomologies, exists (cf [FM], [Mo], [HO], [Co3]). The so called full group, or Dye group, consisting of all automorphism T for which the trajectory partition $\sigma(T)$ is finer than some given trajectory partition, say, $\tau(G)$, was intensely studied. It seems useful to keep in mind that the algebra $\mathfrak{M} = \{M_\varphi :$ $\varphi \in L^\infty(M)$ is an analog of the Cartan algebra in $W(G, M)$, while the Dye group (denoted by $[G]$) is an analog of the Weil group in the theory of semisimple Lie algebras. This analogy plays a substantial role in the study of W^*-algebras ([Ve3], [FM]).

On the other hand, many results in trajectory theory (for example the Feldman-Connes-Ornstein-Weiss theorem, i.e. Theorem 1.2 of this chapter) were obtained only after their analogs in the theory of W^*-algebras had been proved (in our example, after the Connes theorem [Co2]).

---

associate with a given (not necessarily measurable) partition the linear (but, generally, non-closed) subspace of functions, such that their integral over the elements of some measurable partition which is finer than the given one, vanish. So, we have the correspondence between the tame partitions and some, generally, non-closed subspaces of L^2. The orthogonal complements to these subspaces are the subrings corresponding to measurable hulls of the partitions considered.

Bibliography

The relationship between ergodic theory on the one hand and classical mechanics and theory of ordinary differential equations on the other hand is discussed in [Ar1], [Ar2]. The book BVGN contains the exposition of the theory of Lyapunov characteristic exponents for the systems of ordinary differential equations. An important bibliography of works in ergodic theory up to 1975 may be found in the review article [KSS] Approximation theory of dynamical systems by periodic ones with some its applications is the topic of [KS].

Spectral theory, as well as numerous examples of dynamical systems are discussed in [CFS]. The review articles [Ro1], [Ro2], [Ro3] are devoted to the detailed exposition of measure theory and to the principal problems of general metric and entropy theory of dynamical systems. The book [Si5] contains a brief and clear exposition of the elements of ergodic theory. Some chapters of [Bi] are also devoted to entropy theory. One of the first systematic expositions of ergodic theory is [H]. The book [Or] deals with the problem of metric isomorphism and contains the important results due mainly to its author, in particular, it contains the complete solution to the problem of metric isomorphism for Bernoulli automorphisms. The results of recent investigations concerning the problem of the equivalence of dynamical systems in the sense of Kakutani are presented in [ORW]. The relationship between ergodic theory and statistical mechanics of the lattice systems is considered in [Rue]. The main area of applications of thermodynamic formalism is the theory of hyperbolic dynamical systems.

For the convenience of the reader, references to reviews in Zentralblatt für Mathematik (Zbl.), complied using the MATH database, have, as far as possible, been included in this bibliography.

[A]   Abramov, L.M.: Metric automorphisms with quasi-discrete spectra. Izv. Akad. Nauk SSSR, Ser. Mat. 26, 513–530 (1962) [Russian]. Zbl. 132.359

[AK]   Ambrose, W., Kakutani, S.: Structure and continuity of measurable flows. Duke Math. J. 9, 25–42 (1942)

[An]   Anosov, D.V.: On the additive homologic equation related to an ergodic rotation of the circle. Izv. Akad. Nauk SSSR. Ser. Mat. 37, 1259–1274 (1973) [Russian]. Zbl. 298.28016. English transl.: Math. USSR, Izv. 7, 1257–1271 (1975)

[AW]   Araki, H., Woods, E.: A classification of factors. Publ. Res. Inst. Math. Sci., Kyoto Univ., Ser. A 4, 51–130 (1968). Zbl. 206.129

[Ar1]   Arnold, V.I.: Mathematical methods in classical mechanics. Nauka, Moscow 1974 [Russian]; English transl.: Springer-Verlag, New York–Berlin–Heidelberg 1978. 462 p. Zbl. 386.70001

[Ar2]   Arnold, V.I.: Additional chapters of the theory of ordinary differential equations. Nauka, Moscow 1978 [Russian]

[AGH]   Auslander, L., Green, L., Hahn, F.: Flows on homogeneous spaces. Princeton Univ. Press, Princeton N.J., 1963. 107 p. Zbl. 106.368

[Av]   Avez, A.: Entropie des groupes de type fini. C.R. Acad. Sci., Paris, Ser. A 275, 1363–1366 (1972). Zbl. 252.94013

[B]   Belinskaya, R.M.: Partitions of (the) Lebesgue space in(to) (the) trajectories of ergodic automorphisms. Funkts. Anal. Prilozh. 2, No. 3, 4–16 (1968) [Russian]. Zbl. 176.447. English transl.: Funct. Anal. Appl. 2 (1968) 190–199 (1969)

[Be]   Bewely, T.: Extensions of the Birkhoff and von Neumann ergodic theorems to semigroups actions. Ann. Inst. Heni Poincaré, Sect. B 7, 283–291 (1971). Zbl. 226.28009

[BG]   Bezuglyi, S.I., Golodets, V.Ya.: Hyperfinite and II_1-actions for non-amenable groups. J. Funct. Anal. 40, 30–44 (1981). Zbl. 496.22011

[Bi]   Billingsley, P.: Ergodic theory and information. J. Wiley and Sons, New York–London–Sydney 1965. Zbl. 141.167

Bibliography

[Bl] Blanchard, F.: Partitions extrémales de flots d'entropie infinie. Z. Wahrscheinlichkeitstheor. Verw. Geb. *36*, 129–136 (1976). Zbl, 319.28012 (Zbl. 327.28012)

[Blo] Blokhin, A.A.: Smooth ergodic flows on surfaces. Tr. Mosk. Mat. O.-va *27*, 113–128 (1972) [Russian]. Zbl. 252.28008

[BK] Bogolyubov, N.N., Krylov, N.M.: La théorie générale de la mesure dans son application à l'étude de systèmes dynamiques de la mécanique non-linéaire. Ann. Math. II. Ser. *38*, 65–113 (1937). Zbl. 16.86

[Bo1] Bowen, R.: Equilibrium states and the ergodic theory of Anosov diffeomorphisms. Lect. Notes Math. 470 (1975). Zbl. 308.28010

[Bo2] Bowen, R.: Anosov foliations are hyperfinite. Ann. Math., II. Ser. *106*, 549–565 (1977). Zbl. 374.58008

[BVGN] Bylov, B.F., Vinograd, R.E., Grobman, D.M., Nemitsky V.V.: Theory of Lyapunov exponents and applications to stability problems. Nauka, Moscow, 1966. 576 p. [Russian]. Zbl. 144.107

[C] Calderon, A.: A general ergodic theorem. Ann. Math., II. Ser. *58*, 182–191 (1953). Zbl. 52.119

[Ca] Cassels, J.W.S.: Zn introduction to diophantine approximation. Cambridge Univ. Press, 1957. 166 p. Zbl. 77.48

[CO] Chacon, R.V., Ornstein, D.S.: A general ergodic theorem. Ill. J. Math. *4*, 153–160 (1960). Zbl. 134.121

[Ch] Chulaevsky, B.A.: Cyclic approximations of the interval exchange transformations. Usp. Mat. Nauk *34*, 215–216 (1979) [Russian]. Zbl. 415.28014

[Co1] Connes, A.: Une classification des facteurs de type III. Ann. Sci. Ec. Norm. Supér. IV. Ser. *6*, 133–252 (1973). Zbl. 274.46050

[Co2] Connes, A.: Classification of injective factors of types II_1, II_∞, III_λ, $\lambda \neq 1$. Ann. Math., II. Ser. *104*, 73–115 (1976). Zbl. 343.46042

[Co3] Connes, A.: Classification des facteurs. Proc. Symp. Pure Math. *38*, 43–102 (1982). Zbl. 503.46043

[CFW] Connes, A., Feldman, J., Weiss, B.: An amenable equivalence relation is generated by single transformation. Ergodic Theory Dyn. Syst. *1*, 431–450 (1981). Zbl. 491.28018

[Con] Conze, J.P.: Entropie d'un groupe abélien de transformations. Z. Wahrscheinlichkeitstheor. Verw. Geb. *25*, 11–30 (1972). Zbl. 261.28015

[CFS] Cornfeld, I.P., Fomin, S.V., Sinai, Ya.G.: Ergodic Theory. Springer-Verlag, Berlin-Heidelberg—New York 1982. Zbl. 493.28007

[D] Dye, H.: On a group of measure preserving transformations. I, II. Am. J. Math. *81*, 119–159 (1959); *85*, 551–576 (1963). Zbl. 87.115 Zbl. 191.428

[FV] Fedorov, A.L., Vershik, A.M.: Trajectory theory. In: Modern problems of mathematics. New achievements, vol. 26, 121–212. VINITI, Moscow

[FM] Feldman, J., Moore, C.C.: Ergodic equivalence relations, cohomology and von Neumann algebras. I, II. Trans. Am. Math. Soc. *234*, 289–324 (1977). Zbl. 369.22009

[F] Furstenberg, H.: Strict ergodicity and transformations of the torus. Am. J. Math. *83*, 573–601 (1961). Zbl. 178.384

[FKO] Furstenberg, H., Katznelson, Y., Ornstein, D.: The theoretical proof of Szemeredi's theorem. Bull. Am. Math. Soc., New Ser. *7*, 527–552 (1982). Zbl. 523.28017

[FK] Furstenberg, H., Kesten, H.: Products of random matrices. Ann. Math. Stat. *31*, 457–469 (1960). Zbl. 137.355

[GF] Gelfand, I.M., Fomin, S.V.: Geodesic flows on manifolds of constant negative curvature. Usp. Mat. Nauk *7*, 118–137 (1952) [Russian]. Zbl. 48.92

[G] Grigorchuk, R.I.: On Milnor's problem concerning group growth. Dokl. Akad. Nauk SSSR *271*, 30–33 (1983). Zbl. 547.20025. English transl.: Sov. Math., Dokl. *28*, 23–26 (1983)

[Gu] Gurevich, B.M.: Perfect partitions for ergodic flows. Funkts. Anal. Prilozh. *11*, No. 3, 20–23 (1977) [Russian]. Zbl. 366.28006 English transl.: Funct. Anal. Appl. *11*, 179–182 (1978)

Bibliography

[HIK] Hajian, A., Ito, Y., Kakutani, S.: Full groups and a theorem of Dye. Adv. Math. *17*, 48–59 (1975). Zbl. 303.28017

[H] Halmos, P.R.: Lectures on ergodic theory. Math. Soc. Jap. Tokyo, 1956. 99p. Zbl. 73.93

[HO] Hamachi, T., Osikawa, M.: Ergodic groups of automorphisms and Krieger's theorem. Semin. Math. Sci., Keio University, *3*, 113 p. (1981). Zbl. 472.28015

[J] Jewett, R.: The prevalence of uniquely ergodic systems. J. Math. Mech. *19*, 717–729 (1970). Zbl. 192.406

[KV] Kaimanovich, V., Vershik, A.: Random walks on discrete groups boundary and entropy. Ann. Probab. *11*, 457–490 (1983)

[K] Kalikow, S.A.: T, T^{-1}-transformation is not loosely Bernoulli. Ann. Math., II. Ser. *115*, 393–409 (1982). Zbl. 523.28018

[Kal] Kaminski, B.: Mixing properties of two-dimensional dynamical systems with completely positive entropy. Bull. Acad. Pol. Sci., Ser. Sci. Math. *28*, 453–463 (1980). Zbl. 469.28013

[Ka2] Kaminski, B.: The theory of invariant partitions for actions. Bull. Acad. Pol. Sci., Ser. Sci. Math. *29*, 349–362 (1981). Zbl. 479.28016

[Kat1] Katok, A.B.: Spectral properties of dynamical systems with an integral invariant on the torus. Funkts. Anal. Prilozh. *1*, No. 4, 46–56 (1967) [Russian]. Zbl. 172.121. English transl.: Funct. Anal. Appl. *1*, 296–305 (1967)

[Kat2] Katok, A.B.: Invariant measures for the flows on oriented surfaces. Dokl. Akad. Nauk SSSR *211*, 775–778 (1973) [Russian]. Zbl. 298.28013. English transl.: Sov. Math., Dokl. *14*, 1104–1108 (1974)

[KSS] Katok, A.B., Sinai, Ya.G., Stepin, A.M.: Theory of dynamical systems and general transformation groups with invariant measure. Itogi Nauki Tekh., Ser. Mat. Anal. *13*, 129–262. Moscow, (1975) [Russian]. Zbl. 399.28011. English transl.: J. Sov. Math. 7, 974–1065 (1977)

[KS] Katok, A.B., Stepin, A.M.: Approximations in ergodic theory. Usp. Mat. Nauk 22, No. 5, 81–106 (1967) [Russian]. Zbl. 172.72. English transl.: Russ. Math. Surv. 22, No. 5, 77–102 (1967)

[KW1] Katznelson, Y., Weiss, B.: Commuting measure-preserving transformations. Isr. J. Math. *12*, 161–173 (1972). Zbl. 239.28014

[KW2] Katznelson, Y., Weiss, B.: A simple proof of some ergodic theorems. Isr. J. Math. *42*, 291–296 (1982). Zbl. 546.28013

[Ke1] Keane, M.: Interval exchange transformations. Math. Z. *141*, 25–31 (1975). Zbl. 278.28010

[Ke2] Keane, M.: Nonergodic interval exchange transformations. Isr. J. Math. *26*, 188–196 (1977). Zbl. 351.28012

[KeS] Keane, M., Smorodinsky, M.: Bernoulli schemes of the same entropy are finitarily isomorphic. Ann. Math., II. Ser. *109*, 397–406 (1979). Zbl. 405.28017

[Ki] Kiefer, J.C.: A generalized Shannon-McMillan theorem for the actions of an amenable group on a probability space. Ann. Probab. *3*, 1031–1037 (1975). Zbl. 322.60032

[Kin] Kingman, J.F.C.: Subadditive ergodic theory. Ann. Probab. *1*, 883–909 (1973). Zbl. 311.60018

[Kir] Kirillov, A.A.: Dynamical systems, factors and group representations. Usp. Mat. Nauk. 22, No. 5, 67–80 (1967) [Russian]. Zbl. 169.466. English transl.: Russ. Math. Surv. No. 5, 63–75 (1967)

[Ko1] Kochergin, A.V.: On the absence of mixing for special flows over rotations of the circle and for flows on the two-dimensional torus. Dokl. Akad. Nauk SSSR, Ser. Mat. *205*, 515–518 (1972) [Russian]. Zbl. 262.28015; English transl.: Sov. Math., Dokl. *13*, 949–952 (1972)

[Ko2] Kochergin, A.V.: Change of time in flows and mixing. Izv. Akad. Nauk SSSR, Ser. Mat. *37*, 1275–1298 (1973). [Russian]; Zbl. 286.28013. English transl.: Math. USSR, Izv. 7, 1273–1294 (1975)

[Kol1] Kolmogorov, A.N.: A new metric invariant of transient dynamical systems and auto-

	morphisms of Lebesgue spaces. Dokl. Akad. Nauk SSSR, Ser. Mat. *119*, 861–864 (1958) [Russian]. Zbl. 83.106
[Ko12]	Kolmogorov, A.N.: On dynamical systems with an integral invariant on the torus. Dokl. Akad. Nauk SSSR, Ser. Mat. *93*, 763–766 (1953) [Russian]. Zbl. 52.319
[Kr]	Krengel, U.: Recent progress in ergodic theorems. Astérisque *50*, 151–192 (1977). Zbl. 376.28016
[Kri1]	Krieger, W.: On entropy and generators of measure preserving transformations. Trans. Am. Math. Soc. *149*, 453–464 (1970). Zbl. 204.79
[Kri2]	Krieger, W.: On unique ergodicity. Proc. Sixth Berkeley Symp. Math. Stat. Prob. *2*, 337–346 (1970). Zbl. 262.28013
[Kri3]	Krieger, W.: On ergodic flows and the isomorphism of factors. Math. Ann. *223*, 19–30 (1976). Zbl. 332.46045
[Kry]	Krygin, A.B.: An example of continuous flow on the torus with mixed spectrum. Mat. Zametki *15*, 235–240 (1974) [Russian]. Zbl. 295.58011. English transl.: Math. Notes *15*, 133–136 (1974)
[M1]	Mackey, G.W.: Point realizations of transformation groups. Ill. J. Math. *6*, 327–335 (1962). Zbl. 178.172
[M2]	Mackey, G.W.: Ergodic transformations groups with a pure point spectrum. Ill. J. Math. *8*, 593–600 (1964). Zbl. 255.22014
[M3]	Mackey, G.W.: Ergodic theory and virtual groups. Math. Ann. *166*, 187–207 (1966). Zbl. 178.388
[Ma]	Maruyama, G.: Transformations of flows. J. Math. Soc. Japan *18*, 303–330 (1966). Zbl. 166.404
[Mas]	Masur, H.: Interval exchange transformations and measured foliations. Ann. Math. II. Ser. *115*, 169–200 (1982). Zbl. 497.28012
[Mau]	Mautner, F.: Geodesic flows on symmetric Riemann spaces. Ann. Math., II. Ser. *65*, 416–431 (1957). Zbl. 84.375
[Me]	Meshalkin, L.D.: A case of isomorphism of Bernoulli schemes. Dokl. Akad. Nauk SSSR, Ser. Mat. *128*, 41–44 (1959) [Russian]. Zbl. 99.123
[Mi1]	Millionshchikov, V.M.: Metric theory of linear systems of differential equations. Mat. Sb., Nov. Ser. *77*, 163–173 (1968) [Russian]. Zbl. 176.46
[Mi2]	Millionshchikov, V.M.: A criterion of stability of probability spectrum for linear systems of differential equations with reccurrent coefficients and criterion of quasi-reducibility for systems with quasi-periodic coefficients. Mat. Sb., Nov. Ser. *78*, 179–201 (1969) [Russian]. Zbl. 186.147
[Mo]	Moore, C.C.: Ergodic theory and von Neumann algebras. Proc. Symp. Pure Math. *38*, 179–226 (1982). Zbl. 524.46045
[N]	Neumann, J. von: Zur Operatorenmethode in der klassischen Mechanik. Ann. Math., II. Ser. *33*, 587–642 (1932). Zbl. 5.122
[NM]	Neumann J. von, Murrey, F.: On rings of operators IV. Ann. Math., II. Ser. *44*, 716–808 (1943). Zbl. 60.269
[O]	Olshansky, A.Yu.: On the problem of existence of an invariant mean of a group. Usp. Mat. Nauk *35*, No. 4, 199–200 (1980) [Russian]. Zbl. 452.20032. English transl: Russ. Math. Surv. *35*, No. 4, 180–181 (1980)
[Or]	Ornstein, D.S.: Ergodic theory, randomness and dynamical systems. Yale Univ. Press, New Haven and London, 141 p. 1974. Zbl. 296.28016
[OW]	Ornstein, D.S., Weiss, B.: Entropy and isomorphism theorems for actions of amenable groups. Isr. J. Math. 1987
[ORW]	Ornstein, D.S., Rudolph, D.J., Weiss, B.: Equivalence of measure preserving transformations. Mem. Am. Math. Soc. *262*, 116 p. (1982). Zbl. 504.28019
[Os1]	Oseledets, V.I.: Markov chains, skew products and ergodic theorems for "general" dynamical systems. Teor. Veroyatn. Primen. *10*, 551–557 (1965) [Russian]. Zbl. 142.144. English transl.: Theor. Probab. Appl. *10*, 499–504 (1965)
[Os2]	Oseledets, V.I.: A multiplicative ergodic theorem. Lyapunov characteristic exponents of

dynamical systems. Tr. Mosk. Mat. O.-va *19*, 179–210 (1968) [Russian]. Zbl. 236.93034. English transl.: Trans. Mosc. Math. Soc. *19*, 197–231 (1969)

[P] Pitskel, B.S. On informational "futures" of amenable groups. Dokl. Akad. Nauk SSSR, Ser. Mat. *223*, 1067–1070 (1975) [Russian]. Zbl. 326.28027. English transl.: Sov. Math., Dokl. *16*, 1037–1041 (1976)

[PS] Pitskel, B.S., Stepin, A.M.: On the equidistribution property of entropy of commutative groups of automorphisms. Dokl. Akad. Nauk SSSR, Ser. Mat. *198*, 1021–1024 (1971) [Russian]. Zbl. 232.28017. English transl.: Sov. Math., Dokl. *12*, 938–942 (1971)

[R] Ramsay, A.: Virtual groups and group actions. Adv. Math. *6*, 253–322 (1971). Zbl. 216.149

[Re] Renault, J.: A groupoid approach to C*-algebras. Lect. Notes Math. *793* 160 p. (1980). Zbl. 433.46049

[Ro1] Rokhlin, V.A.: On the principal notions of measure theory. Mat. Sb., Nov. Ser. *67*, 107–150 (1949) [Russian]. Zbl. 33.169

[Ro2] Rokhlin, V.A.: Selected topics of metric theory of dynamical systems. Usp. Mat. Nauk *4*, No. 2, 57–128 (1949) [Russian]. Zbl. 32.284

[Ro3] Rokhlin, V.A.: Lectures on the entropy theory of measure preserving transformations. Usp. Mat. Nauk *22*, No. 5, 3–56 (1967) [Russian]. Zbl. 174.455. English transl.: Russ. Math. Surv. *22*, No. 5, 1–52 (1967)

[RS] Rokhlin, V.A., Sinai, Ya.G.: Construction and properties of invariant measurable partitions. Dokl. Akad. Nauk SSSR, Ser. Mat. *141*, 1038–1041 (1961) [Russian]. Zbl. 161,343. English transl.: Sov. Math., Dokl. *2*, 1611–1614 (1961)

[Ru1] Rudolph, D.J.: A two-valued step-coding for ergodic flows. Math. Z. *150*, 201–220 (1976). Zbl. 318.28010 (Zbl. 325.28019).

[Ru2] Rudolph, D.J.: An example of a measure preserving map with minimal self-joinings and applications. J. Anal. Math. *35*, 97–122 (1979). Zbl. 446.28018

[Rue] Ruelle, D.: Thermodynamic formalism. Addison–Wesley, London–Amsterdam–Don Mills, Ontario–Sydney–Tokyo 1978. 183 p. Zbl. 401.28016

[S] Sataev, E.A.: On the number of invariant measures for the flows on oriented surfaces. Izv. Akad. Nauk SSSR, Ser. Mat. *39*, 860–878 (1975) [Russian]. Zbl. 323.28012. Math. USSR, Izv. 9, 813–830 (1976)

[Sh] Shklover, M.D.: On classical dynamical systems on the torus with continuous spectrum. Izv. Vyssh. Uchebn. Zaved. Mat. 1967, No. *10*, 113–124 (1967) [Russian]. Zbl. 153.126

[Si1] Sinai, Ya.G.: On the concept of entropy for a dynamical system. Dokl. Akad. Nauk SSSR, Ser. Mat. *124*, 768–771 (1959) [Russian]. Zbl. 86.101

[Si2] Sinai, Ya.G.: Dynamical systems with countable Lebesgue spectrum I. Izv. Akad. Nauk SSSR, Ser. Mat. *25*, 899–924 (1961) [Russian]. Zbl. 109.112

[Si3] Sinai, Ya.G.: On weak isomorphism on measure preserving transformations. Mat. Sb. *63*, 23–42 (1964) [Russian]. English transl.: Transl., II. Ser., Am. Math. Soc. *57*, 123–143 (1966).

[Si4] Sinai, Ya.G.: Gibbs measures in ergodic theory. [Usp., Nov. Ser. Mat. Nauk *27*, 21–64] (1972) [Russian]. Russ. Math. Surv. *27*, No. 4, 21–69 (1972). Zbl. 255.28016

[Si5] Sinai, Ya.G.: Introduction to ergodic theory. Erevan Univ. Press, Erevan 1973. English transl.: Math. Notes *18*, Princeton, Princeton Univ. Press (1976). Zbl. 375.28011

[Sm] Smorodinsky, M.: A partition on a Bernoulli shift which is not weak Bernoulli. Math. Syst. Theory. *5*, 201–203 (1971). Zbl. 226.60066

[St] Stepin, A.M.: Bernoulli shifts on groups. Dokl. Akad. Nauk SSSR, Ser. Mat. *223*, 300–302 (1975) [Russian]. Zbl. 326.28026. English transl.: Sov. Math., Dokl. *16*, 886–889 (1976)

[Su] Sujan, S.: Generators of an Abelian group of invertible measure preserving transformations. Monatsh. Math. *90*, 67–79 (1980). Zbl. 432.28016

[T1] Tempelman, A.A.: Ergodic theorems for general dynamical systems. Tr. Mosk. Mat. O.-va *26*, 95–132 (1972) [Russian]. Zbl. 249.28015. English transl.: Trans. Mosc. Math. Soc. *26*, 94–132 (1974)

[T2] Tempelman, A.A.: Ergodic theorems on groups. Vilnius 1986

[V1]	Veech, W.A.: Interval exchange transformations. J. Anal. Math. *33*, 222–272 (1978). Zbl. 455.28006
[V2]	Veech, W.A.: Gauss measures for transformations on the space of interval exchange maps. Ann. Math. II. Ser. *115*, 201–242 (1982). Zbl. 486.28014
[V3]	Veech, W.A.: The metric theory of interval exchange transformations I. Generic spectral properties. Am. J. Math. 1983
[Ve1]	Vershik, A.M.: A measurable realization of continuous groups of automorphisms of a unitary ring. Izv. Akad. Nauk SSSR, Ser. Mat. *29*, 127–136 (1965) [Russian]. Zbl. 194.163. English transl.: Transl., II. Ser. Am. Math. Soc. 84, 69–81 (1969)
[Ve2]	Vershik, A.M.: On the lacunary isomorphism of sequences of measurable partitions. Funkts Anal. Prilozh. *2*, No. 3, 17–21 (1968) [Russian]. Zbl. 186.202. English transl.: Funct. Anal. Appl. 2 (1968), 200–203 (1969)
[Ve3]	Vershik, A.M.: Non-measurable partitions, trajectory theory and operator algebras. Dokl. Akad. Nauk SSSR *199*, 1004–1007 (1971) [Russian]. Zbl. 228.28013. English transl.: Sov. Math., Dokl. *12*, 1218–1222 (1971)
[Ve4]	Vershik, A.M.: Decreasing sequences of σ-algebras. Teor. Veroyatn. Primen. *19*, 657–658 (1974) [Russian].
[Ve5]	Vershik, A.M.: The action of PSL $(2, \mathbb{R})$ on $P_1 R$ is approximable. Usp. Mat. Nauk *33*, No. 1, 209–210 (1978) [Russian]. Zbl. 391.28008. English transl.: Russ. Math. Surv. *33*, No. 1, 221–222 (1978)
[Ve6]	Vershik, A.M.: Measurable realizations of groups of automorphisms and integral representations of positive operators. Siberian Math. J. *28*, N 1, 52–60
[Ve6F]	Vershik, A.M., Fedorov, A.A.: Trajectory theory. Itogi Nauki Tekh., Ser. Sovrem. Probl. Mat., vol. *26*, 171–211. (1985) [Russian]. Zbl. 614.28018
[VG]	Vinokurov, V.G., Ganikhodgaev, N.: Conditional functions in trajectory theory of dynamical systems. Izv. Akad. Nauk SSSR, Ser. Mat. *42*, 927–964 (1978) [Russian]. Zbl. 408.28018. English transl.: Math. USSR, Izv. *13*, 221–251 (1979)
[Z]	Zimmer, R.: Ergodic theory, group representations and rigidity. Bull. Am. Math. Soc. New Ser. *6*, 383–416 (1982). Zbl. 532.22009

II. Ergodic Theory of Smooth Dynamical Systems

Contents

Chapter 6. Stochasticity of Smooth Dynamical Systems. The Elements of
KAM-Theory (*Ya.G. Sinai*) 101
§ 1. Integrable and Nonintegrable Smooth Dynamical Systems.
    The Hierarchy of Stochastic Properties of Deterministic Dynamics .. 101
§ 2. The Kolmogorov-Arnold-Moser Theory (KAM-Theory) 104
Chapter 7. General Theory of Smooth Hyperbolic Dynamical Systems
(*Ya.B. Pesin*) 108
§ 1. Hyperbolicity of Individual Trajectories 108
    1.1. Introductory Remarks 108
    1.2. Uniform Hyperbolicity 109
    1.3. Nonuniform Hyperbolicity 110
    1.4. Local Manifolds 111
    1.5. Global Manifolds 113
§ 2. Basic Classes of Smooth Hyperbolic Dynamical Systems. Definitions
    and Examples 113
    2.1. Anosov Systems 113
    2.2. Hyperbolic Sets 115
    2.3. Locally Maximal Hyperbolic Sets 118
    2.4. A-Diffeomorphisms 119
    2.5. Hyperbolic Attractors 119
    2.6. Partially Hyperbolic Dynamical Systems 120
    2.7. Mather Theory 121
    2.8. Nonuniformely Hyperbolic Dynamical Systems. Lyapunov
        Exponents 123
§ 3. Ergodic Properties of Smooth Hyperbolic Dynamical Systems 125
    3.1. u-Gibbs Measures 125
    3.2. Symbolic Dynamics 126
    3.3. Measures of Maximal Entropy 128
    3.4. Construction of u-Gibbs Measures 128
    3.5. Topological Pressure and Topological Entropy 129
    3.6. Properties of u-Gibbs Measures 131

	3.7. Small Stochastic Perturbations	132
	3.8. Equilibrium States and Their Ergodic Properties	133
	3.9. Ergodic Properties of Dynamical Systems with Nonzero Lyapunov Exponents	133
	3.10. Ergodic Properties of Anosov Systems and of UPH-Systems	135
	3.11. Continuous Time Dynamical Systems	137
§4.	Hyperbolic Geodesic Flows	137
	4.1. Manifolds with Negative Curvature	138
	4.2. Riemannian Metrics Without Conjugate Points	141
	4.3. Entropy of Geodesic Flow	143
§5.	Geodesic Flows on Manifolds with Constant Negative Curvature	143
§6.	Dimension-like Characteristics of Invariant Sets for Dynamical Systems	146
	6.1. Introductory Remarks	146
	6.2. Hausdorff Dimension	146
	6.3. Dimension with Respect to a Dynamical System	149
	6.4. Capacity and Other Characteristics	149

Chapter 8. Dynamical Systems of Hyperbolic Type with Singularities (*L.A. Bunimovich*) 151

§1.	Billiards	151
	1.1. The General Definition of a Billiard	152
	1.2. Billiards in Polygons and Polyhedrons	153
	1.3. Billiards in Domains with Smooth Convex Boundary	155
	1.4. Dispersing (Sinai) Billiards	156
	1.5. The Lorentz Gas and the Hard Spheres Gas	163
	1.6. Semidispersing Billiards	164
	1.7. Billiards in Domains with Boundary Possessing Focusing Components	165
	1.8. Hyperbolic Dynamical Systems with Singularities (a General Approach)	167
	1.9. The Markov Partition and Symbolic Dynamics for Dispersing Billiards	169
	1.10. Statistical Properties of Dispersing Billiards and of the Lorentz Gas	170
§2.	Strange Attractors	172
	2.1. Definition of a Strange Attractor	172
	2.2. The Lorenz Attractor	173
	2.3. Some Other Examples of Hyperbolic Strange Attractors	178

Chapter 9. Ergodic Theory of One-Dimensional Mappings (*M.V. Jakobson*) 179

§1.	Expanding Maps	179
	1.1. Definitions, Examples, the Entropy Formula	179
	1.2. Walters Theorem	182
§2.	Absolutely Continuous Invariant Measures for Nonexpanding Maps	184

2.1. Some Examples 184
   2.2. Intermittency of Stochastic and Stable Systems 186
   2.3. Ergodic Properties of Absolutely Continuous Invariant Measures 187
§ 3. Feigenbaum Universality Law 190
   3.1. The Phenomenon of Universality 190
   3.2. Doubling Transformation 191
   3.3. Neighborhood of the Fixed Point......................... 193
   3.4. Properties of Maps Belonging to the Stable Manifold of Φ 195
§ 4. Rational Endomorphisms of the Riemann Sphere 196
   4.1. The Julia Set and its Complement 196
   4.2. The Stability Properties of Rational Endomorphisms 197
   4.3. Ergodic and Dimensional Properties of Julia Sets 198
Bibliography 199

Chapter 6
Stochasticity of Smooth Dynamical Systems. The Elements of KAM-Theory

Ya.G. Sinai

§ 1. Integrable and Nonintegrable Smooth Dynamical Systems. The Hierarchy of Stochastic Properties of Deterministic Dynamics

A Hamiltonian system with n degrees of freedom and with a Hamiltonian $H(p_1, \ldots, p_n, q_1, \ldots, q_n)$ is called integrable if it has n first integrals $I_1 = H, I_2, \ldots, I_n$ which are in involution. The famous Liouville theorem states (see [Ar1], [KSF]) that if the n-dimensional manifold obtained by fixing the values of the integrals $I_1 = C_1, I_2 = C_2, \ldots, I_n = C_n$ is compact, and these integrals are functionally independent in some neighbourhood of the point $(C_1, \ldots, C_n)$ then this manifold is the n-dimensional torus. One could introduce cyclic coordinates $\varphi_1, \ldots, \varphi_n$ such that the equations of motion would obtain a simple form $\dot\varphi_i = F_i(I_1, \ldots, I_n) = \text{const}$, $1 \leqslant i \leqslant n$, and the motion would be quasi-periodic with n frequencies.

From the point of ergodic theory this situation means that a flow $\{S^t\}$ which corresponds to a Hamiltonian system with the Hamiltonian $H(p, q)$ and with the invariant measure $dp\,dq$ is non-ergodic. Its ergodic components are (mod 0) n-dimensional tori and in every such torus an ergodic flow with purely point spectrum is obtained. We shall call a dynamical system integrable one also in a more general case when it is non-ergodic and when on almost every of its ergodic components a dynamical system with purely point spectrum is obtained.

There are many examples of integrable systems: the geodesic flows on surfaces of revolution, the geodesic flow on a triaxial ellipsoid, a billiard in an ellipse, the system of three point eddies in two-dimensional hydrodynamics and so on. A large number of new examples of integrable systems were found in the last decade by the inverse scattering method [T].

Nevertheless one has to consider integrability of a dynamical system not as a rule, but instead as an exclusion. The property of complete integrability of a Hamiltonian system has disappeared already under small sufficiently generic type perturbations of a Hamiltonian. The famous KAM-theory stated that invariant tori with "sufficiently incommensurable" frequencies did not disappear but only shifted slightly in the phase space and the corresponding motion remains quasi-periodic. These invariant tori form a set of positive measure which is not a domain. Its complement consists of stochastic layers where the dynamics is essentially more complicated. The main mechanism whereby such layers occur is connected with the appearance of heteroclinic and homoclinic trajectories (see Chap. 7, Sect. 2). These layers have a complicated topological structure, in some directions it is the structure of a Cantor set type, and contain an infinite number of periodic trajectories. We shall describe below rigorous mathematical results concerning a structure and properties of trajectories in stochastic layers. Unfortunately these results are dealing yet with trajectories belonging to some subsets of measure zero and do not refer to "generic" trajectories. Even the theorem on positivity of the measure of the set occupied by stochastic layers is not proved yet, but all investigators generally believe that it is true.

The other general mechanism of non-integrability is connected with the occurence of Smale's horseshoe (see Chap. 7, Sect. 2), i.e. of the subspace of a phase space where dynamics has some special properties of instability. By moving away from integrability, a set occupied by invariant tori shrinks and correspondingly a set filled with trajectories with a complex nonintegrable behavior is extended. One can consider it as limiting such dynamical systems which possess the most strong statistical properties in the whole phase space. The most important examples of such systems form geodesic flows on compact manifolds of negative curvature, billiards in domains with convex inwards boundary (see Chaps. 7 and 8) and some one-dimensional mappings (see Chap. 9). The investigation of ergodic properties of such systems is based on the notion of hyperbolicity which will be considered in detail in Sect. 1 of Chap. 7.

We shall now enumerate those properties of a dynamical system, which will be called stochastic properties.

1) The existence of an invariant measure. In many cases, for instance for Hamiltonian systems or for dynamical systems of the algebraic nature, a priori there exists a natural invariant measure. Nevertheless, there are some important situations when such invariant measure is not known a priori but it could be

Chapter 6. Stochasticity of Smooth Dynamical Systems

found by analysis of properties of the dynamics. Among these systems there are dynamical systems with hyperbolic (see Chap. 7, Sect. 2 and Sect. 3) and strange (see Chap. 8, Sect. 2) attractors. The natural method of constructing an invariant measure for these systems is to take any initial smooth measure and to study problems of the existence and of the properties of a limit, as $t \to \infty$, of its iterates under the action of the dynamics, and of the dependence of this limit on the choice of the initial measure. In some cases this program can be realized.

2) Ergodicity (see Chap. 1, Sect. 3). When the invariant measure is chosen, it is natural to consider ergodic properties of the dynamical system with respect to this measure. The simplest one is the problem of ergodicity.

3) Mixing (see Chap. 1, Sect. 3). From the point of statistical mechanics, mixing means the irreversibility, i.e. that any initial measure absolutely continuous with respect to the invariant measure converges to it weakly, under the action of dynamics.

4) The K-property (see Chap. 1, Sect. 3). If a dynamical system is a K-system, then it has countably-multiple Lebesgue spectrum, positive entropy and is mixing of all degrees. The K-property means that a deterministic dynamical system can be coded into a regular stationary random process of probability theory.

5) The Bernoulli property. In the case of discrete time, if a code can be constructed so that the resulting process is a sequence of independent random variables, then the dynamical system is called a Bernoullian one (see Chap. 3, Sect. 4). In the case of continuous time, this definition refers to any automorphism from the flow. It is worthwhile mentioning that usually a code is given by complicated nonsmooth functions and therefore a connection with the smooth structure in the phase space would be lost.

6) The central limit theorem. Let (M, μ, T) be an ergodic automorphism. In view of the Birkhoff-Khinchin ergodic theorem (see Chap. 1, Sect. 2) for any $f \in L^1(M, \mu)$ almost everywhere

$$\lim_{n \to \infty} \left[\frac{1}{n} \sum_{k=1}^{n} f(T^k(x)) - \bar{f} \right] = 0$$

where $\bar{f} = \int f d\mu$. The difference $\left[\frac{1}{n} \sum_{k=1}^{n} f(T^k(x)) - \bar{f} \right]$ is called a time fluctuation with respect to the mean. We shall say that f satisfies the central limit theorem (CLT) of probability theory if there exists a number $\sigma = \sigma(f)$ such that

$$\lim_{n \to \infty} \mu \left\{ x: \sqrt{n} \left[\frac{1}{n} \sum_{k=1}^{n} f(T^k(x)) - \bar{f} \right] < a \right\} = \frac{1}{\sqrt{2\pi\sigma}} \int_{-\infty}^{a} e^{-u^2/2\sigma} du$$

C.L.T. states that time fluctuations have the order $O(1/\sqrt{n})$. In some cases one has to prove CLT for a possibly wider class of phase functions. It is more difficult as a rule, than to prove the K-property.

7) The rate of correlations decay. A function $f \in L^2(M, \mu)$ with zero mean satisfies the property of exponential decay of correlations if there exist positive numbers $C, \rho < 1$ such that

$$\left| \int f(T^n(x)) f(x) \, d\mu \right| < C \rho^{|n|}$$

This property can be verified easily enough when T is a Markov automorphism and f is a function with a sufficiently simple structure. In the general case, in order to prove the exponential decay of correlations or for an analysis of their decay rate for smooth functions f, one has to construct approximations of the dynamical system under consideration by Markov automorphisms. It is often very complicated to construct such approximations.

In what follows, when referring to stochastic properties of dynamical systems we shall mean the hierarchy 1)–7) enumerated above.

§2. The Kolmogorov-Arnold-Moser Theory (KAM-Theory)

We begin to discuss the KAM-theory by considering one important example, and then more general results will be studied. Let us consider the difference equation of stationary sin-Gordon equation type

$$u_{n+1} - 2u_n + u_{n-1} = \lambda V'(u_n) \qquad (6.1)$$

where V is a periodic function with period 1 and λ is a parameter. If $V(u) = 1 - \cos 2\pi u$ then we obtain precisely a stationary sin-Gordon difference equation. We shall assume that $V \in C^\infty(R^1)$, while this assumption can be taken sufficiently milder. Let us introduce new variables $z_n = u_n - u_{n-1}$, $\varphi_n = \{u_n\}$ where as usual the fractional part of a number is denoted by $\{\cdot\}$. Then (6.1) takes the form

$$z_{n+1} = z_n + \lambda V'(\varphi_n), \qquad \varphi_{n+1} = \varphi_n + z_{n+1} \quad (\text{mod } 1) \qquad (6.2)$$

The space of pairs (z, φ) where $-\infty < z < \infty$, $\varphi \in S^1$, forms a two-dimensional cylinder C. The relations (6.2) show that sequences (z_n, φ_n) are trajectories of the transformation of the cylinder C into itself, where

$$T(z, \varphi) = (z', \varphi'), \quad z' = z + \lambda V'(\varphi), \quad \varphi' = \varphi + z' \quad (\text{mod } 1) \qquad (6.3)$$

The transformation T in the case of $V(u) = 1 - \cos 2\pi u$ was introduced by B.V. Chirikov (cf [Chi]) and since then it has been called the Chirikov transformation. A general transformation of type (6.3) will be called a standard transformation.

Standard transformations arise in many problems of the theory of nonlinear oscillations, in plasma physics, in solid state physics (the Frenkel-Kontorova model). Much physical and mathematical literature is devoted to investigations of these transformations. We want to mention [Chi] from the corresponding physical review papers, and from the mathematical works besides the books by V.I. Arnold [Ar1], [Ar2] and by J. Moser [Mos] we should like to mention papers by S. Aubry (see for instance [AP]), by J. Mather [Mat1] and by I. Parseval [P].

In the three last papers a more general situation is also considered. In order to discuss it we shall consider a Lagrangian $L(u, u')$ which satisfies the following conditions

a) $L(u + 1, u' + 1) = L(u, u')$ (periodicity).

b) $-\dfrac{\partial^2 L}{\partial u\, \partial u'} \geq \text{const} > 0$ and is continuous (twist condition).

For any set of points $u_0, u_1, \ldots, u_m$ on a line we should define the Lagrangian $L(u_0, u_1, \ldots, u_m) = \sum_{i=1}^{m} L(u_i, u_{i-1})$. Variational principles of mechanics and of statistical physics force to construct chains $\{u_i\}_{-\infty}^{\infty}$ such that $L(u_s, u_{s+1}, \ldots, u_{s+m})$ takes its minimal value for any $s, m > 0$ and for fixed u_s, u_{s+m}. The corresponding necessary condition takes the following form

$$\frac{\partial L(u_s, \ldots, u_{s+m})}{\partial u_n} = \frac{\partial L(u_{n+1}, u_n)}{\partial u_n} + \frac{\partial L(u_n, u_{n-1})}{\partial u_n} = 0, \qquad -\infty < n < \infty.$$

Let us introduce a momentum $p = \dfrac{\partial L(u, u')}{\partial u}$ conjugated to u. Then the last chain of equations changes to

$$p_n + \frac{\partial L(u_{n+1}, u_n)}{\partial u_n} = 0, \qquad p_{n+1} - \frac{\partial L(u_{n+1}, u_n)}{\partial u_{n+1}} = 0, \qquad -\infty < n < \infty. \tag{6.4}$$

It follows from b) that the first equation can be uniquely solved, and therefore u_{n+1} can be considered as a function of p and u_n. Then p_{n+1} can be found from the second equation. So the transformation T which maps (u_n, p_n) into (u_{n+1}, p_{n+1}) is defined. The condition a) allows us to pass to the phases $\varphi_n\{u_n\}$ and to consider the action of T in the same cylinder C. One could obtain a standard transformation taking $L(u, u') = \tfrac{1}{2}(u - u')^2 + \lambda V(u')$. The useful example of the other type arises in the following way. Let $\Gamma \subset C^\infty$ be a convex curve and it forms a boundary of a convex domain on a plane. We introduce a cyclic variable u on Γ and set

$$L(u, u') = \|\mathbf{r}(u) - \mathbf{r}(u')\|^2$$

where $\mathbf{r}(u)$ is a radius-vector of the point with coordinate u. Then it is easy to demonstrate that sequence $\{u_n\}$ satisfying (6.4), correspond to billiard trajectories inside Γ.

Let us come back to the transformation (6.2). We shall consider it formally in case $\lambda = 0$. The corresponding solutions are formed by those sequences $\{z_n =$

$I, \varphi_n = \varphi_0 + nI\}$, where I is a constant. In geometrical language it means that every circle $\Gamma_{I,0} = \{(z, \varphi): z = I\}$ is invariant with respect to T and T can be reduced on Γ to the rotation on the angle I. The KAM-theory states that if λ is sufficiently small, the majority of these circles are preserved and suffer only small variation in form on the cylinder C. More precisely, the following theorem is valid (see [Ar1], [Ar2], [Mos]).

Theorem 2.1. *Let a number I be poorly approximated by rationals, i.e. $\min_p |I - p/q| > 1/q^{2+\varepsilon}$ for some $\varepsilon > 0$ and for all sufficiently large q. Then one can find $\lambda_0(I) = \lambda_0$ such that for all λ, $|\lambda| \leq \lambda_0$, a standard transformation T has an invariant curve $\Gamma_{I,\lambda}$ which is C^0-close to $\Gamma_{I,0}$ and T is reduced on $\Gamma_{I,0}$ to the rotation. If λ is sufficiently small, then a set of curves $\Gamma_{I,\lambda}$ has a positive measure.*

Surely this theorem of the KAM-theory is a theorem of perturbation theory. Its essential peculiarity is connected with the fact that constructed invariant curves do not form any domain. According to one of the Birkhoff theorems (cf [Bi]) in any neighbourhood of such a curve, there are periodic trajectories. In a "generic" case, i.e. when a function V belongs to an everywhere dense subset of the second category among these periodic trajectories there are hyperbolic ones and their separatrices intersect transversally. In the case mentioned above in Section 1, stochastic layers arise, which are situated between invariant curves given by the KAM-theory. One can find some information on these layers in Chap. 7, Sect. 2.

The new approach to the KAM-theory considered in the papers by Aubry [AD], Mather [Mat1] and Parseval [P] mentioned above which is developed yet again in the two-dimensional case only. Aubry introduces the following definition.

Definition 2.1. A sequence $\{u_n\}$ is called a *configuration with a minimal energy* (CME) if the following property holds: if we consider for any s, t, $s < t$, the sequence $\{u'_n\}$ such that $u'_n = u_n$ if $n \leq s, n \geq t$, then

$$\sum_{n=s+1}^{t} L(u'_n, u'_{n-1}) \geq \sum_{n=s+1}^{t} L(u_n, u_{n-1}).$$

In other words, if we consider L as an energy, then any finite perturbation of CME does not decrease the energy of the configuration. Aubry proved that for any CME $\{u_n\}$ there exists $\lim_{n \to \infty} \frac{1}{n} u_n = \omega$ which can be considered as the analogy of the rotation number of a homeomorphism of the circle. Moreover for any ω there exist CME with this ω. If ω is a rational number, $\omega = p/q$, then in a generic case among CME there are hyperbolic periodic trajectories of period q. If ω is an irrational number, then there exists a measurable (mod 0) injective mapping f_ω of the unit circle S^1 into C such that $f_\omega(S^1)$ is invariant with respect to T and $Tf_\omega(x) = f(x + \omega)$, i.e. $T|f_\omega(S^1)$ is the rotation on the angle ω. If the mapping f_ω is continuous, then $f_\omega(S^1)$ is a continuous curve which envelops the cylinder C.

Chapter 6. Stochasticity of Smooth Dynamical Systems

If ω is poorly approximable by rational numbers and λ is sufficiently small, then the curve $f_\omega(S^1)$ coincides with a curve given by the KAM theory. If f_ω is discontinuous then $f_\omega(S^1)$ is a Cantor perfect set which is called a cantorus. It is invariant under the transformation T and as before T reduces on it to the rotation. It was proposed in [AD] the geometrical approach to constructing of such invariant sets, which is based on one of the Birkhoff theorems.

For any standard transformation there exists a critical value λ_{cr}, or such that for all $\lambda > \lambda_{cr}$ there are no invariant curves enveloping C. Therefore for such values of λ, all invariant sets $f_\omega(S^1)$ are cantori. Apparently for each value of I mentioned in Theorem 2.1 there exists $\lambda_{cr}(I)$ such that for all $\lambda < \lambda_{cr}(I)$ there is an invariant curve $\Gamma_{I,\lambda}$, and for $\lambda > \lambda_{cr}(I)$ it destroys and transforms into a cantorus, but the nature of this bifurcation is not completely clear yet. If λ increases further, then a projection of any cantorus is concentrated in a small neighbourhood of the minimum of the function V. Aubry proved that for such λ a projection on the φ-axis of a cantorus has zero measure (cf [AD]). Apparently a Hausdorff dimension of this projection is also equal to zero.

The above given description shows how separate ergodic components of a standard transformation can be constructed. Invariant curves $\Gamma_{I,\lambda}$ form a set of positive measure if λ is small, and therefore T has invariant sets of positive measure where it is nonergodic. From the other side in the case of large λ, numerical simulations and considerations on the physical level, using Chirikov's criterion of resonanses overlapping, show that T has invariant sets of large measure where it obeys some properties of stochasticity. Nevertheless there are no corresponding rigorous mathematical results.

We are now going to formulate the main statement of the KAM-theory. Not only concrete results are significant in the KAM-theory but also a method for obtaining these results, because it could be applied to a large variety of problems. Unfortunately we have no opportunity to discuss this question in more detail. The KAM-theory will be considered explicitly in Volume 3 of the given series. We discuss only the main theorem of this theory in its application to the Hamiltonian systems. Let us consider an integrable Hamiltonian system with n degrees of freedom. Then in a phase space there exists an open set $\mathcal{O} = U \times \text{Tor}^n$ where U is a neighbourhood in the n-dimensional space, Tor^n is a n-dimensional torus with coordinates $\varphi_1, \ldots, \varphi_n$, and a Hamiltonian H_0 has the form $H_0 = H_0(I_1, \ldots, I_n)$ where $(I_1, \ldots, I_n)$ are coordinates in U. Let the Hamiltonian $H(I, \varphi) = H_0(I) + \varepsilon H_1(I, \varphi, \varepsilon)$ be a small perturbation of the Hamiltonian H_0. We assume that H is analytic in the domain $\mathcal{O}' = U' \times S$ where U' is a complex neighbourhood of U in $\mathbb{C}^n$ and S is a complex neighbourhood of Tor^n in $\mathbb{C}^n$.

The main theorem of the KAM-theory (see [Ar1], [Ar2], [Mos]). *Let* $\det \left\| \frac{\partial^2 H}{\partial I_i \partial I_j} \right\| \neq 0$ *in* U. *Then for all sufficiently small* $\varepsilon > 0$, *one can find a subset* $K \subset \mathcal{O}$ *such that* $\text{mes } K \to \text{mes } \mathcal{O}$ *as* $\varepsilon \to 0$ *and there exists a measurable partition of* K *into invariant n-dimensional tori. The Hamiltonian dynamical system with* the Hamiltonian H reduces on every such torus to the quasi-periodic motion with pure point spectrum and n basic frequencies.

This theorem shows that a small perturbation of an integrable system is nonergodic and has an invariant subset of positive measure. Ergodic components belonging to this subset have pure point spectrum. In particular it disproved completely the hypothesis which appeared often in physical works, that a generic multi-dimensional nonlinear Hamiltonian system is ergodic. We would like to mention additionally that tori which are constructed in the KAM-theory depend smoothly on a parameter $u \in K$.

There exist different modifications of this theorem which allow to extend the KAM-theory onto non-Hamiltonian systems, or to consider the milder conditions of smoothness of the system, or to admit some separate degeneracies when $\det \left\| \dfrac{\partial^2 H}{\partial I_i \partial I_j} \right\| \equiv 0$. We shall not consider these questions here.

Chapter 7
General Theory of Smooth Hyperbolic Dynamical Systems

Ya.B. Pesin

§ 1. Hyperbolicity of Individual Trajectories

1.1. Introductory Remarks. The idea of studying the global behavior of dynamical systems using their local properties seems to be very attractive. The theory of hyperbolic dynamical systems may be considered as a partial realization of this idea.

The hyperbolicity of individual trajectory will now be described, proceeding from the behavior of near-by trajectories or more exactly of infinitely close trajectories. There are three rough types of behavior in a neighbourhood of a given trajectory $S^t x$.

a) The trajectory $S^t x$ attracts all near-by trajectories when $t \to +\infty$ (complete stability); b) $S^t x$ attracts all near-by trajectories when $t \to -\infty$ (complete instability); c) $S^t x$ attracts some trajectories when $t \to +\infty$ and attracts some others when $t \to -\infty$. The definition of hyperbolicity is based on this last type of behavior.

In this chapter we shall typically consider the case when the phase space M is a compact n-dimensional Riemannian C^∞-manifold, and transformations T (or flows S^t) are also supposed to be C^∞ unless stated otherwise. Let S^t be a flow generated by a smooth vector field $X(x)$. Then it corresponds to $X(x)$, a vector field on the tangent space $\mathcal{T}M$, which is called the vector field of variation of X, and it is denoted by $(X(x), X_x(x))$. The corresponding flow is denoted by $S^t(x, v) = (S^t x, dS^t_x v)$. Each transformation $\tilde{S}^t$ maps the tangent space $(\mathcal{T}M)_x$ into $(\mathcal{T}M)_{S^t(x)}$. The corresponding system of differential equations is called the *system of variational equations*

$$\dot{x} = X(x), \qquad \dot{v} = X_x(x)v. \tag{7.1}$$

In discrete time we deal with a diffeomorphism T, and instead of variational equations we consider the differential dT which maps the tangent space $\mathcal{T}_x$ to the tangent space $\mathcal{T}_{T^t(x)}$.

The solutions of (7.1) define a cocycle $\mathfrak{U}(x, t)$ which induces a map $\mathcal{T}_x \to \mathcal{T}_{S^t(x)}(\mathcal{T}_{T^t(x)}$ when the time is discrete). Suppose that the dynamical system under consideration admits an ergodic invariant measure v. Then the multiplicative ergodic theorem is applicable (see Chap. 1, Sect. 2) and it implies the existence of Lyapunov exponents $\chi_1 \leq \chi_2 \leq \ldots \leq \chi_n$. This theorem shows in particular that the v typical trajectories of variational equations behave in a sufficiently regular way.

In this chapter we deal essentially with a case when there is no zero exponents (when the time is discrete) or there is only one zero exponent corresponding to the flow direction (when the time is continuous). It is a generally assumed hypothesis that such a situation is "typical", although the corresponding strong results are still not proved.

In many problems there is no natural invariant measure. In this chapter we expose a theory based on certain properties of solutions of variational equations, which are related in some way or another to the multiplicative ergodic theorem. We shall see below that these properties (which can be checked by analyzing local behavior of dynamical systems) often allow us to construct invariant measures.

1.2. Uniform Hyperbolicity. We begin by considering discrete time dynamical systems. In order to standardize the notation we shall denote by S^1 a diffeomorphism and by $\{S^t\}$, $t \in \mathbb{Z}$ the group generated by S^1.

Definition 1.1. A trajectory $\{S^t(x)\}$ is called *uniformly completely hyperbolic*[1] if there exist subspaces $E^s(S^t(x))$ and $E^u(S^t(x))$[2] and constants $C > 0$, λ, μ such that

$$\mathcal{T}_{S^t(x)} = E^s(S^t(x)) \oplus E^u(S^t(x)), \tag{7.2}$$

$$dS^\tau E^s(S^t(x)) = E^s(S^{t+\tau}(x)), dS^\tau E^u(S^t(x)) = E^u(S^{t+\tau}(x)), \tag{7.3}$$

$$0 < \lambda < 1 < \mu,$$

---

[1] We shall also call it *uniformly hyperbolic* if there is no misunderstanding.
[2] The indices "s", "u" mean "stable" and "unstable" respectively.

and for all t and $\tau \geq 0$

$$\|dS^\tau v\| \leq C\lambda^\tau \|v\|, \qquad v \in E^s(S^t(x)), \tag{7.4}$$

$$\|dS^\tau v\| \geq C^{-1}\mu^\tau \|v\|, \qquad v \in E^u(S^t(x)), \tag{7.5}$$

$$\gamma(S^t(x)) \geq \text{const},$$

where $\gamma(S^t(x))$ is the angle between the subspaces $E^s(S^t(x))$ and $E^u(S^t(x))$.

When x is a periodic point (in particular a fixed point) the trajectory $\{S^t(x)\}$ is uniformely completely hyperbolic if and only if x is a hyperbolic periodic (fixed) point.

Thus Definition 1.1 suggests that the tangent space $\mathcal{T}_x$ is split into two subspaces $E^s(x)$, $E^u(x)$, such that infinitely close trajectories corresponding to $v \in E^s(x)$ ($v \in E^u(x)$) approach each other with an exponential rate when $t \to +\infty$ ($t \to -\infty$). We shall see below that such behavior of the solutions of variational equations implies a similar behavior for the trajectories of S^t.

1.3. Nonuniform Hyperbolicity. One can weaken the conditions of uniform hyperbolicity in two ways. Firstly, the hyperbolicity may be nonuniform (but complete). Secondly, it may be only partial, i.e. holds only for a part of the tangent space.

Definition 1.2. A trajectory $\{S^t(x)\}$ is said to be *nonuniformely hyperbolic* if there exist subspaces $E^s(S^t(x))$, $E^u(S^t(x))$, $t \in \mathbb{Z}$ satisfying (7.2.) and (7.3) and some numbers λ, μ satisfying (7.4), and a number $0 < \alpha < 1$, and for any ε, $0 < \varepsilon < \alpha \cdot \min\{\ln \mu, (-\ln \lambda)\}$ there exists a function $C(x, \varepsilon) > 0$ such that

$$C(S^t(x), \varepsilon) \leq C(x, \varepsilon)e^{\varepsilon|t|} \quad (t \in \mathbb{Z}) \tag{7.6}$$

and for all t and $\tau \geq 0$

$$\|dS^\tau v\| \leq C(S^{t+\tau}(x), \varepsilon)\lambda^\tau \|v\|, \qquad v \in E^s(S^t(x)),$$

$$\|dS^\tau v\| \geq C^{-1}(S^{t+\tau}(x), \varepsilon)\mu^\tau \|v\|, \qquad v \in E^u(S^t(x)), \tag{7.7}$$

$$\gamma(S^t(x)) \geq C^{-1}(S^t(x), \varepsilon)\gamma(x).$$

The inequality (7.6) has the following sense: the estimates (7.7) get worse along the trajectory (when t is increasing) but relatively slowly (i.e. with an exponential rate which is small comparing with $\max\{\ln u, \ln \lambda^{-1}\}$. These conditions are not merely technical, but play a basic role in the study of nonuniform hyperbolicity. At first sight one can consider the condition (7.6) to be rather artificial and restrictive; but we shall see below that this is not so and that typical trajectories (with respect to some invariant measure) often satisfy it.

Notice that for a periodic trajectory $\{S^t(x)\}$ the conditions of uniform and nonuniform hyperbolicity coincide.

We obtain the definition of *partial hyperbolicity* if we replace (7.4) in the definition of complete hyperbolicity by

Chapter 7. General Theory of Smooth Hyperbolic Dynamical Systems

$$0 < \lambda < \min(1, \mu) \tag{7.8}$$

(i.e. without the assumption $\mu > 1$). Similarly we differ the concepts of *uniform partial hyperbolicity* and of *nonuniform partial hyperbolicity*. In examples, the conditions of partial hyperbolicity are usually realized in the following way. The tangent space $\mathcal{T}_{S^t(x)}$ can be split into a direct sum of three subspaces invariant under dS^t

$$\mathcal{T}_{S^t(x)} = E^s(S^t(x)) \oplus E^0(S^t(x)) \oplus E^u(S^t(x)), \tag{7.9}$$

where $dS^t|E^s(S^t(x))$ is a contraction, $dS^t|E^u(S^t(x))$ is an expansion ((7.5) or (7.7) holds depending on whether hyperbolicity is uniform or not). The subspace $E^0(S^t(x))$ is called *neutral*. Vectors lying in this subspace may contract and expand but not too fast.

More exactly (we restrict ourselves to the case of uniform partial hyperbolicity) there exist C_1, λ_1, μ_1 such that

$$C_1 > 0, \quad 0 < \lambda < \lambda_1 \leqslant 1 \leqslant \mu_1 \leqslant \mu < \infty, \tag{7.10}$$

and for all t and $\tau > 0$

$$C_1^{-1} \lambda_1^\tau \|v\| \leqslant \|dS^\tau v\| \leqslant C_1 \mu_1^\tau \|v\|, \quad v \in E^0(S^t(x)). \tag{7.11}$$

We shall call this variant of hyperbolicity *partial hyperbolicity in a narrow sense*.

The conditions of hyperbolicity for dynamical systems with continuous time (uniform, nonuniform, partial) are defined similarly; the only difference consists in another (comparatively to (7.2)) splitting of the tangent space $\mathcal{T}_{S^t(x)}$, $(t \in \mathbb{R})$, namely

$$\mathcal{T}_{S^t(x)} = E^s(S^t(x)) \oplus E^u(S^t(x)) \oplus X(S^t(x)), \tag{7.12}$$

where $X(S^t(x))$ is a one-dimensional subspace generated by the flow (obviously invariant under dS^t). Notice that comparing (7.9) and (7.12) one sees that any diffeomorphism S^t (which acts as the time t translation along the trajectories of the flow) satisfies the conditions of partial (uniform or non-uniform) hyperbolicity.

1.4. Local Manifolds. First of all hyperbolicity conditions allow us to study the assymptotic behavior in a neighbourhood of a hyperbolic trajectory. The exact description is given by the following theorem which is one of the key facts in the hyperbolic theory.

Theorem 1.1 (on local manifold, see [Pe2]). *Let $\{S^t(x)\}, (t \in \mathbb{Z})$ be a nonuniform hyperbolic trajectory. Then there exists a local stable manifold (LSM) $V^s(x)$ such that for $y \in V^s(x)$ the distance between $\{S^t(x)\}$ and $\{S^t(y)\}$ decreases with an exponential rate, i.e. for any t and $\tau \geqslant 0$*

$$d(S^{t+\tau}(x), S^{t+\tau}(y)) \leqslant KC(S^t(x), \varepsilon) \lambda^\tau e^{\varepsilon \tau} d(S^t(x), S^t(y)), \tag{7.13}$$

where d is the distance induced by the Riemannian metric and $K > 0$ is a constant.

Let us make some remarks.

Remark 1.1. In our case a local manifold is defined using a smooth map

$$\psi(x): B^s(r) \to E^u(x)$$

satisfying

$$\psi(0) = 0 \quad \text{and} \quad d\psi(0) = 0. \tag{7.14}$$

Here $B^s(r)$ is a ball of radius r in $E^s(x)$ centered at the origin: $r = r(x)$ is said to be the size of LSM. One gets LSM by projecting the graph of ψ into M by exp.

It follows from (7.14) that

$$x \in V^s(x), \qquad \mathcal{T}_x V^s(x) = E^s(x). \tag{7.15}$$

Remark 1.2. One can prove the existence of LSM if $\{S^t\} \in C^{1+\alpha}, 0 < \alpha \leq 1$ (i.e. the first derivatives satisfy Hölder condition of degree α). If $\{S^t\} \in C^{r+\alpha}, r \geq 1$, $0 < \alpha \leq 1$, then $V^s(x) \in C^r$ (cf [Pe2]).

Remark 1.3. Any point $y = S^t(x)$ of a nonuniform hyperbolic trajectory may be considered as an initial point. Thus LSM $V^s(y)$ can be constructed for any such y. The sizes of LSM at x and at $S^t(x)$ are related by

$$r(S^t(x)) \geq K e^{-\varepsilon|t|} r(x), \tag{7.16}$$

where $K > 0$ is some constant. In a typical situation $r(S^t(x))$ is an oscillating function of t and for many values of t, $r(S^t(x))$ is of the same order as $r(x)$. Nevertheless, for some t, $r(S^t(x))$ may become as small as (7.16) allows.

Remark 1.4. Comparing 7.13 and 7.16 one can see that the rate of decreasing of LSM is small comparatively to the rate of convergence of $\{S^t(x)\}$ and $\{S^t(y)\}$.

Remark 1.5. If a trajectory $\{S^t(x)\}$ is uniformely hyperbolic (completely or partially) then it certainly satisfies the conditions of Theorem 1 (uniform hyperbolicity is a particular case of nonuniform). The corresponding statement is known as Hadamard-Perron theorem (cf [An1], [PE2]). It holds for $\{S^t\} \in C^1$, and if $\{S^t\} \in C^r, r \geq 1$ then $V^s(x) \in C^r$. In addition, one can prove in this case that $r(S^t(x)) \geq $ const and $C(S^t(x), \varepsilon) \leq $ const (see (7.6)) for all t.

Remark 1.6. There is a symmetry in hyperbolic theory between the objects with indices "s" and "u". Namely when time direction is reversed the statements concerning some object with index "s" go over into the statements about the corresponding object with index "u". In particular this allows us to define a *local unstable manifold* (LUM) $V^u(x)$ of the point as a LSM for S^{-1}. Of course it has all the properties similar to those of $V^s(x)$.

Remark 1.7. For a dynamical system with continuous time, LSM and LUM are defined as being constructed for the diffeomorphism S^1, a unit time translation along the trajectories of the flow. We shall see below that LSM and LUM play a crucial role in the analysis of ergodic and topological properties of

Chapter 7. General Theory of Smooth Hyperbolic Dynamical Systems   113

hyperbolic dynamical systems. Only in a very specific situation when a system has additional symmetry can local manifolds be defined explicitely (see below). Typically they are constructed using a variant of the contractive transformation theorem (cf [Pe2]).

1.5. Global Manifolds. The *global stable manifold* (GSM) at a point $x \in M$ is the smoothly immersed submanifold

$$W^s(x) = \bigcup_{-\infty < t < \infty} S^{-t}(V^s(S^t(x))) \quad (t \in \mathbb{Z}) \tag{7.17}$$

This manifold has the same class of smoothness as LSM and $y \in W^s(x)$ if and only if

$$d(S^t(x), S^t(y)) \to 0 \quad \text{when } t \to \infty.$$

The *global unstable manifold* (GUM) is defined similarly.

For dynamical systems with continuous time, GSM is defined by (7.17) where $t \in \mathbb{R}$. An additional structure related to the continuity of the time allows to construct a smooth *global weakly stable* immersed *manifold* as

$$W^{so}(x) = \bigcup_{-\infty < t < \infty} W^s(S^t(x)). \tag{7.18}$$

For any $y \in W^{so}(x)$ every trajectory $\{S^t(y)\}$ is in $W^{so}(x)$. The *global unstable manifold* and the *global weakly unstable manifolds* for flows are defined similarly.[3]

§2. Basic Classes of Smooth Hyperbolic Dynamical Systems. Definitions and Examples

2.1. Anosov Systems. Definition 2.1. A dynamical system is called *Anosov* (*Anosov diffeomorphism* or *Anosov flow* respectively) if any trajectory of this system is uniformely completely hyperbolic and the constants C and λ can be chosen independent of the point.[4]

Theorem 2.1. (cf [An1], [AS]). *Let $\{S^t\}$ be a C^1-Anosov system. Then*
1) *the subspaces $E^s(x)$, $E^u(x)$, $x \in M$ generate two continuous invariant distributions (subbundles) of the tangent space $\mathcal{T}$ (denoted respectively by E^s and E^u). These distributions are transversal (i.e. $E^s(x) \cap E^u(x) = 0$ for any $x \in M$).*

---
[3] One uses also a different system of notation and different terminology for continuous time systems. GSM is denoted by $W^{ss}(s)$ and is called the strong stable manifold, and the global weakly stable manifold is denoted by $W^s(x)$ and is called the stable manifold. A similar system of notation and terminology is used with respect to unstable manifolds.
[4] There is an interesting question: If any trajectory of a system is uniformely completely hyperbolic, is this an Anosov system? [Pe2] gives a positive answer to this question for C^2-systems which conserve a measure equivalent to the Riemannian volume.

2) *if $\{S^t\}$ is a $C^{1+\alpha}$ system, $\alpha > 0$ then the distribution E^s satisfies the Hölder condition: $\tilde{d}(E^s(x), E^s(y)) \leq C\rho(x,y)^\beta$ where $C > 0$, $\beta \in (0, \alpha]$ are constants independent of x and y, $\tilde{d}$ is the distance in the Grassman bundle induced by the Riemannian metric; the similar assertion is true for E^u.*

3) *the distribution E^s is integrable; the corresponding maximal integral submanifolds form a continuous C^1-foliation on M (see [An1] for definition) denoted by W^s. The leaf of W^s passing through x coincides with GSM $W^s(x)$. The foliation W^s is S^t-invariant (i.e. $S^t(W^s(x)) = W^s(S^t(x))$ for any $x \in M$).*

The foliation W^s is contracting, i.e. for any $x \in M$; $y \in W^s(x)$, $t \geq 0$

$$d^s(S^t(x), S^t(y)) \leq C\lambda^t d^s(x, y),$$

where $C > 0$ is a constant (independent of x and y) and d^s is the distance measured along the leaves of $W^s(x)$ induced by the Riemannian metric on the leaf which is considered as a smooth submanifold in M.

4) *similar assertions are true for the distribution E^u. The corresponding distribution is denoted by W^u, it is invariant under $\{S^t\}$ and is contracting for $t \leq 0$. The foliations W^s and W^u are transversal at each point $x \in M$ (i.e. $V^s(x) \cap V^u(x) \cap U(x) = x$, where $U(x)$ is a small neighbourhood of x).*

5) *If the time is continuous, the distributions $E^s \oplus X$ and $E^u \oplus X$ (see 7.12) are continuous, invariant under dS^t and integrable; their maximal integrable submanifolds form continuous S^t-invariant C^1-foliations on M denoted respectively by W^{so} and W^{uo}, the leaf of W^{so} (respectively of W^{uo}) passing through x coincides with $W^{so}(x)$ (respectively with $W^{uo}(x)$).*

We shall formulate two important properties concerning the topological structure of Anosov diffeomorphisms (a more detailed description of topological properties is given in [An1], [PS1], the case of continuous time is also considered there).

A point $x \in M$ is called *nonwandering* with respect to a homeomorphism S if for any neighbourhood U of **x** there exists an $n \neq 0$ such that $S^n(U) \cap U \neq \emptyset$. A homeomorphism S is called *topologically transitive* if it has an everywhere dense trajectory.

Theorem 2.2 (cf [An1], [PS1]). 1) *If S is a topologically transitive C^2-Anosov diffeomorphism, then $\overline{W^s(x)} = M$, $\overline{W^u(x)} = M$ for any $x \in M$.*

2) *Periodic points of a C^2-Anosov diffeomorphism S are dense in the set of nonwandering points $\Omega(S)$[5]. The number P_n of periodic points with period $\leq n$ is finite and*

$$\lim_{n \to \infty} \frac{\log P_n}{n} = h,$$

where h is the topological entropy of $S|\Omega(S)$ (for the definition of topological entropy see Sect. 3).

---

[5] The set $\Omega(S)$ is closed and S-invariant.

We now consider some examples of Anosov systems. We shall use Tor^n to denote n-torus identified with the factor space $\mathbb{R}^n/\mathbb{Z}^n$. Let A: Tor$^n \to$ Tor^n be an algebraic automorphism given by an integer matrix $A = (a_{ij})$ and let $\lambda_1, \ldots, \lambda_n$ be the complex eigenvalues of A.

Theorem 2.3 (cf [An1]). *An algebraic automorphism is Anosov if and only if $|\lambda_i| \neq 1$ for all $i \in [1, n]$ (in this case A is called a hyperbolic toral automorphism).*

Since det $A = \pm 1$, A conserves the Lebesgue measure on Tor^n. GSM and GUM coincide with the projection of k-dimensional and of $(n - k)$-dimensional hyperplanes on Tor^n. Here k is the number of i: $|\lambda_i| < 1$, k-dimensional hyperplanes are parallel to the subspace spanned by the corresponding eigenvectors and $(n - k)$-dimensional hyperplanes are parallel to the subspace spanned by eigenvectors corresponding to $|\lambda_i| > 1$. A little C^1-perturbation of a hyperbolic toral automorphism[6] is an Anosov diffeomorphism on the torus (see Theorem 2.11 below). Moreover any Anosov diffeomorphism on the torus is topologically conjugate to some hyperbolic automorphism. There exist examples of Anosov automorphisms on nilmanifolds (nilmanifold is a factor of some nilpotent Lie group by a discrete subgroup) different from tori.

The most simple way to obtain an Anosov flow is to construct a special flow over an Anosov diffeomorphism (see Chap. 1, Sect. 4). Geodesic flows on Riemannian manifolds of negative curvature give another class of interesting examples. Their properties are discussed in details in Sect. 4.

2.2. Hyperbolic Sets

Definition 2.2. A set Λ is called a *hyperbolic set* with respect to the dynamical system $\{S^t\}$ if it is closed and consists of trajectories satisfying the condition of uniform complete hyperbolicity with the same constants C and λ.

If Λ coincides with M then $\{S^t\}$ is an Anosov system. If $\Lambda \neq M$ then Λ contains "holes" i.e. has a structure of a Cantor set. Nevertheless local properties of trajectories in Λ are similar to those of Anosov systems.

Similarly to Anosov systems, the subspaces $E^s(x)$ and $E^u(x)$ form two continuous distributions (subbundles) of the tangent bundle $\mathcal{T}\Lambda$ (denoted respectively by E^s and E^u). They are invariant under dS^t, transversal to each other and satisfy Hölder condition (if $\{S^t\} \in C^{1+\alpha}$). The intersections of Λ with GSM i.e. the sets $\Lambda \cap W^s(x)$ and similarly the sets $\Lambda \cap W^u(x)$ form two C^1-*laminations of* Λ[7] (cf [PS1]) tangent to E^s and to E^u respectively. In the case of continuous time there exist two additional C^1-laminations of Λ formed by $\Lambda \cap W^{so}(x)$ and $\Lambda \cap W^{uo}(x)$. The sets

---

[6] i.e. a map $A + \varepsilon S$ where S is a diffeomorphism of the torus $S(x_i, \ldots, x_n) = (f_1(x_1, \ldots, x_n), \ldots, f_n(x_1, \ldots, x_n))$ and $f_i(x_1, \ldots, x_n)$ are periodic in $x_1, \ldots, x_n$ with the period 1.
[7] A C^1-lamination of Λ is a partition of Λ induced by the restriction on Λ of a C^1-foliation defined in a neighbourhood of Λ.

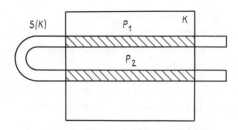

Fig. 1

$$W^s(\Lambda) = \bigcup_{x \in \Lambda} W^s(x), \quad W^u(\Lambda) = \bigcup_{x \in \Lambda} W^u(x)$$

are called respectively *stable* and *unstable manifolds of hyperbolic set* Λ.

For $C^{1+\alpha}$ diffeomorphisms $\alpha \in (0, 1]$ it is known that the Riemannian volume of Λ equals 0 or 1 (cf [BO2]); in the last case we dealt with some Anosov diffeomorphisms. It is not the same for C^1-diffeomorphisms: a counterexample is given in [BO2].

An example of a hyperbolic set is the so-called *Smale horseshoe*. In order to have an image of it we consider a map S of the square $K = [0, 1] \times [0, 1] \subset \mathbb{R}^2$ with the following properties. First, the square is strongly stretched in the horizontal direction and compressed in the vertical direction, then it is bent in the form of a horseshoe, and this horseshoe is finitely mapped back on the square so that $K \cap S(K)$ consists of two strips $P_i = [0, 1] \times [a_i, b_i]$, $i = 1, 2$, $0 < a_1 < b_1 < a_2 < b_2 < 1$ (see Fig. 1). Each strip P_i contains two substrips $P_{i_1 i_2}$, $i_1, i_2 = 1, 2$ which are parts of $S^2(K)$ and so on. When $n \to \infty$ the width of $P_{i_1 i_2 \cdots i_n}$ tends to zero. Under the action of S^{-1}, horizontal and vertical directions are interchanged, otherwise the situation is similar. The existence of a C^2-map with the above properties is proved for example in [N]. It is also proved that the set $\Lambda = \bigcap_{-\infty < n < \infty} S^n(K)$ is hyperbolic.

One can generalize the above construction by forming at the first step any finite number of strips, say l, instead of 2. Then at the second step l^2 strips are formed, and so on (the number of vertical strips corresponding to the map S^{-1} may differ from l). One can finitely generalize this construction for the multi-dimensional case (cf [C]). There are other modifications of the above construction. We wish to point out a smooth realization of Smale horseshoe from [N]: there exists a C^2-diffeomorphism $\tilde{S}$ of two-dimensional sphere which extends the map S constructed above so that $\Omega(S) = \Lambda \cup p \cup q$ where p is a stable fixed point and q is an unstable one[8].

Other examples of hyperbolic sets give invariant sets in a neighbourhood of the trajectory of a homoclinic point. Let p be a hyperbolic periodic point of a

---

[8] First a map $S_0 : \mathscr{D}^2 \to \mathscr{D}^2$ of a disk $\mathscr{D}^2$ containing K is constructed which extends S and then one extends S_0 to the desired map S.

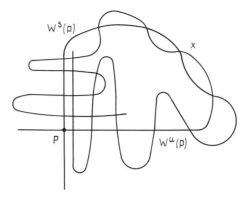

Fig. 2

diffeomorphism S. A point x of intersection between the stable manifold $W^s(p)$ and the unstable manifold $W^u(p)$ which is different from p is called a *homoclinic point*. If the intersection of $W^s(p)$ with $W^u(p)$ is transversal at x (i.e. $\mathcal{T}_x W^s(p) \oplus \mathcal{T}_x W^u(p) = \mathcal{T}_x$ and $\mathcal{T}_x W^s(p) \cap \mathcal{T}_x W^u(p) = 0$) then x is called a *transversal homoclinic point*. The existence of one transversal homoclinic point implies an infinite number of homoclinic points with different trajectories. Roughly speaking we have the following situation. Since $W^s(p)$ and $W^u(p)$ are invariant, the trajectory of a homoclinic point consists of homoclinic points. This gives rise to strong oscillations of $W^u(p)$ which stretches along itself when the iterates of x converge to p along $W^s(p)$. A similar picture is valid for $W^s(p)$. Joining the two pictures generates a grid of homoclinic points (see Fig. 2). Now let us consider a compact set Γ which is the union of p with the orbit of some transversal homoclinic point x and choose some open sets $U_0, U_1, \ldots, U_n$, $U_i \cap U_j \neq \emptyset$ in the following way: U_0 is a neighborhood of p which contains all but a finite number of points of the orbit $\{S^n(x)\}$ and any of the remainder points of orb x is covered by one of U_i.

Theorem 2.4 (cf [A2]). *For any neighborhood V of the set Γ there exist neighborhoods $U_i \in V$, $i = 0, 1, \ldots, n$ such that*

$$\Lambda = \bigcap_{m=-\infty}^{\infty} S^m \left(\bigcup_{i=1}^{n} U_i \right)$$

is an invariant hyperbolic set[9].

Thus Λ consists of points y whose orbits $\{S^n y\}$ $-\infty < n < \infty$ are contained in U. When U diminishes Λ tends to Γ.

Let us give a simple explanation of how the appearance of a homoclinic point gives rise to "stochastic trajectories". We restrict ourselves to the case of a

---

[9] One can show that Λ is a Cantor set, i.e. a perfect nowhere dense set.

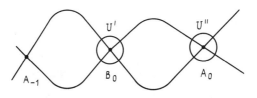

Fig. 3

C^2-diffeomorphism of two-dimensional cylinder M which has a fixed hyperbolic point $\mathcal{O}$. Let γ^u and γ^s denote the pieces of $W^u(\mathcal{O})$ and of $W^s(\mathcal{O})$ bounded by θ from one side and by a transversal homoclinic point A_0 from another side. The point $S^{-1}A_0 = A_{-1}$ obviously is also homoclinic. It is clear that there must be at least one other homoclinic point B_0 between A_{-1} and A_0 (see Fig. 3). Consider two sufficiently small neighbourhoods $U' \ni B_0$ and $U'' \ni A_0$ and for a given point x let 1 correspond to an iterate of x if this iterate falls in U', and 0 if it falls in U''. Let $\varepsilon = \{\varepsilon_1, \varepsilon_2, \ldots, \varepsilon_n\}$ be a sequence of 0 and 1.

Then there exists a point $x \in \Lambda$ such that the iterates of x fall consequently in U' and in U'' according to the sequence ε. As the major part of sequences ε are "random", we see that random sequences are "realized" by trajectories of S'.

Homoclinic points are a particular case of the so-called *heteroclinic points* (a point x is called heteroclinic if there exist two periodic points p_1 and p_2 such that $x \in W^s(p_1) \cap W^u(p_2)$). The preceding constructions and results are still valid for cycles of heteroclinic orbits.

2.3. Locally Maximal Hyperbolic Sets

A given hyperbolic set may have a very complicated structure. We shall therefore restrict ourselves to a special though sufficiently large class of hyperbolic sets introduced by V.M. Alekseev [A1] and D.V. Anosov [An2].

Definition 2.3. A compact set Λ invariant under a diffeomorphism S is called *locally maximal* (or isolated) if there exists a neighborhood $U(\Lambda) \supset \Lambda$ such that for any S-invariant Λ', with $\Lambda \subset \Lambda' \subset U(\Lambda)$, Λ' coincides with Λ.

Let Λ be a locally maximal hyperbolic set (LMHS) of a diffeomorphism S. Let us denote by $\Omega(S|\Lambda)$ the set of non-wandering points and by $\text{Per}(S|\Lambda)$ the set of periodic points of S restricted on Λ.

Theorem 2.5 (*Spectral decomposition theorem for LMHS cf* [Bo2], [K1]).
1) The set $\Omega(S|\Lambda)$ may be uniquely represented as a disjoint union of closed invariant subsets $\Omega_1, \ldots, \Omega_k$, such that the action of S on any Ω_i is topologically transitive.
2) Any Ω_i may be uniquely represented as a disjoint union of closed sets $\Omega_i^1, \ldots, \Omega_i^{n_i}$ which are cyclically permuted by S and $S^{n_i}|\Omega_i^1$ is topological mixing (i.e. for any $U, V \subset \Omega_i^1$ open in the topology of Ω_i^1 there exists an N such that $(S^{n_i})^n(U) \cap V \neq 0$ for any $n \geq N$).

3) $\overline{W^s(x)} \supseteq \Omega_i$, $\overline{W^u(x)} \supseteq \Omega_i$ for any $x \in \Omega_i$.
4) $\Omega(S|\Lambda) = \overline{\text{Per}(S|\Lambda)}$.

A similar spectral decomposition theorem of LMHS is true for a dynamical system $\{S^t\}$ with continuous time. In this case the following alternative is an analog of statements 2), 3) about a component of topological transitivity Ω_i: either for any $x \in \Omega_i$ $\overline{W^s(x)} \supset \Omega_i$ and $\overline{W^u(x)} \supset \Omega_i$, or $\{S^t\}$ may be represented as a special flow (see Chap. 1, Sect. 4) over a homeomorphism T of some compact X(cf [Pla]). Although the set X may have a complicated topological structure (it may contain "holes"), it can be shown that $T|X$ possesses some kind of hyperbolic properties which are expressed by the so-called axiom A^* (cf [PS1]) or its other variant axiom $A^{\#}$ (cf [AJ]). A similar alternative is certainly valid for topologically transitive Anosov flows. In this case when $\{S^t\}$ is a special flow over the map $T: X \to X$ one can prove that X is a smooth compact Riemannian manifold and T is an Anosov diffeomorphism ([An]).

2.4. A-Diffeomorphisms. One says that a diffeomorphism S satisfies axiom A (or is an A-diffeomorphism (cf [Bo2])) if its non-wandering set $\Omega(S)$ is hyperbolic and its periodic points are dense in $\Omega(S)$. One can show that if S satisfies axiom A then $\Omega(S)$ is locally maximal. The components of topological transitivity of $\Omega(S)$ (i.e. the sets Ω_i of Theorem 2.5.), are called basic sets.

2.5. Hyperbolic Attractors

Definition 2.4. A set Λ is called an *attractor* of a dynamical system $\{S^t\}$ ($t \in \mathbb{Z}$ or $t \in \mathbb{R}$) if there exists a neighborhood $U \supset \Lambda$ such that $S^t(U) \subset U$ for $t > 0$ and $\Lambda = \bigcap_{t \geq 0} S^t(U)$.[10]

One can easily see that Λ is closed, locally maximal and invariant under $\{S^t\}$.

Definition 2.5. A set Λ is called a *hyperbolic attractor* if it is an attractor and simultaneously a hyperbolic set.

Although hyperbolic attractors do not appear in simple physical systems they are good models for real situations.

Theorem 2.6 (cf [PS1]). *A LMHS Λ is a hyperbolic attractor if and only if $W^u(x) \subset \Lambda$ for any $x \in \Lambda$.*

Thus a hyperbolic attractor is a union of unstable manifolds of its points and the complexity of topological structure of such an attractor is related to the fact that the intersections of unstable manifolds with some transversal submanifold constitute a set of Cantor type.

The first example of a hyperbolic attractor, the *solenoid*, was constructed by Smale and Williams.

---

[10] If we require beforehand that Λ be closed, then the condition $\overline{S^t(U)} \subset U$ may be rejected.

The corresponding general construction is given for example in [N] and here we shall consider only the most simple variant. Let g be a map of the circle S^1 given by $g(x) = 2x \pmod 1$ and let ψ_0 be the immersion of S^1 into $M = \mathscr{D}^2 \times S^1$ given by $\psi(x) = (0, 0, 2x)$. We approximate this immersion in C^1-topology by an inclusion $\psi(x) = (\psi_1(x), \psi_2(x), 2x)$ and then extend ψ to a map $S: M \to M$ such that any section $\mathscr{D}^2 \times x_0$ of $\mathscr{D}^2 \times S^1$ is uniformely compressed and then is mapped inside the section $\mathscr{D}^2 \times \psi(x_0)$ (an example of such a map S is $S(\alpha, \beta, x) = (10^{-10}\alpha + \frac{1}{2}\cos x, 10^{-10}\beta + \frac{1}{2}\sin x, 2x)$. The set $\Lambda = \bigcap_{n \geq 0} S^n(M)$ is called solenoid. Topologically it is a locally trivial fibre bundle with the base space S^1 whose fibres are perfect Cantor sets. Λ is connected but not locally connected ([N]).

One can obtain another example of a hyperbolic attractor from a hyperbolic automorphism A of the torus Tor^2. More exactly the following holds.

Theorem 2.7 (c.f. [Pl]). *There exists a C^2-diffeomorphism $S: \text{Tor}^2 \to \text{Tor}^2$ and two open neighborhoods $U_1 \subset U_2$ of the origin $\mathcal{O} = (0,0)$ such that*
1) $S(x) = A(x)$ *for* $x \in \text{Tor}^2 \setminus U_2$;
2) $S(x) = B(x)$ *for* $x \in U_1$ *where* $B: U_1 \to \text{Tor}^2$
has three fixed points: an unstable fixed point $\mathcal{O}$ and two hyperbolic points $\mathcal{O}_1$, $\mathcal{O}_2 \in W^u(\mathcal{O})$ such that the stable and the unstable manifolds of $\mathcal{O}_1$ and $\mathcal{O}_2$ are subsets of the corresponding GSM and GUM of these points with respect to A.

S has a hyperbolic attractor which coincides with the closure of global unstable manifolds of $\mathcal{O}_1$ and $\mathcal{O}_2$.

In a small neighborhood of a hyperbolic set (in particular of a hyperbolic attractor) the stochastic properties of a dynamical system are displayed the most evidently. In many cases the origin of stochastic behavior and of some or another kind of instability is the existence in the phase space of a dynamical system of invariant sets which may be modelled approximately by the Smale horseshoe or by the solenoid of Smale and Williams (or by their modifications).

2.6. Partially Hyperbolic Dynamical Systems

Definition 2.6. A dynamical system is said to be *uniformely partially hyperbolic* (UPH-system) if any trajectory of this system satisfies the condition of partial hyperbolicity and one can choose the same constants C, λ, μ for all trajectories.

An invariant set Λ which consists of uniformely partially hyperbolic trajectories with the same constants C, λ, μ is called a *uniformely partially hyperbolic set* (UPH-set). If an UPH-set is an attractor, then it is called a *uniformely partially hyperbolic attractor* (UPH-attractor). Under UPH-system (respectively UPH-set of UPH-attractor) in the narrow sense we mean an UPH-system whose trajectories all satisfy the condition of partial hyperbolicity in the narrow sense.

Perhaps the simplest example of an UPH-system is constructed in the following way. Let T be an Anosov diffeomorphism of a smooth manifold M, and let N be a smooth manifold. Then $S: M \times N \to M \times N$ defined by $S(x, y) = (T(x), y)$, $x \in M$, $y \in N$ is an UPH-diffeomorphism. The neutral distribution of S is integrable, the corresponding leaves are compact and diffeomorphic to N. Let $\pi\colon M \times N \to M$ be the projection. Then using natural notation we have $\pi(W_S^s(z)) = W_T^s(\pi(z))$, $\pi(W_S^u(z)) = W_T^u(\pi(z))$.

A more general construction is a skew-product over an Anosov diffeomorphism. Any small perturbation of S defined above is an UPH-diffeomorphism (see Theorem 2.11). Moreover it is conjugate to a skew-product over an Anosov diffeomorphism.

One can similarly construct examples of UPH-sets and of UPH-attractors: if a map T has a hyperbolic set (respectively a hyperbolic attractor) then $\Lambda \times N$ is an UPH-set (respectively an UPH-attractor) for S and any small perturbation of S has an UPH-set (respectively an UPH-attractor).

2.7. Mather Theory. A dynamical system $\{S^t\}$ on a smooth compact Riemannian manifold M generates a group of continuous linear operators $\{S_*^t\}$ acting on a Banach space $\Gamma^0(\mathcal{T}M)$ of continuous vector fields $v(x)$ on M by the formula

$$(S_*^t v)(x) = dx^t v(S^{-t}(x)).$$

The spectrum Q of the complexification of S_*^1 is called the *Mather spectrum* of the dynamical system $\{S^t\}$.

Theorem 2.8 (on the structure of Mather spectrum, Mather [Mat1], [KSS]). *Let $\{S^t\}$ be a C^1-dynamical system on a smooth compact Riemannian manifold M. Suppose in addition that nonperiodic trajectories of S^t are dense in M. Then:*

1) *Any connected component of the spectrum Q coincides with some ring Q_i with radii λ_i, μ_i, the number of such rings p is less than or equal to $\dim M$, and $0 < \lambda_1 \leqslant \mu_1 < \cdots < \lambda_l \leqslant 1 \leqslant \mu_l < \cdots < \lambda_p \leqslant \mu_p$;*

2) *The invariant subspace $\mathcal{E}_i \in \Gamma^0(\mathcal{T}M)$ of S_*^1 corresponding to the component Q_i of the spectrum is a module over the ring of continuous functions; the union of the subspaces $E_i(x) = \{v(x) \in \mathcal{T}_x : v \in \mathcal{E}_i\}$ constitutes a dS^1-invariant continuous distribution on $\mathcal{T}M$ and $\mathcal{T}_x = \bigoplus_{i=1}^p E_i(x)$, $x \in M$;*

3) *If the time t is continuous then the unit circle lies in Q;*

4) *Similar assertions about the spectrum of the complexification of S_*^t are true for any $t \neq 0$.*

We add to this the following.

Theorem 2.9 (cf [BP]). *Under the conditions of Theorem 2.8,*

1) *if $\mu_k < 1$ then the distribution $F_k^s = \bigoplus_{i=1}^k E_i$ is integrable and the maximal integral manifolds of this distribution generate a continuous C^1-foliation of M which we denote by W_k^s. A leaf $W_k^s(x)$ of W_k^s at some point $x \in M$ is a C^1-immersed submanifold of M[11];*

2) *if $\lambda_k > 1$ then a similar statement holds about the distribution $F_k^u = \bigoplus_{i=k}^p E_i$ (we shall denote by W_k^u the corresponding foliation and by $W_k^u(x)$ its leaf at some point $x \in M$);*

---

[11] It can be shown that for $\{S^t\} \in C^r$, $W_k^s(x) \in C^r$.

3) *the foliation W_k^s is S^t-invariant and contracting, i.e. for any $x \in M$, $y \in W_k^s(x)$, $t \geq 0$*

$$d^{(s)}(S^t(x), S^t(y)) \leq C(\lambda_k + \varepsilon)^t d^{(s)}(x, y),$$

where $C > 0$ is some constant independent of x, y, t; ε is any number satisfying $0 < \varepsilon \leq \min\{\lambda_{k+1} - \mu_k, 1 - \mu_k\}$; $d^{(s)}$ is a distance induced on $W_k^s(x)$ by the Riemannian metric on the leaf considered as a smooth submanifold of M;

4) *the foliation W_k^u is S^t-invariant and contracting for $t \leq 0$.*

Theorems 2.8, 2.9 imply the existence of two filtrations of the distributions

$$F_1^s \subset F_2^s \subset \cdots \subset F_{l-1}^s, \qquad F_{l+1}^u \subset \cdots \subset F_p^u$$

and the existence of two corresponding filtrations of the foliations

$$W_1^s \subset W_2^s \subset \cdots \subset W_{l-1}^s, \qquad W_{l+1}^u \subset \cdots \subset W_p^u.$$

Of course for an aribtrary dynamical system, filtrations may happen to be trivial (in that case the spectrum S consists of one ring, which usually contains the unit circle).

If instead of the whole manifold M, some S^t-invariant set Λ is considered, then Theorems 2.8, 2.9 remain true (in this case $\Gamma^0(\mathcal{T}\Lambda)$ must be considered instead of $\Lambda^0(\mathcal{T}M)$).

The classes of dynamical systems considered above are characterized by their Mather spectrum.

Theorem 2.10 (cf [Mat]). 1) *A diffeomorphism S^1 is Anosov if and only if its Mather spectrum Q is a disjoint union $Q = Q_1 \cup Q_2$ where Q_1 is strictly inside and Q_2 is strictly outside the unit circle.*

2) *A flow $\{S^t\}$ is Anosov if and only if its Mather spectrum Q is a disjoint union $Q = Q_1 \cup Q_2 \cup Q_3$ where Q_1, Q_2 are as above and Q_3 coincides with the unit circle.*

3) *A dynamical system is an UPH-system (respectively an UPH-system in the narrow sense) if its Mather spectrum may be represented an a disjoint union $Q = Q_1 \cup Q_2$ with $Q_1 \neq \emptyset$, $Q_2 \neq \emptyset$ (respectively $Q = Q_1 \cup Q_2 \cup Q_3$ where $Q_i \cap Q_j = \emptyset$ for $i \neq j$ and $Q_i \neq \emptyset$ for $i = 1, 2, 3$).*

4) *An invariant set Λ is hyperbolic (respectively an UPH-set) if the Mather spectrum on Λ satisfies 1) or 2) (respectively 3)).*

Let us denote by $\text{Diff}^r(M)$ (respectively by $\Gamma^r(\mathcal{T}M)$) the space of C^r-diffeomorphisms (respectively C^r-vector fields) equipped with C^r-topology. The Mather spectrum is stable under small perturbations of dynamical systems in the following sense.

Theorem 2.11 (cf [KSS]). *Let $Q = \bigcup_{i=1}^p Q_i$ be the decomposition of the Mather spectrum of a C^r-dynamical system, where $r \geq 1$, Q_i are the rings with the radii λ_i, μ_i and $0 < \lambda_1 \leq \mu_1 < \cdots < \lambda_p \leq \mu_p$, where $p \leq \dim M$. Let $\mathcal{T}M = \bigoplus_{i=1}^p E_i$ be the corresponding decomposition of the tangent bundle into dS^1-invariant subbundles E_i, $i = 1, \ldots, p$. Then for any sufficiently small $\varepsilon > 0$ there exists a neighborhood*

Chapter 7. General Theory of Smooth Hyperbolic Dynamical Systems   123

U of $\{S^t\}$ (respectively in $\text{Diff}^r(M)$ if $t \in \mathbb{Z}$ and in $\Gamma^r(\mathcal{T}M)$ if $t \in \mathbb{R}$) such that for any system $\{\tilde{S}^t\} \in U$ its Mather spectrum Q is a union of components $\tilde{Q}_i$ which are contained in the rings wih the radii $\tilde{\lambda}_i$, $\tilde{\mu}_i$ satisfying $|\lambda_i - \tilde{\lambda}_i| \leq \varepsilon$, $|\mu_i - \tilde{\mu}_i| \leq \varepsilon$. Moreover the distribution $\tilde{E}_i$ corresponding to the component Q_i satisfies $\tilde{d}(E_i, \tilde{E}_i) \leq \gamma$.

Remark 2.1. While the component Q_i of the spectrum $\{S^t\}$ is a ring, the corresponding component Q_i of the spectrum of $\{\tilde{S}^t\}$ may be a union of several rings.

It follows from Theorem 2.11 that Anosov systems and UPH-systems form open subsets respectively in $\text{Diff}^r(M)$ (in the case of diffeomorphisms) and in $\Gamma^r(\mathcal{T}M)$ (in the case of flows), $r \geq 1$. A stronger result may be proved about Anosov systems: they are structurally stable (cf [An1]). If Λ is a LMHS of a C^r-system $\{S^t\}$, then it can be proved (cf [KSS]) that any C^r-system $\{\tilde{S}^t\}$ sufficiently closed to $\{S^t\}$ in C^r-topology has a LMHS $\tilde{\Lambda}$ and there exists a homeomorphism $h: \Lambda \to \tilde{\Lambda}$ close to identity in C^0-topology (in particular $\tilde{\Lambda}$ is in a small neighborhood of Λ) such that $h \circ S^t|\Lambda = \tilde{S}^t \circ h|\Lambda$.

2.8. Nonuniformely Hyperbolic Dynamical Systems. Lyapunov Exponents. We fix a measure v invariant under a dynamical system $\{S^t\}$.

Definition 2.7. A system $\{S^t\}$ is said to be *nonuniformly completely hyperbolic* (NCH), respectively *nonuniformly partially hyperbolic* (NPH), if there exists an invariant set Λ of positive v-measure, consisting of trajectories satisfying the conditions of nonuniform complete hyperbolicity (respectively nonuniform partial hyperbolicity). In this case the functions $C(S^t(x), \varepsilon), \gamma(S^t(x))$ and constants $\lambda, \mu, \varepsilon$ (see (7.6), (7.7)) are the restrictions on a given trajectory of some measurable functions $C(x), \gamma(x), \lambda(x), \mu(x), \varepsilon(x)$.

We shall now give another definition of NCH and of NPH-systems. As it was mentioned in the beginning of this chapter, to any smooth dynamical system there is a corresponding co-cycle $\mathcal{U}(X, t)$ which describes the behavior of solutions of variational equations. This cocycle satisfies the conditions formulated in Sect. 2 of Chap. 1, thus it defines a Lyapunov characteristic exponent $\chi(x, v)$ on the tangent bundle $\mathcal{T}M(x \in M, v \in \mathcal{T}_x)$.

Suppose that $\{S^t\}$ is a regular trajectory (see Chap. 1, Sect. 2).

Theorem 2.12 (cf [Pe2]). *A trajectory $\{S^t(x)\}$ is nonuniformely (completely) hyperbolic if*

$$\chi(x, v) \neq 0 \begin{cases} \text{for any } v \neq \alpha X(x) & \text{if } t \in \mathbb{R} \\ \text{for any } v \in \mathcal{T}_x & \text{if } t \in \mathbb{Z} \end{cases} \quad (7.19)$$

(here $X(x)$ is the vector field corresponding to the flow $\{S^t\}$, $\alpha \in \mathbb{R}$). A trajectory $\{S^t(x)\}$ is nonuniformely partially hyperbolic, if

$$\chi(x, v) < 0 \quad \text{for some } v \in \mathcal{T}_x. \quad (7.20)$$

It follows from this theorem and from the multiplicative ergodic theorem (see Chap. 1, Sect. 2) that the definition of an NCH (respectively of an NPH)-system is equivalent to the following: A set Λ consisting of trajectories satisfying 7.19 (respectively 7.20) is the support of some invariant Borel measure (for this reason NCH-systems are called the *systems with non-zero Lyapunov exponents*).

For NCH and NPH-systems the distribution E^s is only measurable. In order to have more exact information we have to consider the sets

$$\Lambda_l = \left\{ x \in M : C(x) \leqslant l, \gamma^{-1}(x) \leqslant l, \lambda(x) \leqslant \frac{l-1}{l} < \frac{l+1}{l} \leqslant \mu(x) \right\}. \quad (7.21)$$

The sets Λ_l are obviously measurable, $\Lambda_l \subset \Lambda_{l+1}$ and $\bigcup_l \Lambda_l = \Lambda \pmod{0}$. The estimates (7.6), (7.7) are uniform for $x \in \Lambda_l$ but become worse with the growth of l. As shows in [Pe2], the sets Λ_l are closed and the distribution E^s is continuous on Λ_l. Moreover the sizes of LSM are uniformly bounded from below on Λ_l and LSM $V^s(x)$ continuously depend on $x \in \Lambda_l$ (for this reason Λ_l is said to be "a set with uniform estimates").

We shall use mes to denote the Riemannian volume on M. When mes $\Lambda > 0$ (we do not suppose here that mes is an invariant measure) LSM on Λ_l have an important property of *absolute continuity*. Consider an $x \in \Lambda$, fix an $l > 0$ and choose a small neighborhood $U(x)$ of x with the size depending on l. For $y \in \Lambda_l \cap U(x)$ consider the LSM $V^s(y)$. Take two smooth submanifolds W_1, $W_2 \subset U(x)$ transversal to $V^s(y)$. We define $A_i = \{z : z = W_i \cap V^s(y) \text{ for some } y \in \Lambda_l \cap U(x)\}$, $i = 1, 2$, and let $p: A_1 \to A_2$ be defined as a map which lets the point $p(z)$ correspond to $z \in V^s(y)$, lying on the same LSM (p is called the *projection map*).

Let μ_{W_i} denote the measure induced on W_i, $i = 1, 2$ by the restriction of the Riemannian measure. If l is sufficiently large we have $\mu_{W_1}(A_1) > 0$.

Theorem 2.13 (on the absolute continuity of LSM, see [Pe2]). *The measure $p^* \mu_{W_1}$ is absolutely continuous with respect to the measure μ_{W_2} (for any choice of x, $U(x)$, l, W_1, W_2).*

Let ξ_l be the partition of $U \cap \Lambda_l$ generated by $V^s(y)$ where $y \in \Lambda_l \cap U$. It follows from Theorem 2.13 that for almost every element $C_{\xi_l}(y) = V^s(y) \cap U \cap \Lambda_l$ of ξ_l the conditional measure mes $(\cdot | C_{\xi_l})$ is absolutely continuous with respect to $\mu_{V^s(y)}$. This fact plays a crucial role in the study of ergodic properties of hyperbolic dynamical systems. One can say that it is a bridge which allows us to pass from local properties of dynamics to the study of global behavior. The property of absolute continuity and the proof of Theorem 2.13 for Anosov systems was given in [An1] (see also [AS]). For UPH-systems it was formulated in [BP]. The case of NPH-systems was considered in [Pe2]. In the case of Anosov systems and of UPH-systems the formulation of this property may be simplified: in this case the sizes of LSM are "equal" and the projection map is defined for all $z \in W_1$.

There is a construction which lets an NCH-flow of class C^2 without fixed points correspond to a C^2-Anosov flow (cf [Pe1]). There exists an NCH-diffeomorphism of class C^2 on any two dimensional manifold, and one can construct a C^2-diffeomorphism with all but one non-zero Lyapunov exponents on any n-dimensional manifold for $n > 2$ (cf [BFK], [K2]). In these examples the Lebesgue measure is invariant and the corresponding diffeomorphisms are isomorphic to Bernoulli shifts with respect to this measure.

§3. Ergodic Properties of Smooth Hyperbolic Dynamical Systems

3.1. u-Gibbs Measures. The aim of this section is to show that hyperbolic dynamical systems have invariant measures which define statistical properties of typical trajectories and stochastic properties of the whole system. If a system has some smooth invariant measure it coincides with the measure which is constructed in the general case.

In order to simplify the exposition we restrict ourselves to the case of diffeomorphisms; the modifications concerning the case of continuous time will be done at the end of this section. We shall always suggest that the dynamical system under consideration is of class C^2. Consider an $x \in M$ such that the trajectory $\{S^n(x)\}$ is nonuniformely completely hyperbolic (see Sect. 1) and let $W^u(y)$ be the GUM for $y = S^n(y)$. We shall construct on every $W^u(y)$ a measure which is absolutely continuous with respect to the Riemannian volume on $W^u(y)$ and is uniquely defined (up to a constant factor). Let $z \in W^u(y)$, $S(z) \in W^u(S(y))$. Let $J^{(u)}(z)$ define the Jacobian of the map $W^u(S(y)) \xrightarrow{S^{-1}} W^u(y)$ which evaluates the contraction of $W^u(S(y))$ under the action of S^{-1}. Consider the fraction

$$\psi_n(z, y) = \prod_{k=1}^{n} \frac{J^{(u)}(S^k(z))}{J^{(u)}(S^k(y))}. \tag{7.22}$$

Since $\mathrm{dist}(S^k(y), S^k(z)) \to 0$ exponentially as $k \to \infty$ the factors corresponding to large k tend to 1 with exponential rate. Thus there exists a positive limit $\lim_{n\to\infty} \psi_n(z, y) = \psi(z, y)$. It follows from the construction that

$$\psi(z, w) \cdot \psi(w, y) = \psi(z, y).$$

Let $v^u(y)$ denote the Lebesgue measure on $W^u(y)$ considered as a smooth submanifold of M. Consider a measure μ^u on $W^u(y)$ given by $d\mu^u(z)/dv^u(z) = \psi(z, y)$. Thus we have intrinsically defined a measure μ^u on every $W^u(y)$. When we vary y, the measure μ^u is multiplied by a constant factor.

Suppose that a dynamical system has a hyperbolic attractor Λ (see Sect. 2). Then for any $x \in \Lambda$ $W^u(x)$ is contained in Λ. Fix a small neighborhood $U(x)$ of x in M and consider LUM $V^u(y)$ for $y \in U(x) \cap \Lambda$. These manifolds generate the partition $\xi^u = \xi^u_{U(x) \cap \Lambda}$ of the set $U(x) \cap \Lambda$. Consider some Borel measure λ on Λ and the restriction of this measure on $U(x) \cap \Lambda$. Since the partition ξ^u is measurable λ induces on λ-almost every C_{ξ^u} a conditional measure $\lambda(\cdot | C_{\xi^u})$.

Definition 3.1. A measure λ is said to be an u-Gibbs measure if $\lambda(\cdot|C_{\xi^u})$ may be written as $d\lambda(z|C_{\xi^u}) = \Xi \cdot \psi(z, x) dv^u(z)$, $x \in C_{\xi^u}$ where Ξ is a normalizing factor

$$\Xi = \int_{C_{\xi u}} \psi(z,x)\, dv^u(z).$$

It is clear that $\lambda(\cdot | C_{\xi u})$ is uniquely defined, i.e. it does not depend on the choice of ψ. We chose the notation Ξ following the statistical mechanics, where it is used to denote the statistical sum.

Theorem 3.1. *An u-Gibbs measure exists and is invariant under S; if $S|\Lambda$ is topologically transitive, then u-Gibbs measure is unique.*

We shall see below that this measure has other nice properties. There is a natural way to construct such a measure. Let $A \in M$ be an open set with the piecewise smooth boundary. In order to obtain $\lambda(A)$ consider $V^u(x)$ and its images under S^n. The leaf $S^n(V^u(x))$ will be approximatively equidistributed along the attractor. Let

$$\lambda_n(A) = \frac{\int_{A \cap S^n(V^u(x))} \psi(z, S^n(x))\, dv^u(z)}{\int_{S^n(V^u(x))} \psi(z, S^n(x))\, dv^u(z)}. \tag{7.23}$$

In a good situation when $n \to \infty$, $\lambda_n(A)$ tend to a limit which does not depend on the initial leaf $V^u(x)$. This limit is the value of $\lambda(A)$. The proof of this assertion and the analysis of the properties of λ is based on the special symbolic representation of $S|\Lambda$ which also reveals the profound connection between the theory of hyperbolic dynamical systems and the equilibrium statistical mechanics of lattice systems.

3.2. Symbolic Dynamics. By a symbolic representation of a smooth dynamical system we mean that almost every trajectory (in topological sense or with respect to some invariant measure) is encoded by means of a finite or a countable alphabet, so as the dynamical system turns out to be associated with the shift acting on some subset of the space of two-sided sequences. If the coding is good then the structure of the corresponding subset in the space of sequences is relatively simple. Good coding requires special partitions.

Let Λ be a LMHS. We say $R \subset \Lambda$ is a *rectangle* if the diameter of R is sufficiently small, $R = \overline{\operatorname{Int} R}$, and for any $x, y \in R$, $V^s(x) \cap V^u(y) \in R$.

This definition does not exclude that rectangles are full of holes like a direct product of Cantor sets. Any rectangle may be obtained using the following procedure. Take some $z \in \Lambda$ and choose open subsets $U^s \subset V^s(z) \cap \Lambda$, $U^u \subset V^u(z) \cap \Lambda$. If we construct $V^s(y)$ for any $y \in U^u$ and $V^u(x)$ for any $x \in U^s$, (where it can be done) then $R = \bigcup_{x,y} (V^u(x) \cap V^s(y))$ is a rectangle.

Definition 3.2. We define *Markov partition* as a finite cover of Λ by rectangles $R_1, R_2, \ldots, R_k$ such that
1) $\operatorname{Int} R_i \cap \operatorname{Int} R_j = \emptyset$ for $i \neq j$ and
2) if $x \in \operatorname{Int} R_i$, $S(x) \in \operatorname{Int} R_j$, then

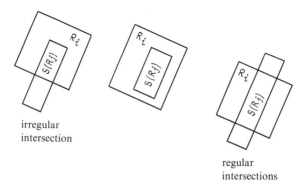

Fig. 4

$$S(V^u(x) \cap R_i) \supset V^u(S(x)) \cap R_j,$$
$$S^{-1}(V^s(x) \cap R_i) \supset V^s(S^{-1}(x)) \cap R_j.$$

The Markov condition imposes very strong restrictions on the agreement between the "boundaries" of elements of Markov partition. For example, if we consider a hyperbolic automorphism of the two-torus, then the intersection $R_i \cap S(R_j)$ must be right (see Fig. 4).

Theorem 3.2. *For any $\varepsilon > 0$ there exists a Markov partition of the LMHS Λ whose elements have diameters less than ε.*

The idea of Markov partition and the first approach to the construction of such partitions for Anosov systems appeared in [Si2], [Si3]. Important improvements and generalizations to the case of hyperbolic sets are due to Bowen [Bo2]. A simple concrete example of the Markov partition for hyperbolic toral automorphism was constructed by Adler and Weiss [AW]. This partition includes two elements which coincide with the projections on the torus of two "real" parallelograms constructed on the plane.

Any finite partition $\xi = \{R_1, \ldots, R_k\}$ of the set Λ allows us to construct the map ψ from Λ into the space $\Sigma_{(k)}$ of two-sided sequences of k symbols, namely $\psi(x) = \{\omega_n\}$, where ω_n is such that $S^n(x) \in R_{\omega_n}$. The map ψ conjugates S with a shift σ on some subset of $\Sigma_{(k)}$. The triple $(\psi, \psi(\Lambda), \sigma)$ is called a *symbolic representation* of the map S restricted on Λ. If ξ is a Markov partition then the map ψ constructed above is one-to-one everywhere except for the trajectories of the boundaries of rectangles, i.e. on the set $\Lambda \setminus \bigcup_n S^n(\bigcup_{i=1}^k \partial^i R_i)$. The main advantage of Markov partitions is that the image $\psi(\Lambda)$ turns out to be a topological Markov chain (TMC) which we denote by (Σ_A, σ) where $A = (a_{ij})$ is the transition matrix:

$$a_{ij} = \begin{cases} 1, & \text{if Int } R_i \cap S^{-1}(\text{Int } R_j) \neq \emptyset \\ 0, & \text{otherwise} \end{cases}$$

(see Chap. 3, Sect. 6). More exactly the following holds.

Theorem 3.3 (cf [Bo2]). *Let* $\omega = \{\omega_n\} \in \Sigma_A$. *Then the intersection* $\bigcap_{n \in \mathbb{Z}} S^n(R_{\omega_n})$ *is nonempty and consists of exactly one point. Thus a map* ψ^{-1} *is defined on* Σ_A *by the formula* $\psi^{-1}(\omega) = \bigcap_{n \in \mathbb{Z}} S^n(R_{\omega_n})$; ψ^{-1} *is continuous and one-to-one with the exception of some set of the first category in the sense of Bair.*

3.3. Measures of Maximal Entropy.
Theorem 3.3 allows us to reduce several questions about topological and ergodic properties of hyperbolic attractors to some problems of statistical mechanics of u-Gibbs measures.

Let Λ be a LMHS of a diffeomorphism S, and suppose that $S|\Lambda$ is topological transitive. Let (Σ_A, σ) be a symbolic representation of Λ constructed using some Markov partition ξ. Consider a stationary Markov chain with the transition probabilities $p_{ij} = a_{ij} z_i / \lambda(A) z_i$, where $\lambda(A)$ is the maximal positive eigenvalue of A and $Z = \{z_i\}$ is the corresponding eigenvector (see [AJ]). We denote by μ_0^* the Markov measure corresponding to this Markov chain and by μ_0 the preimage of μ_0^* under ψ. As shown in [AJ] μ_0 is the measure of maximal entropy for $S|\Lambda$ (for the definition see Sect. 3.5 below). We will formulate some important properties of μ_0.

We use P_k to denote the number of periodic points of period k lying in Λ and $P_k(B)$ to denote the number of such points lying in some subset $B \subset \Lambda$.

Theorem 3.4 (cf [AJ]). *If* $B \subset \Lambda$ *satisfies* $\mu_0(B \setminus \operatorname{Int} B) = 0$, *then*

$$\mu_o(B) = \lim_{n \to \infty} \frac{P_n(B)}{P_n}.$$

This theorem shows that periodic points form a kind of "uniform lattice" inside any LMHS. Some concrete applications of Theorem 3.4 will be formulated in Section 4. For an $x \in \Lambda$ we shall use μ_0^s and μ_0^u to denote the conditional measure induced by μ_0 on $W^s(x)$ and on $W^u(x)$ respectively.

Theorem 3.5 (cf [Mar]). *The following hold*

$$\mu_0^u(S(A)) = K\mu_0^u(A) \quad \text{for any } A \in W^u(x),$$

$$\mu_0^s(S(A)) = K^{-1}\mu_0^s(A) \quad \text{for any } A \in W^s(x),$$

where $K = \exp h > 1$ *and* $h = h(S)$ *is the topological entropy of* $S|\Lambda$ (*for the definition see Sect. 3.5 below*).

Thus the measure of maximal entropy has a nice property: it induces conditional measures on GSM and on GUM which are uniformly contracted (respectively expanded) under the action of S. It can be shown that μ_0 is the unique measure with this property.

3.4. Construction of u-Gibbs Measures.
Here we present a method to construct an u-Gibbs measure arising from statistical physics and developed in [Si]. In consequence of Theorem 2.5 we may suppose without loss of generality that $S|\Lambda$ is topologically transitive (otherwise we shall consider a component of topo-

logical transitivity). Let $\xi = \{R_1, \ldots, R_k\}$ be a Markov partition of Λ and ψ: $\Lambda \to \Sigma_A$ the corresponding symbolic representation. For $x \in M$, consider

$$\varphi^u(x) = -\ln|\text{Jac}(dS|E^u(x)| \tag{7.24}$$

Then $\varphi_*^u(x) = \varphi^u(\psi^{-1}(x))$ is a continuous function on Σ_A. Consider a sequence of measures $\mu_n^* = \rho_n \mu_0^*$ on Σ_A where μ_0^* is the measure with maximal entropy considered above and the densities ρ_n are given by

$$\rho_n = \frac{\exp \sum_{k=0}^{n} \varphi_*^u(\sigma^k(\omega))}{\int_{\Sigma_A} \exp \sum_{k=0}^{n} \varphi_*^u(\sigma^k(\omega)) \, d\mu_0^*}.$$

The function $\varphi^u(x)$ satisfies a Hölder condition because it is true for the distribution $E^u(x)$. This simply implies that

$$|\varphi_*^u(\omega^1) - \varphi_*^u(\omega^2)| \leq d_\beta(\omega^1, \omega^2)^\alpha \, {}^{12} \tag{7.25}$$

for any $\omega^1, \omega^2 \in \Sigma_A$. Now it follows from Sect. 6 of Chap. 3 that there exists a unique weak limit of the measures which we denote by μ^*.

Theorem 3.6 (cf [Si6]). *The preimage of μ^* under ψ coincides with the weak limit of the measures λ_n (cf. (7.23)) and is an u-Gibbs measure on the attractor Λ.*

3.5. Topological Pressure and Topological Entropy. In this section we describe an another approach to the construction of u-Gibbs measures due to Bowen [Bo2]. But beforehand, we shall introduce some general concepts and formulate some related results which are also self-important for the theory of dynamical systems.

Consider some $T: X \to X$ which is a continuous map of the compact metric space X, some $\varphi \in C(X)$ and some finite open cover $\mathcal{U}$ of X. Denote by $\mathcal{W}_m(\mathcal{U})$ the set of all m-tuples $\mathbb{U} = U_{i_1} U_{i_2} \ldots U_{i_n}$ consisting of elements of $\mathcal{U}$. We define also

$$X(\mathbb{U}) = \{x \in X: T^k(x) \in U_{i_k}, k = 0, 1, \ldots, m-1\},$$

$$S_m \varphi(\mathbb{U}) = \sup \left\{ \sum_{k=0}^{m-1} \varphi(T^k(x)): x \in X(\mathbb{U}) \right\}.$$

If $X(\mathbb{U}) \neq \emptyset$ then we set $S_m \varphi(\mathbb{U}) = -\infty$. We say that $\Gamma \in \mathcal{W}_m(\mathcal{U})$ covers X, if $X = \bigcup_{\mathbb{U} \in \Gamma} X(\mathbb{U})$. Consider

$$Z_m(\varphi, \mathcal{U}) = \inf_{\Gamma} \sum_{\mathbb{U} \in \Gamma} \exp S_m \varphi(\mathbb{U}),$$

where infinum is taken over all elements Γ of $\mathcal{W}_m(\mathcal{U})$ which cover X. It can be

---

[12] Recall that the distance in Σ_A is defined by some β, $0 < \beta < 1$ as $d_\beta(\omega^1, \omega^2) = \beta^N$ where N is the maximal nonnegative integer satisfying $\omega_n^1 = \omega_n^2$ for $|n| < N$.

shown (cf [Bo2]) that the finite limit

$$P(\varphi, \mathcal{U}) = \lim_{m \to \infty} \frac{1}{m} \log Z_m(\varphi, \mathcal{U})$$

exists.

Topological pressure of the function φ defined on X with respect to the map T is defined by

$$P(\varphi) = \lim_{\text{diam}\, \mathcal{U} \to 0} P(\varphi, \mathcal{U}).$$

It can be proved that the limit in the right-handside which is taken over all finite open covers of X exists (cf [Bo2]). The notation $P_T(\varphi)$ is also used for this limit.

The following statement has the name of the variational principle for topological pressure.

Theorem 3.7 (Ruelle [Ru1], Walters [W1], Bowen [Bo2]).

$$P_T(\varphi) = \sup_{\mu} \left(h_\mu(T) + \int \varphi \, d\mu \right),$$

where supremum is taken over all T-invariant Borel measures.

Let $\varphi \in C(X)$. A T-invariant Borel measure μ_φ is called an *equilibrium state* of a function φ if

$$h_{\mu_\varphi}(T) + \int \varphi \, d\mu_\varphi = P_T(\varphi).$$

Equilibrium states may not exist (cf [PS1] for more details). Thus the following general criterion for the existence of equilibrium states is important.

A homeomorphism T is said to be expensive if, for arbitrary small $\varepsilon > 0$ and for any $x, y \in X$, there exists an integer n such that $\rho(T^n x, T^n y) \geq \varepsilon$ (here ρ is the distance in X).

Theorem 3.8 (cf [Bo2]). *If T is expansive, then for any $\varphi \in C(X)$ there exists an equilibrium state μ_φ.*

A given function φ may have several equilibrium states; this situation has a slight likeness to the theory of phase transitions (cf [PS1]).

If we consider $\varphi(x) \equiv 0$, then $P_T(0)$ is called the *topological entropy* of the map T and is denoted by $h(T)$. There is an equivalent definition of the topological entropy. Fix $h > 0$ and consider a new distance $\rho_n(x, y)$ in X defined as

$$\rho_n(x, y) = \max_{0 \leq i \leq n} \rho(T^i(x), T^i(y)).$$

For an $\varepsilon > 0$ let $N(n, T, \varepsilon)$ be the minimal number of ε-balls (with respect to ρ_n) necessary to cover X. It can be shown (cf [K3]) that

$$h(T) = \varlimsup_{\varepsilon \to 0} \varlimsup_{n \to \infty} \frac{\log N(n, T, \varepsilon)}{n} = \lim_{\varepsilon \to 0} \lim_{n \to \infty} \frac{\log N(n, T, \varepsilon)}{n}.$$

Theorem 3.7 implies the following *variational principle for topological entropy* (cf [Bo2]):

$$h(T) = \sup_\mu h_\mu(T),$$

where supremum is taken over all T-invariant Borel probability measures. *Measures with maximal entropy* are those which give the supremum in the above formula. They are equilibrium states of $\varphi = 0$.

Remark 3.1. The above definitions may be generalized to the case of maps which are defined and continuous on some noncompact subsets of compact metric spaces. This approach allows to consider discontinuous maps which are continuous on noncompact sets consisting of the points whose trajectories do not fall into the set of discontinuity. In this case the definition of topological entropy is given in [Bo2] and the definition of topological pressure is given in [PP].

We are now able to describe Bowen's approach to the construction of u-Gibbs measures. Consider in Σ_A the equilibrium state of the function $\varphi_*^u(\omega) = \varphi^u(\psi^{-1}(\omega))$ which is uniquely defined because of (7.25). We denote it by μ^*.

Theorem 3.9 (cf [Bo2]). *The image $\psi^{-1}\mu^*$ is an u-Gibbs measure on a hyperbolic attractor Λ.*

It follows from this theorem that u-Gibbs measure is the single equilibrium state of the function $\varphi^u(x)$.

3.6. Properties of u-Gibbs Measures. They play a crucial role in the study of statistical properties of "typical" trajectories in the neighbourhood of a hyperbolic attractor. More exactly the following holds.

Theorem 3.10 (cf [Bo2], [PS1]). *If Λ is a hyperbolic attractor of a diffeomorphism S and $S|\Lambda$ is topologically transitive, then there exists a neighbourhood U of Λ, such that for any function φ continuous on U and for almost every point x (with respect to the Riemannian volume)*

$$\lim_{n\to\infty} \frac{1}{n} \sum_{k=0}^{n-1} \varphi(S^k(x)) = \int \varphi \, d\mu,$$

where μ is the unique u-Gibbs measure on Λ.

Such μ are called *Sinai-Ruelle-Bowen measures* (S-R-B measures).

Consider some measure ν absolutely continuous with respect to the Riemannian volume with the support inside the above neighbourhood U of Λ. According to the general approach of Bogolyubov-Krilov let us consider a sequence of measures (See Chap. 1, Sect. 1).

$$\mu_n = \frac{1}{n} \sum_{k=0}^{n-1} S_*^k \nu. \tag{7.26}$$

It follows from Theorem 3.10 that

$$\mu_n \to \mu \text{ as } n \to \infty \tag{7.27}$$

in the weak topology. A usual interpretation of this fact is that in a neighbourhood of a hyperbolic attractor the system "forgets an initial distribution of trajectories" which is one of the most important indices of stochasticity.

For hyperbolic attractors, a stronger statement than (7.27) holds. Namely the sequence of measures $S_*^n v$ converges weakly to μ. If we want however to consider not only hyperbolic but also more general attractors, it is natural to expect only the convergence in the mean.

Another approach is developed in [PS2]. It does not use Markov partitions and allows to prove directly that the measures μ_n converge weakly to an u-Gibbs measure on Λ. This approach turns out to be effective (and the only possible one) for studying UPH-attractors.

Theorem 3.11 (cf [PS2]). *Let Λ be an UPH-attractor of a C^2-diffeomorphism S. Then any limit measure for the sequence μ_n is some u-Gibbs measure on Λ.*

Since the topological structure of UPH-attractors is generally speaking more complicated than that of hyperbolic attractors the sequence μ_n may not converge even in the case when S is topologically transitive.

In conclusion we point out that Bowen's approach to the construction of equilibrium states may be generalized to arbitrary LMHS. Although in this case we cannot define an u-Gibbs measure, the equilibrium state corresponding to φ^u is of some interest (cf [Bo2]). An analog of u-Gibbs measure for Smale horseshoe was constructed in [C].

u-Gibbs measures have rich ergodic properties which determine together with Theorem 3.10 their role in hyperbolic theory. Let μ be an u-Gibbs measure on a hyperbolic attractor Λ of a diffeomorphism S. Then the results of Section 3.4 and Theorem 3.12 stated below imply that on any component of topological transitivity on Λ the dynamical system (S, μ) is ergodic, some power of S is Bernoulli and has exponential decay of correlations; for a wide class of smooth or piecewise smooth functions the central limit theorem of probability theory holds. The entropy of $S|\Lambda$ with respect to μ may be calculated by formula (7.28) below.

3.7. Small Stochastic Perturbations. We consider a family of probability distributions $q(\cdot|x,\varepsilon)$ on M where ε is a parameter. We suppose that $q(\cdot|x,\varepsilon)$ continuously depends on x and for any fixed ρ

$$\min q(U_\rho(x)|x,\varepsilon) \to 1 \text{ for } \varepsilon \to 0$$

where $U_\rho(x)$ is a ball of radius ρ centered in x. Let S be a homeomorphism of M. We construct a family of Markov chains Π_ε where a random point x moves according to the following rule: first x is mapped in $S(x)$ and then in a random point y chosen according to the distribution $q(\cdot|S(x),\varepsilon)$. A family of Markov chains Π_ε defined as above is called small stochastic perturbation of the homeomorphism S. It is not difficult to prove that if $I_\varepsilon = \{\pi_\varepsilon\}$ is a family of Π_ε-invariant measures, then for $\varepsilon \to 0$ any limit measure (in the sense of weak convergence) for the family I_ε coincides with some S-invariant measure.

In the case when S is a C^2-diffeomorphism of a smooth manifold M and Λ is a hyperbolic attractor, Kifer found some conditions which imply the sequence π_ε to converge to the unique u-Gibbs measure on Λ. In the particular case when S is an Anosov diffeomorphism of a smooth compact manifold, this result was obtained earlier in [Si6].

3.8. Equilibrium States and Their Ergodic Properties. Apart from the u-Gibbs measures described above, some other invariant measures are of interest for the hyperbolic theory. A method to construct such measures was given in [Bo2]. Let Λ be a LMHS of a C^2 diffeomorphism S, and let φ be a LMHS function continuous on Λ. It is easy to show that S is expansive on Λ and hence there exists an equilibrium state μ_φ (see Sect. 3.5). In general, μ_φ is not unique and in order to ensure the uniqueness it is enough to impose on φ some sufficiently weak conditions of smoothness. Let us suppose for example that φ is a Hölder function on Λ. Then $\phi(\omega) = \varphi(\psi^{-1}(\omega))$ is continuous on Σ_A and satisfies (7.25)[13]. Thus (cf [AJ]) there exists a unique equilibrium state μ_ϕ on Σ_A and its preimage under ψ is the unique equilibrium state of φ.

Since our knowledge of ergodic properties of equilibrium states corresponding to the functions satisfying (7.25) on TMC with finite alphabet is sufficiently good (see Chap. 3 Sect. 6), this approach allows us to obtain sufficiently detailed results about ergodic properties of μ_φ.

Theorem 3.12 (cf [Bo2]). *Let φ be a Hölder function on Λ. Suppose that S is topologically transitive. Then there exists a unique equilibrium state μ_φ with the following properties:*

1) *the measure μ_φ is positive on open subsets of Λ;*
2) *S is ergodic with respect to μ_φ;*
3) *there exist $m \geq 1$ and a set $\Lambda_0 \subset M$ such that $\Lambda = \bigcup_{k=0}^{m-1} S^k(\Lambda_0)$, $S^m(\Lambda_0) = \Lambda_0$, $S^i(\Lambda_0) \cap S^j(\Lambda_0) = \emptyset$ for $i \neq j$, $0 \leq i, j \leq m$ the system $(S^m | \Lambda_0, \mu_\varphi)$ is isomorphic to some Bernoulli automorphism (in particular is mixing and has K-property);*
4) *the system $(S^m | \Lambda_0, \mu_\varphi)$ is characterized by exponential decay of correlations and for a function satisfying some Hölder condition the central limit theorem holds (see Chap. 6, Sect. 1).*

If S is not topologically transitive then the above statements hold on any component of topological transitivity (see Theorem 2.5). Several results about the uniqueness and the properties of equilibrium states for functions satisfying some weaker assumptions than the Hölder ones were obtained in [Si6].

3.9. Ergodic Properties of Dynamical Systems with Nonzero Lyapunov Exponents. The definitive reason for the existence of u-Gibbs measures and for the possibility of studying their ergodic properties is the high degree of instability of

---

[13] Recall that (ψ, Σ_A, σ) is a symbolic representation of S on Λ.

the trajectories on hyperbolic attractors. Here we shall consider a situation when the instability of the trajectories of a dynamical system is sufficiently small.

Let S be a C^2-diffeomorphism of a manifold M. We shall denote by Λ the set of biregular points[14] which have nonzero values $\chi_i(x)$ of characteristic exponents $\chi^+(x,v)$. The diffeomorphism $S|\Lambda$ is an NCH-diffeomorphism.

Definition 3.3. A measure μ is said to be a *measure with nonzero exponents*, if μ is S-invariant and $\mu(\Lambda) = 1$.

We will introduce a class of measures on Λ which are u-Gibbs measure with nonzero exponents.

Definition 3.4. A measure μ is called *Sinai measure* on Λ if μ has nonzero Lyapunov exponents and the family of LUM (or the family of LSM) has the property of absolute continuity with respect to μ (see Sect. 2.8).

Generally speaking Λ is not an attractor, but if it admits a Sinai measure, then for μ-almost every x the LUM $V^u(x)$ (or the LSM $V^s(x)$) is contained in Λ.

The question about the existence of Sinai measures is difficult and still unsolved. The only known result in this direction proved in [Pe2] is the following: if μ is a Liouville measure (see Chap. 1, Sect. 1) with nonzero exponents, then μ is a Sinai measure.

Sinai measures have rich ergodic properties described below. For NCH systems with Liouville measure they were proved in [Pe2] the general case was considered in [L3].

We begin by formulating a general result which gives an upper estimate for the entropy of any C^1-diffeomorphism.

Theorem 3.13 (cf [K3]). *The entropy of a C^1-diffeomorphism S with respect to some Borel invariant measure μ satisfies the following inequality*

$$h_\mu(S) \leq \int_M \sum_{i=1}^{k(x)} q_i(x)\chi_i(x)\,d\mu(x),$$

where $\chi_1(x) \geq \cdots \geq \chi_{k(x)}(x) \geq 0 \geq \chi_{k+1}(x) \geq \cdots \geq \chi_n(x)$ *are the ordered values of characteristic Lyapunov exponents at the point x, $n = \dim M$, and $q_i(x)$ are the multiplicities of the corresponding values.*

In particular if for almost every point (with respect to the measure μ) Lyapunov exponents are zero, then $h_\mu(S) = 0$. If we consider arbitrary Borel measures μ, then the preceding inequality may be strict (cf [K3]). It can be proved that it is an equality for Sinai measures.

Theorem 3.14 (*the entropy formula*; [Pe2] [K3] [L3]). *For a $C^{1+\varepsilon}$-diffeomorphism S, $\varepsilon > 0$ the entropy with respect to some Sinai measure (in particular with respect to some Liouville measure with nonzero Lyapunov exponents)*

---

[14] For the definition of biregular points and of characteristic Lyapunov exponents, see Sect. 2 of Chap. 1.

is given by the formula

$$h_\mu(S) = \int_M \sum_{i=1}^{k(x)} q_i(x)\chi_i(x)\,d\mu(x). \tag{7.28}$$

For a generalization of (7.28) to any Borel probability measure invariant under a $C^{1+\varepsilon}$-diffeomorphism, refer to the work of F. Ledrappier and L.S. Young [LY].

The following describes the partition of Sinai measures into ergodic components.

Theorem 3.15 (cf [Pe2], [L3]). *Let μ be a Sinai measure for a $C^{1+\varepsilon}$-diffeomorphism, $\varepsilon > 0$. Then the invariant sets Λ_i, $i = 0, 1, 2, \ldots$ exist, such that*
1) $\bigcup_{i \geq 0} \Lambda_i = \Lambda$, $\Lambda_i \cap \Lambda_j = \emptyset$ *for* $i \neq j$;
2) $\mu(\Lambda_0) = 0$, $\mu(\Lambda_i) > 0$ *for* $i > 0$;
3) $S|\Lambda_i$ *is ergodic for* $i > 0$.

Some conditions sufficient for the ergodicity of $S|\Lambda$ are given in [Pe2]. It turns out that if μ is a smooth measure and W^u (or W^s) is locally continuous on Λ (cf [Pe2] for definition), then any ergodic component Λ_i is (mod 0) and open set. The topological transitivity of $S|\Lambda$ therefore implies the ergodicity of $S|\Lambda$. For smooth systems the local continuity of W^u is similar to a property formulated in the main theorem for dispersed billiards (see Theorem 1.5 of Chap. 8, Sect. 1). The following statement describes the partition into K-components.

Theorem 3.16 (cf [Pe2], [L3]). *Any set Λ is a disjoint union of sets $\Lambda_i^{(j)}$, $j = 1, \ldots, n_i$ such that*
1) $S(\Lambda_i^{(j)}) = \Lambda_i^{(j+1)}$ *for* $j = 1, \ldots, n_i - 1$, $S(\Lambda_i^{(n_i)}) = \Lambda_i^{(1)}$,
2) $S^{n_i}|\Lambda_i$ *is isomorphic to some Bernoulli automorphism.*

Let $\pi(S)$ be a π partition for S (i.e. the maximal partition with zero entropy (cf [Ro]). We shall use ξ^- (respectively ξ^+) to denote the partition of M into $W^s(x)$ (respectively into $W^u(x)$). For $x \in M \setminus \Lambda$ we set $W^s(x) = W^u(x) = x$.

Theorem 3.17 (cf [Pe2]). *There exists a measurable partition η of M with the following properties:*
1) *For almost every $x \in \Lambda$ the element $C_\eta(x)$ is an open (mod 0) subset of $W^s(x)$;*
2) $S\eta \geq \eta$, $\bigvee_0^\infty S^k \eta = \varepsilon$;
3) $\bigwedge_{-\infty}^0 S^k \eta = v(\xi^-) = v(\xi^+) = v(\xi^-) \wedge v(\xi^+) = \pi(S)$[15]
4) $h_\mu(S) = h_\mu(S, \eta)$.

3.10. Ergodic Properties of Anosov Systems and of UPH-Systems. The theorems stated above are fully applicable to Anosov systems. We only have to set $\Lambda = M$. The following assertion is a consequence of Theorem 3.10.

Theorem 3.18. *Let S be a C^2-Anosov diffeomorphism. Suppose that S is topologically transitive. Then*

---

[15] Recall that the measurable hull of ξ is denoted by $v(\xi)$.

1) *the sequence of measures μ_n from (7.26) weakly converges as $n \to +\infty$ to a u-Gibbs measure μ_1 which is the unique equilibrium state of the function φ^u* (see (7.24));

2) *the sequence of measures μ_n weakly converges as $n \to -\infty$ to an s-Gibbs measure*[16] *μ_2 which is the unique equilibrium state of the function*

$$\varphi^s(x) = \ln|\text{Jac}(dS|E^s(x))|.$$

Ya.G. Sinai and A.N. Lifschitz proved (cf [Si6]) that $\mu_1 = \mu_2$ if an only if for any periodic x of period p

$$|\text{Jac}(dS^p(x))| = 1.$$

It is a very uncommon property: Anosov diffeomorphisms with $\mu_1 = \mu_2$ lie in the complement in $\text{Diff}^2(M)$ to some set of the second category in the sense of Bair (cf [Si6]). Nevertheless this class of diffeomorphisms include some interesting examples of automorphisms with algebraic origin. Thus we shall suppose that $\mu = \mu_1 = \mu_2$ which implies that μ is absolutely continuous. Then it follows from Theorem 3.12.

Theorem 3.19 (cf [An1], [AS]). *An Anosov diffeomorphism which admits a C^2-Liouville measure is isomorphic to a Bernoulli automorphism (in particular it is ergodic, mixing, has K-property, has positive entropy); moreover it has exponential decay of correlations and satisfies the central limit theorem for Hölder functions.*

Notice that the major part of Theorem 3.19 (up to the K-property) was proved earlier (cf [Au1]).

The existence of u-Gibbs measures and of s-Gibbs measures for UPH-diffeomorphisms follows from Theorem 3.11 (cf [PS2]). These measures may be obtained as limit measures for the sequence μ_n (see (7.26)). They have positive entropy (cf [PS2]). When studying ergodic properties of an UPH-system with Liouville measure the notion of transitivity for a pair of foliations W^u, W^s is important.

Consider two points $x, y \in M$. A sequence of points $x_1, \ldots, x_n$ is called an (R, N)-Hopf chain if $x_1 = x$, $x_N = y$ and for $i = 2, \ldots, N$, x_i lies on the stable or on the unstable manifold of x_{i-1} and the distance (measured along this manifold) between x_i and x_{i-1} does not exceed R. For an $x \in M$ we define the (R, N)-set of x as the set of all $y \in M$ which are connected with y by an (R, N)-Hopf chain. ForAnosov diffeomorphisms there exist some R and N such that for any $x \in M$ the (R, N)-set of x coincides with the whole manifold. The proof of ergodicity for Anosov diffeomorphisms is based on this fact. In the case of UPH-diffeomorphisms the sum of dimensions of stable and unstable manifolds is less than the dimension of M. Therefore it is quite possible that for any $x \in M(R, N)$-

---

[16] The definition of an s-Gibbs measure is similar to the definition of an u-Gibbs measure.

set of x has measure zero. For example if a pair W^u, W^s is integrable[17] and generates a foliation W, then the (R, N)-set of x coincides with $W(x)$. Certainly in such a case S is not ergodic.

A pair of foliations W^u, W^s is called *locally transitive* if for any $\varepsilon > 0$, there exists an N such that (ε, N)-set of any $x \in M$ contains some neighbourhood of x. UPH-*diffeomorphism* is called *locally transitive* if the pair of foliations W^u, W^s is locally transitive. A small perturbation of a sufficiently smooth locally transitive UPH-diffeomorphism is again locally transitive and UPH (cf [BP]). It is also proved in [BP] that if S is a C^2-locally transitive UPH-diffeomorphism with Liouville measure, then under some additional suppositions about W^u and W^s S has K-property.

3.11. Continuous Time Dynamical Systems. The definitions of u-Gibbs measures, of measures with nonzero Lyapunov exponents, of Sinai measures may be transferred to the continuous time dynamical systems (the zero exponent corresponding to the flow direction must be excluded). The definition and the construction of Markov partition and the corresponding symbolic model for flows on hyperbolic sets require some modifications (cf [Bo2]). Theorems 3.1, 3.2, 3.10–3.12 (with exception of statements 3) and 4)), 3.13–3.15, 3.17 and the corresponding consequences of these theorems are literally transferred to this case (one has only to suppose that $n \in \mathbb{R}$). Theorems 3.4 and 3.5 may be transferred with obvious modifications (cf [A5]). The analog for statements 3) and 4) of Theorem 3.12 is: if the first possibility of the alternative (see 2.3) is realized, then the flow $\{S^t\}$ is Bernoulli with respect to μ_φ (cf [Bo1]).

§4. Hyperbolic Geodesic Flows

During a long time geodesic flows played an important stimulating role in the development of hyperbolic theory. Thus, for example, when Hadamard and Morse were studying in the beginning of the 20th century the statistics of behavior of geodesics on surfaces with negative curvature, they pointed out that the local instability of trajectories gave rise to some global properties of dynamical systems: such as ergodicity, topological transitivity and so on. Later the investigations related to geodesic flows provoked the introducing of different classes of hyperbolic dynamical systems (Anosov systems, UPH-systems and NCH-systems with Liouville measures). At that same time, geodesic flows always were a suitable object for applying dynamical methods, which in particular allowed to obtain some interesting results of differential geometric character.

---

[17] I.e. the distribution $\mathcal{T}_x W^u(x) \oplus \mathcal{T}_x W^s(x)$ is integrable; some examples are given by diffeomorphisms S^t (t is fixed) which are the time t translations of an Anosov flow; another example is the direct product of an Anosov diffeomorphism and the identic map (see Sect. 2).

The connection of geodesic flows with classical mechanics was considered in Chap. 1, Sect. 1[18].

4.1. Manifolds with Negative Curvature. Let Q be a compact P-dimensional Riemannian *manifold with negative curvature*, i.e. for any $x \in Q$ and for any two linearly independent vectors $v_1, v_2 \in \mathcal{T}_x$ the sectional curvature $K_x(v_1, v_2)$ satisfies

$$K_x(v_1, v_2) \leqslant -k, \qquad k > 0. \tag{7.29}$$

A geodesic flow $\{S^t\}$ on the unit tangent bundle $M = SQ$ was defined in Chap. 1, Sect. 1. The key to the study of topological and ergodic properties of this flow is the following.

Theorem 4.1. *A geodesic flow on a compact Riemannian manifold with negative curvature is Anosov.*

The proof of this theorem is based on the study of the solutions of variational equations corresponding to the flow $\{S^t\}$ i.e. of the Jacobi equation (cf [Pe3]).

$$Y''(t) + R_{XY}Y = 0, \tag{7.30}$$

where $Y(t)$ is a vector field along $\gamma(t)$, $X(t) = \dot{\gamma}(t)$ and R_{XY} is the curvature tensor.

More exactly let $v \in M$. We define by $\pi: M \to Q$ the projection, and by $K: M \to \mathcal{T}_{\pi(v)}Q$ the connection map induced by the Riemannian metric (i.e. K is Levi-Civita connection). Consider some $\xi \in \mathcal{T}_v M$ and let $Y_\xi(t)$ be the solution of the Jacobi equation with the initial conditions $Y_\xi(0) = d\pi \xi$, $Y'_\xi(0) = K\xi$ correspond to ξ. Then the map $\xi \xrightarrow{\chi} Y_\xi(t)$ is an isomorphism and $d\pi \, dS^t \xi = Y_\xi(t)$, $K dS^t \xi = Y'_\xi(t)$. The above identification χ allows us to say that (7.30) are the variational equations corresponding to the flow S^t. Fix some $v \in M$ and denote by $\gamma_v(t)$ the geodesic given by the vector v. Using Fermi coordinates $\{e_i(t)\}$, $i = 1, \ldots, p$, along $\gamma_v(t)$[19] we can rewrite (7.30) in the matrix form

$$\frac{d^2}{dt^2} A(t) + K(t)A(t) = 0, \tag{7.31}$$

where $A(t) = (a_{ij}(t))$, $K(t) = (k_{ij}(t))$ are matrix functions and $K_{ij}(t) = K_{\gamma_v(t)}(e_i(t), e_j(t))$. Using (7.29) it can be shown (cf [An1]) that the boundary value problem for (7.31) may be uniquely solved. Thus for given s there exists a unique solution $A_s(t)$ satisfying the boundary conditions $A_s(0) = I$ (identic matrix), $A_s(s) = 0$. Further there exists a limit $\lim_{\to \infty} \dot{A}_s(t)|_{t=0} = A^+$. A positive limit solution $D^+(t)$ of (7.31) is defined by its initial conditions $D^+(t)|_{t=0} = I$, $\dot{D}^+(t)|_{t=0}$ A^+. It is nondegenerate (i.e. $\det(D^+(t)) \neq 0, t \in \mathbb{R}$) and $D^+(t) = \lim_{s \to +\infty} A_s(t)$. Moreover

$$D^+(t) = C(t) \int_t^\infty C^{-1}(s)(C^{-1}(s))^* \, ds, \tag{7.32}$$

---

[18] For detailed references see [Pe3]
[19] $\{e_i(t)\}$ is obtained by the time t parallel translation along $\gamma_v(t)$ of an orthonormal basis $\{e_i(0)\}$ in $\mathcal{T}_{\gamma(0)}M$; $e_1(t) = \dot{\gamma}(t)$.

Chapter 7. General Theory of Smooth Hyperbolic Dynamical Systems   139

where $C(t)$ is the solution of (7.31) corresponding to the initial conditions $C(0) = 0$, $\left.\frac{d}{dt}C(t)\right|_{t=0} = I$. Similarly considering $s \to -\infty$ we can define a *negative limit solution* of (7.31). The vectors $\chi^{-1}(D^{\pm}(0)e_i(0))$, $i = 1, \ldots, p$ are linearly independent and span a subspaces $E^{\pm}(v) \subset \mathcal{T}_v M$. It can be shown that the subspaces $E^{+}(v)$, $E^{-}(v)$ generate two S^t-invariant distributions satisfying the conditions of uniform hyperbolicity. This proves Theorem 4.1.

It can be shown that for geodesic flows on manifolds with negative curvature, the second possibility of the alternative for Anosov flows is realized (see 2.3), namely $\{S^t\}$ is mixing. This fact may be deduced from a theorem due to Arnold, or may be proved directly (cf [Si1]). Now we can obtain topological and ergodic properties of geodesic flows as a consequence of general results about Anosov flows.

Theorem 4.2. *Let $\{S^t\}$ be a geodesic flow on a compact Riemannian manifold with negative curvature. Then*

1) *$\{S^t\}$ is isomorphic to a Bernoulli flow, in particular $\{S^t\}$ is ergodic, mixing, has continuous spectrum, positive entropy and K-property;*

2) *$\{S^t\}$ is topologically mixing, in particular topologically transitive;*

3) *periodic orbits of $\{S^t\}$ are dense in M the number $P(T)$ of periodic orbits of period $\leq T$ is finite and*

$$\lim_{T \to \infty} \frac{hTP(T)}{e^{hT}} = 1,$$

where $h = h(S^1)$ is the topological entropy of S^1;

4) *$\{S^t\}$ admits a uniquely defined measure with maximal entropy μ_0. If for a given Borel set $B \subset M$, we define*

$$N_{t,\varepsilon}(B) = \sum_{\gamma \in Q(\varepsilon, t)} \omega_\gamma(B) \bigg/ \sum_{\gamma \in Q(\varepsilon, t)} \tau(\gamma)$$

where $Q(\varepsilon, t)$ is the set of periodic orbits γ with the period (not necessarily the minimal one) lying between $t - \varepsilon$ and $t + \varepsilon$, $\tau(\gamma)$ is the minimal period of γ, $\omega_\gamma(\cdot)$ is the measure on M which coincides with the image of the Lebesgue measure on $[0, \tau(\gamma))$ under the map $t \to S^t(x)$, $x \in \gamma$. Then for B satisfying $\mu_0(B \setminus \text{int } B) = 0$ we have

$$\lim_{t \to \infty} N_{t,\varepsilon}(B) = \mu_0(B)$$

Consider the stable foliation W^s and the unstable foliation W^u for the flow $\{S^t\}$. The leaf $W^s(v)$ (respectively $W^u(v)$) passing through some linear element v is called the *stable* (respectively *unstable*) *horosphere*.[20] It can be shown that for

---

[20] The modern terminology is somewhat different from the traditional one which uses horospheres (horocycles in two dimensional case) to denote the projections of $W^s(v)$ and $W^u(v)$ in Q. Nowadays the latter are called limit spheres (respectively limit circles) and horospheres are the framing of limit spheres (see below).

any $v \in M$ the corresponding stable and unstable horospheres are everywhere dense in M.

We see that the methods of the dynamical systems theory give us important information about geometric properties of compact manifolds with negative curvatures. We obtain namely the density of closed geodesics in M, their asymptotic number, and their distribution, the existence of everywhere dense geodesics, the density in M of stable and unstable horospheres, and some other properties.

Now we shall consider another approach to the construction of stable and unstable horospheres. This approach firstly is useful for the description of topological and geometrical structure of horospheres, and secondly it allows to transfer part of the above results to a more wide class of metrics (see 4.2).

We denote by H the universal Riemannian covering of Q. According to the Hadamard-Cartan theorem, any two points in H are joint by the single geodesic and for any $x \in H$ the map $\exp_x: \mathbb{R}^p \to H$ is a diffeomorphism. Hence the map

$$\varphi_x(y) = \exp_x[(1 - \|v_y\|^{-1}]$$

is a homeomorphism of the open unit ball B on H (v_y is a vector in $\mathcal{T}_x Q$ connecting the origin with $y \in B$). Let G be the group of motions (i.e. of isometries) of H. Then Q coincides with H/Γ where Γ is a discrete subgroup of G.

The geodesics $\gamma_1(t)$, $\gamma_2(t)$ in H are *asymptotic* for $t > 0$ if

$$\rho(\gamma_1(t), \gamma_2(t)) \leqslant \text{const}$$

for all $t > 0$ (ρ is the distance in H induced by the Riemannian metric). It can be shown that geodesics which are asymptotic for $t > 0$ do not intersect and that for any point x there is a single geodesic passing through x and asymptotic for $t > 0$ to the given geodesic. Thus we obtain an equivalence relation. We use $H(\infty)$ to denote the set of corresponding equivalence classes. Any such class coincides with a bundle of geodesics which are asymptotic for $t > 0$ (these classes are also referred to as *points at infinity*); $H(\infty)$ is called the *absolute*. One can define a topology τ on $H \cup H(\infty)$ so that $H \cup H(\infty)$ becomes a compact metric space and the restriction of τ on H coincides with the topology induced in H by the Riemannian metric.

The map φ_x may be extended to a homeomorphism (still denote by φ_x) of the closed ball $\bar{B} = B \cup S^{p-1}$ (S^{p-1} is the $(p-1)$-dimensional sphere in $\mathbb{R}^p$) on $H \cup H(\infty)$ by the equality

$$\varphi_x(y) = \gamma_{v_y}(+\infty), \qquad y \in S.$$

In particular φ maps S homeomorphically on $H(\infty)$. The image of an asymptotic bundle under φ generates the partition of B into disjoint curves "tending" to the same point in S^{p-1}. A *limit sphere* is a submanifold in H orthogonal to some bundle of asymptotic geodesics. More exactly it means that through any point of this submanifold passes a single geodesic contained in the bundle which is directed orthogonally to the submanifold. It can be shown that limit spheres exist and have the following properties:

1) any limit sphere is uniquely defined by a point q corresponding to the bundle under consideration and by $x \in M$ (for this reason it will be denoted by $L(q, x)$);

2) $\bigcup_{-\infty < t < \infty} L(q, \gamma(t)) = H$;

3) $L(q, \gamma(t_1)) \cap L(q, \gamma(t_2)) = \emptyset, t_1 \neq t_2$;

4) $\rho(L(q, \gamma(t_1)), L(q, \gamma(t_2))) = |t_2 - t_1|$.

Here $\gamma(t)$ is a geodesic satisfying $\gamma(+\infty) = q$. The last property expresses the equidistance of limit spheres.

The set $\varphi^{-1}(L(q, x)) \cup \varphi^{-1}(q) \subset \bar{B}$ is homeomorphic to the $(p-1)$-dimensional sphere in $\bar{B}$ which is tangent to S^{p-1} in the single point $\varphi^{-1}(q)$. There is an another method to construct a limit sphere as a limit of some spheres. More exactly, let us consider a geodesic $\gamma(t)$ joining q and x and for any t the sphere in H centered in $\gamma(t)$ and passing through x. The limit of these spheres for $t \to +\infty$ coincides with $L(q, x)$[21].

The framing of the limit sphere $L(q, x)$ with the orthogonal vectors pointing inside (respectively outside) coincides with the stable horosphere $W^s(v)$ (respectively with the unstable horosphere $W^u(v)$) where v is defined by $\pi(v) = x$, $\gamma_v(+\infty) = q$.

4.2. Riemannian Metrics Without Conjugate Points.
Two points x, y are *conjugate* along a geodesic $\gamma(t)$ if for some $t_1, t_2 \in \mathbb{R}$, $x = \gamma(t_1)$, $y = \gamma(t_2)$ and there exists a nonzero solution $Y(t)$ of the Jacobi equation (7.30) along $\gamma(t)$ satisfying $Y(t_1) = Y(t_2) = 0$[22]. A Riemannian metric has no conjugate points if there are no conjugate points on any geodesic. If there are no conjugate points, then, similarly to 4.1, one can construct positive and negative limit solutions of (7.31) along $\gamma(t)$ and then two distributions of subspaces $E^+(v)$, $E^-(v)$, $v \in SH$. With some additional geometric supposition the fields E^+, E^- are integrable. To explain this let us consider the universal Riemannian covering manifold H (so that $M = H/\Gamma$ where Γ is a discrete subgroup of isometries of H) and lift the fields E^+, E^- on SH. It can be shown that in this case as in 4.1, limit spheres (defined as the limit of spheres) orthogonal to the bundles of asymptotic geodesics exist. Their framing (called horospheres) are the integral manifolds for the distributions E^+ and E^- (cf [Pe3] for more details).

If for any $v \in M$

$$E^-(v) + E^+(v) + X(v) = \mathcal{T}_v M \qquad (7.33)$$

($X(v)$ is the one-dimensional subspace corresponding to the flow direction), then $\{S^t\}$ is an Anosov flow. The condition (7.33) holds for metrics with negative curvature, but it may also hold for metrics which have some domains with zero or even with positive curvature. For this reason one considers the *manifolds of Anosov type* which admit some Riemannian metric such that the corresponding

---
[21] This explains the name "limit sphere".
[22] For a more geometric definition of conjugate points see [Pe3].

geodesic flow is Anosov (an important subclass consists of *manifolds of hyperbolic type* which admit a Riemannian metric with negative curvature; in particular this class contains any surface of genus >0). For these manifolds one gets information about the properties of geodesic flows generated by any metric without conjugate points.

Now we give a sufficient condition which implies that (7.33) holds along a geodesic, defined by some vector $v \in M$. Let $w \in M$ be orthogonal to v and $Y_w(t)$ a positive limit solution of (7.31) defined by w (i.e. $Y_w(0) = D^+(0)w$, $Y'_w(0) = D^+(0)w$). We use $K_w(t)$ to denote the sectional curvature at $\gamma_v(t)$ defined by $Y_w(t)\gamma_v(t)$. Now the condition is: for any w

$$\overline{\lim_{t \to +\infty}} \frac{1}{t} \int_0^t K_w(s)\,ds < 0. \tag{7.34}$$

It can be shown that (7.34) implies (7.33) along $\gamma_v(t)$. Moreover for any $\xi \in E^+(v)$ the characteristic Lyapunov exponent satisfies $\chi^+(\xi, v) > 0$ and for $\xi \in E^-(v)$ we have $\chi^+(\xi, v) < 0$. If (7.34) holds for vectors v constituting a set $\Lambda \subset M$ of positive measure, then $\{S^t\}|\Lambda$ is an NCH-flow. It turns out that with some geometrical restrictions on the Riemannian metric (expressed by the so-called axiom of visibility) and other conditions $\{S^t\}$ is isomorphic to a Bernoulli flow. It is realized for geodesic flows on the surfaces of genus >0 with a Riemannian metric without conjugate points, on the manifolds with a Riemannian metric without focal points satisfying the axiom of visibility, and in some other cases.

We see that hyperbolic properties characterizing geodesic flows on manifolds with negative curvature may take place (may be in a somewhat weaker form) for more general metrics. Notice however that while the property of having negative curvature is local and a rough one[23], the absence of conjugate points is a global property which is not rough and usually it is difficult to verify it. Nevertheless the above results are essential for metrics with nonpositive curvature which lie on the boundary of the set of metrics with negative curvature. A simple example of the geodesic flow on n-torus with standard metric, which is not ergodic and has zero entropy, shows that this situation is rather different. This example is not typical however. It turns out that if the universal covering H of a compact manifold M with nonpositive curvature does not admit any geodesically isometric embedding of the plane (for metrics with nonpositive curvature the last condition is equivalent to the axiom of visibility), then the geodesic flow on M is an NCH-flow and it is isomorphic to a Bernoulli flow. Roughly speaking the zero curvature reveals the possibility of a geodesically isometric embedding in H of an infinite strip of zero curvature which consists of geodesics joining two given points on the absolute. When the above condition holds the total measure of all such strips is equal to zero. Progress has recently been made in the study

---

[23] i.e. any sufficiently close metric also has negative curvature.

of geometrical structure and in the classification of manifolds with non-positive curvature (c.f. the survey [Bk]).

4.3. Entropy of Geodesic Flow. Let $v \in M$. We define v^1 as the set of $w \in M$ orthogonal to v. Consider a linear map $S_v: v^1 \to v^1$ defined by the equality: $S_v w = K_\xi(w)$, where $\xi(w)$ is a vector in $E^-(v)$ such that $d\pi\xi(w) = w$.

Theorem 4.3 (cf [Pe3]). *For a Riemannian metric of class C^4 without focal points S_v is a linear self-adjoint operator which coincides with the operator of the second quadratic form for the limit sphere $L(\pi(v), \gamma_v(+\infty))$ in the point $\pi(v)$.*

Using formula (7.32) we can represent S_v in a form similar to the continued fraction decomposition for the operator $B(x)$ from Sect. 1 of Chap. 8.

Denote by $\{e_i(v)\}$, $i = 1, \ldots, p-1$ the orthonormal basis in v^1 consisting of eigenvectors of S_v and let $K_i(v)$ be the corresponding eigenvalues. The numbers $K_i(v)$ are called the *principal curvatures* and $e_i(v)$ are called the *directions of principal curvatures* for the limit sphere at $\pi(v)$.

Theorem 4.4. *From the conditions of Theorem 4.3 we have the following formula for the entropy of geodesic flow*

$$h_\mu(S^1) = -\int_M \sum_{i=1}^{p-1} K_i(v)\, d\mu(v) = -\int_M \operatorname{tr} S_v\, d\mu(v)$$

(tr—*is the trace of* S_v).

This result is analogous to the result of [SC] for dispersed billiards.

Some estimates from below for the entropy of geodesic flow were recently obtained in [OS].

§5. Geodesic Flows on Manifolds with Constant Negative Curvature

A profound study of ergodic theory of geodesic flows on manifolds with constant negative curvature may be realized using the methods of unitary representations of Lie groups. The idea of the algebraic construction of these flows first appeared in an article by I.M. Gelfand and S.V. Fomin (cf [GGV]) where several important results were obtained. Sometimes dynamical systems to which the Gelfand-Fomin method is applicable are called systems of algebraic origin. Several results concerning these systems are described in the survey [KSS]. Here we shall consider only the geodesic flows on manifolds with constant negative curvature. As a model for Lobachevsky plane we shall use the Poincaré model on the upper semiplane $H = \{z = (x + iy): y > 0\}$. The line $y = 0$ is called absolute (and is denoted by $H(\infty)$), points on this line are called points at infinity.

The straight lines in H coincide with the semicircles[24] centered on the absolute or with the rays orthogonal to the absolute. The Riemannian metric with the curvature $-K$ is given as a scalar product $<,>_L$ at a point $z = x + iy \in H$ by $<,>_L = \dfrac{k}{y^2} <,>$ where $<,>$ is the Euclidean scalar product, $k > 0$. The geodesics of this metric are simply the lines indicated above. An *asymptotic bundle* is the set of all oriented lines in Lobachevsky geometry which are parallel to the given line, i.e. reaching the same point of the absolute. Limit circles orthogonal to some asymptotic bundle are the circles tangent to the absolute and the lines parallel to it.

The *motions* (isometries) in *Lobachevsky geometry* are the fractional linear transformations $z \to \dfrac{a + bz}{c + dz}$ which conserve the upper semiplane. Here a, b, c, d are real and they may be normalized such that $\det \begin{vmatrix} a & b \\ c & d \end{vmatrix} = 1$. The matrices $\begin{vmatrix} a & b \\ c & d \end{vmatrix}$ and $\begin{vmatrix} a' & b' \\ \gamma' & d' \end{vmatrix}$ define the same transformation if $\begin{vmatrix} a & b \\ c & d \end{vmatrix} = \begin{vmatrix} a' & b' \\ c' & d' \end{vmatrix}$. The matrices $\begin{vmatrix} 1 & 0 \\ 0 & 1 \end{vmatrix}$ and $\begin{vmatrix} -1 & 0 \\ 0 & -1 \end{vmatrix}$ form the centre Z of $SL(2, \mathbb{R})$ and the group of isometries of Lobatchevski plane is isomorphic to $SL(2, \mathbb{R})/Z$. A motion $g \in G$ is called

a) *elliptic* if g is conjugate with a transformation $z \to \lambda z$, $|\lambda| = 1$;

b) *hyperbolic* if g is conjugate with a transformation $z \to \lambda z$, $|\lambda| > 1$;

c) *parabolic* if g is conjugate with $z \to z + 1$.

We are especially interested in hyperbolic transformations. Consider some g_0: $g_0 z = \lambda z$, $\lambda > 1$. The line $\{\operatorname{Re} z = 0\} = \lambda$ is invariant under g_0, i.e. $g_0 \gamma = \gamma$ and any point $z \in \gamma$ moves under the action of g_0 by the non-Euclidean distance $\ln \lambda$. If g is conjugate with g_0 then there exists some g-invariant line with the same property. Since $g = g_1 g_0 g_1^{-1}$ for some $g_1 \in G$ we find that λ is the eigenvalue of g, $\lambda > 1$. We shall use M_0 to denote the unit tangent bundle SH. Then $G = M_0 = SL(2, \mathbb{R})/N$, where N consists of two elements $\begin{vmatrix} 1 & 0 \\ 0 & 1 \end{vmatrix}$ and $\begin{vmatrix} -1 & 0 \\ 0 & -1 \end{vmatrix}$, because any motion is uniquely defined by its action on the unit tangent vector based at $z = i$ and directed vertically upwards. A surface Q of constant negative curvature coincides with the factor space $Q = H/\Gamma$ where Γ is some discrete subgroup of G. The phase space of the geodesic flow, i.e. the unit tangent bundle over Q coincides with the factor space $M = \Gamma \backslash G$ of the left cosets, the measure μ on M is induced by the Haar measure on G. The two following observations play an important role in Gelfand-Fomin construction:

1) The right action of G induces in the Hilbert space $\mathscr{L}^2(M, \mu)$ the unitary representation by the formula

---

[24] Here and below semicircles, circles, straight lines are understood in the Euclidean sense.

Chapter 7. General Theory of Smooth Hyperbolic Dynamical Systems

$$U_g f(x) = f(xg), \quad x \in \Gamma \backslash G, \quad g \in G.$$

2) The geodesic flow $\{S^t\}$ is generated by the one-parameter subgroup $g_t = \begin{pmatrix} e^t & 0 \\ 0 & e^{-t} \end{pmatrix}$, i.e. $S^t(x) = xg_t$.

We may find the spectrum of geodesic flow by considering irreducible representations of G and defining the spectrum of $\{U_t\}$ for any irreducible representation (it was first done in [GGV]).

For the surfaces of constant negative curvature there is a close relation between the lengths of closed geodesics and the eigenvalues of hyperbolic motions $g \in \Gamma$. Consider a geodesic γ in H which is invariant under g. Take some $q \in \gamma$ and a linear element x tangent to γ at q. Then gx and x are identified. From the geometric point of view this means that the interval of γ between x and gx is a closed geodesic. It can be shown that all closed geodesics may be obtained in such a way. Thus we see that the lengths of simple closed geodesics are equal to the logarithms of eigenvalues larger than 1 of the transformations $g \in \Gamma$.

There is a remarkable formula due to Selberg (cf [GGV]) which relates the eigenvalues of $g \in \Gamma$ to the eigenvalues of the Laplacian operator on Q. Thus there is a relation between the eigenvalues of the Laplacian operator on the surface with constant negative curvature and the lengths of closed geodesics. To date no generalization of this fact exists for the surfaces with nonconstant negative curvature. This question seems to be one of the most interesting in the subject.

Now let us consider a concrete example of a noncompact surface with constant negative curvature which has a finite volume.

We denote by Γ the *modular subgroup* of $SL(2, R)$ consisting of integer matrices $\begin{vmatrix} a & b \\ c & d \end{vmatrix}$ with determinant 1. Using a nice theorem of Artin we show how to describe the ergodic properties of $\{S^t\}$ on $Q = H/\Gamma$. Consider a geodesic γ in Q and one of its lifting $\tilde{\gamma}$ in H. Denote $\tilde{x} = \tilde{\gamma}(-\infty)$, $\tilde{y} = \tilde{\gamma}(+\infty)$. Suppose, by way of example, that $\tilde{x} > 0$, $\tilde{y} < 0$ and let $\tilde{x} = [\tilde{n}_1, \tilde{n}_2, \ldots]$, $-\tilde{y} = [\tilde{m}_1, \tilde{m}_2, \ldots]$ be the continued fraction expansions of $\tilde{x}$ and $\tilde{y}$ with $\tilde{n}_i > 0$ and $\tilde{m}_i > 0$. Let $\hat{\gamma}$ be an another lifting of γ in H and $\hat{x} = \hat{\gamma}(-\infty)$, $\hat{y} = \hat{\gamma}(+\infty)$. Then $\hat{\gamma} = g\tilde{\gamma}$ for some $g \in \Gamma$. Let $\hat{x} = [\hat{n}_1, \hat{n}_2, \ldots]$, $-\hat{y} = [\hat{m}_1, \hat{m}_2, \ldots]$ be the corresponding continued fraction expansions (again we suppose by way of example that $\hat{x} > 0, \hat{y} < 0$). The Artin Theorem says that $(\tilde{x}, \tilde{y})$ and $(\hat{x}, \hat{y})$ define the same geodesic in M if and only if for some integer k $\sigma^k(\ldots, \tilde{m}_2, \tilde{m}_1, \tilde{n}_1, \tilde{n}_2, \ldots) = (\ldots, \hat{m}_2, \hat{m}_1, \hat{n}_1, \hat{n}_2, \ldots)$ where σ is the shift in the space Σ of doubly infinite sequences of positive integers. Thus we obtain a map $\psi: SQ \to \Sigma$ and it can be shown that the ergodicity of $\{S^t\}$ with respect to the Riemannian volume μ in SQ is equivalent to the ergodicity of σ with respect to the measure $\psi_*\mu$. This is obviously equivalent to the ergodicity of σ in the space Σ_+ of one-sided sequences of positive integers (Σ is the natural extension of Σ_+) with respect to the measure ν which is the projection of $\psi_*\mu$. From the other side let us consider a map T of $[0,1]$ into itself given by $T(x) = \dfrac{1}{x} - \left[\dfrac{1}{x}\right]$.[25] It is easy to see that if $x = [n_1, n_2, \ldots]$, then $T(x) = [n_2, n_3, \ldots]$ so that T is conjugate to the shift σ in Σ_+. Let χ be the corresponding conjugating map. The measure $\chi_*^{-1}\nu$ coincides with the Gauss measure on $[0,1]$ with density $\dfrac{dx}{\log 2(1+x)}$ which is T-invariant. The ergodicity of Gauss measure with respect to T is well-known (cf [KSF]). The generalization of the above construction for some other discrete groups is given in [S].

§6. Dimension-like Characteristics of Invariant Sets for Dynamical Systems

6.1. Introductory Remarks. As it has been mentioned above, the topological structure of hyperbolic sets (including strange attractors) may be rather complicated and irregular (see Sect. 2). It is, in some sense, described by a process similar to the construction of a Cantor set. This allows to characterize the geometrical structure of an invariant hyperbolic set using certain dimension-like characteristics. In many situations we do not need to know the whole complicated topological structure of individual trajectories of dynamical systems, but it is enough to study only the ergodic characteristics of their global behavior. The dimension-like characteristics take an intermediate place between topological and ergodic ones, and according to the recent investigations, they are closely related to the latter. In recent physics literature several dimension-like characteristics for invariant sets of dynamical systems were introduced (these sets are not exactly typically hyperbolic but have much in common with hyperbolic sets; c.f. [FOY]). However, the strict mathematical results known today concern only the Hausdorff dimension, the capacity and the dimension with respect to a map (see below).

6.2. Hausdorff Dimension. Let X be a compact metric space, F the set of open balls in X. The Hausdorff α-measure of some subset $Y \subset X$ is defined by

$$m_H(Y, \alpha) = \lim_{\varepsilon \to 0} \inf_{G \subset F} \left\{ \sum_{S \in G} (\text{diam } S)^\alpha : \bigcup_{S \in G} S \supset Y, \text{ diam } S \leq \varepsilon \right\}. \quad (7.35)$$

Here G is a finite or countable subset of Γ; it is easy to see that the above limit exists. When α is fixed, $m_H(Y, \alpha)$ considered as a function on Y, is a regular, σ-additive, outer Borel measure (cf [Bi1]). When Y is fixed, $m_H(Y, \alpha)$ considered as a function on α, is nonincreasing and has the following property: there exists an α_0 such that $m_H(Y, \alpha) = \infty$ for $\alpha < \alpha_0$ and $m_H(Y, \alpha) = 0$ for $\alpha > \alpha_0$. The

---

[25] $[y]$ is the integer part of y.

Hausdorff dimension of Y is defined by

$$\dim_H Y = \alpha_0 = \inf\{\alpha\colon m_H(Y,\alpha) = 0\} = \sup\{\alpha\colon m_H(Y,\alpha) = \infty\}.$$

Let us consider some properties of Hausdorff dimension:

1) $\dim_H(\bigcup_i Y_i) = \sup_i \dim_H Y_i$ for any countable union of $Y_i \subset X$.

2) $\dim_H(Y \times Z) \geqslant \dim_H Y + \dim_H Z$ for any Borel (in particular for any compact) subsets $Y, Z \subset X$ the strict inequality being possible (more over there exists an A satisfying $\dim_H(A \times A) > 2\dim_H A$).

3) If A is a Borel subset of X and B is a ball in $\mathbb{R}$, then $\dim_H(A \times B) = \dim_H A + n$.

Now we shall consider some results concerning the calculation and the estimates of Hausdorff dimension for invariant sets of dynamical systems. Let Λ be an invariant subset of a C^1-diffeomorphism S of a compact Riemannian manifold M. We denote by $M_S(\Lambda)$ the set of all S-invariant Borel probability ergodic measures on Λ. For some $\mu \in M_S(\Lambda)$ consider a set G_μ which consists of $x \in \Lambda$ satisfying $\frac{1}{n}\sum_{k=0}^{n-1} \varphi(S^k(x)) \to \int_\Lambda \varphi \, d\mu$ for any continuous function φ on Λ. An equivalent definition: G_μ is the union of $x \in \Lambda$ such that the sequence of measure $\frac{1}{n}\sum_{k=0}^{n-1} \delta(S^k(x))$ ($\delta(x)$ is the measure concentrated at x) weakly converges to μ. Consider the ordered characteristic Lyapunov exponents $\chi^1(x) \geqslant \cdots \geqslant \chi^n(x)$ ($n = \dim M$) at x. The functions $\chi^i(x)$ are S-invariant; thus for $\mu \in M_S(\Lambda)$ and for μ almost all $x \in M$ we have $\chi^i_\mu \stackrel{\text{def}}{=} \chi^i(x) = \text{const}$ where $\chi^1_\mu \geqslant \cdots \geqslant \chi^n_\mu$. Set

$$\alpha(\mu) = \begin{cases} 0, & \text{if } \chi^1_\mu < 0 \\ \sup\left\{\alpha\colon 0 \leqslant \alpha \leqslant n, \ \sum_{i=1}^{[\alpha]} \chi^i_\mu + (\alpha - [\alpha])\chi^{[\alpha]+1}_\mu \geqslant 0\right\}, & \text{if } \chi^1_\mu \geqslant 0. \end{cases}$$

Defined in this way, $\alpha(\mu)$ is called *Lyapunov dimension* of μ. A modified version of a hypothesis from [Mor] may be formulated as: $\dim_H \Lambda = \alpha(\mu)$ for some measure μ. This hypothesis was supported in several physics papers (for example see [FOY]). Some modifications of this hypothesis are proved for the two dimensional case (cf [Y2]). For the multidimensional case similar results hold if, instead of Hausdorff dimension, we consider another characteristic (see 6.3 below and [Pe1], [Pe2]). There is no hope for a general formula relating Hausdorff dimension with the characteristic Lyapunov exponents. Therefore the following estimate of Hausdorff dimension obtained by Ledrappier in [L2] is very interesting.

Theorem 6.1.

$$\dim_H \Lambda \leqslant \sup_{\mu \in M_S(\Lambda)} \alpha(\mu).$$

If Λ is a LMHS of a $C^{1+\alpha}$-diffeomorphism S, then it is possible to obtain more exact and more interesting results. Fix an $x \in \Lambda$ and a small neighbourhood $U(x)$ of x in M. As Λ is a locally maximal set, there exist two open sets $U_1 \subset V^s(x)$,

$U_2 \subset V^u(x)$ such that $\Lambda \cap U(x) = (\Lambda \cap U_1) \times (\Lambda \cap U_2)$. It turns out that when $\dim M = 2$, the Hausdorff dimension of Λ equals the Hausdorff dimension of $\Lambda \cap U(x)$ (which does not depend on the choice of $x \in \Lambda$ and of $U(x)$) and that one equals the sum of Hausdorff dimensions of $\Lambda \cap U_1$ and $\Lambda \cap U_2$ (which also do not depend on x). This characterizes the Hausdorff dimension of the intersections of Λ with stable and unstable manifolds. A similar situation arises for $\mu = M_S(\Lambda)$, but instead of Λ we have to consider the support of μ or the set G_μ (remember that $\mu(G_\mu) = 1$). Now we give exact formulations.

Consider the function $\psi^u(t) = P(t\varphi^u)$ where $\varphi^u(x)$ is defined in (7.24) and $P(t\varphi^u)$ is the pressure of $t\varphi^u$ on Λ. It is easy to see that $\varphi^u(0) = P(0) = h(S) \geq 0$. It can also be shown that $\psi^u(t)$ is strictly decreasing (cf [W1]). The results of [Bo2] imply $\psi^u(1) \leq 0$. Therefore there exists a unique root t^u of the equation $P(t^u \varphi^u) = 0$. Similarly considering the function $\psi^s(t) = P(t\varphi^s)$, one can prove the existence of a unique root t^s of the equation $P(t^s \varphi^s) = 0$.

Theorem 6.2 (cf [Ma1]). *Suppose that $\dim M = 2$. Λ is a LMHS for a $C^{1+\alpha}$-diffeomorphism S. $S|\Lambda$ is topologically transitive. Then for any $\mu \in M_S(\Lambda)$ and for any $x \in \Lambda$*

$$\dim_H G_\mu = \dim_H(G_\mu \cap V^u(x)) + \dim_H(G_\mu \cap V^s(x)),$$
$$\dim_H(G_\mu \cap V^s(x)) = h_\mu / \chi_\mu^1, \qquad (7.36)$$
$$\dim_H(G_\mu \cap V^u(x)) = h_\mu / |\chi_\mu^2|,^{26}$$

where h_μ is the metric entropy of S.

We sketch the proof of (7.36). Let ξ be a Markov partition of Λ and $\xi_n = \bigvee_{k=0}^{n-1} S^k \xi$. It follows from the Shannon-McMillan-Breiman theorem (see Chap. 3, Sect. 3) and from the properties of characteristic Lyapunov exponents that for any $\varepsilon > 0$ there exists a set $G_\mu^\varepsilon \subset G_\mu$ satisfying $\mu(G_\mu^\varepsilon) \geq 1 - \varepsilon$ such that 1) the number of elements of ξ_n which cover the set $V^u(x) \cap G_\mu^\varepsilon$ is proportional to $\exp(h_\mu \pm \varepsilon)n$; 2) the diameter of any element of ξ_n intersecting G_μ^ε is proportional to $\exp(\chi_\mu^2 \pm \varepsilon)n$. One can then show that ξ_n is "the best" partition, i.e. it almost realizes inf in (7.55). For this reason $\dim_H(V^u(x) \cap G_\mu^\varepsilon) \approx (h_\mu \pm \varepsilon)/(|\chi_\mu^2| \pm \varepsilon)$. Now take the limit for $\varepsilon \to 0$.

Theorem 6.3 (cf [Ma1]). *In the conditions of Theorem 6.2 for any $x \in \Lambda$ the following hold:*
1) $\dim_H \Lambda = \dim_H(\Lambda \cap V^u(x)) + \dim_H(\Lambda \cap V^s(x))$;
2) $\dim_H(\Lambda \cap V^u(x)) = t^u = \sup_{\mu \in M_S(\Lambda)} \dim_H(G_\mu \cap V^u(x)) = \dim_H(G_{\mu_1} \cap V^u(x))$, *where μ_1 is the Gibbs measure of the function $t^u \varphi^u$;*
3) $\dim_H(\Lambda \cap V^s(x)) = t^s = \sup_{\mu \in M_S(\Lambda)}(G_\mu \cap V^s(x)) = \dim_H(G_{\mu_2} \cap V^s(x))$, *where μ_2 is the Gibbs measure of the function $t^s \varphi^s$.*

---

[26] As $\dim M = 2$ and Λ is a hyperbolic set, we have according to our notation $\chi_\mu^1 > 0 > \chi_\mu^2$.

Let Λ be a hyperbolic attractor. Then $V^u(x) \subset \Lambda$ for any $x \in \Lambda$; thus $\dim_H(V^u(x) \cap \Lambda) = 1$ and consequently $\dim_H \Lambda = 1 + t^s$. Besides if μ is a Sinai-Ruelle-Bowen measure on Λ, then $\dim_H(G_\mu \cap V^s(x)) = \chi^1_\mu/|\chi^2_\mu|$ (see Section 3) which implies $\dim_H G_\mu = 1 + \chi^1_\mu/|\chi^2_\mu|$.

6.3. Dimension with Respect to a Dynamical System. The situation is much more complicated for multidimensional systems where it is difficult to relate Hausdorff dimension to dynamics. In [Pe4] we define another characteristic of invariant hyperbolic set Λ: the so-called *dimension with respect to a dynamical system*. We do not formulate an exact definition but give some reasons for introducing such notion. If $\dim W^s(x) > 1$ then a different rate of contraction in different directions tangent to $W^s(x)$ implies that the image of a ball in $W^s(x)$ under $\{S^t\}$ is a "small, elongated ellipsoid". When calculating the Hausdorff dimension we ignore this fact (we substitute "ellipsoids" with balls of radii equal to the length of the maximal semi-axis of the ellipsoid). Instead, when we calculate the dimension with respect to a dynamical system we simply take into account the "best packing", consisting of these "ellipsoids". Notice that both dimensions coincide when $\dim W^s(x) = 1$. Certainly the dependence of the dimension with respect to a dynamical system not only from Λ but also from dynamics may be considered as a "defect". At the same time Theorems 6.2, 6.3 hold for this dimension (cf [Pe4]).

6.4. Capacity and Other Characteristics. Let X be a compact metric space, $Y \subset X$. We denote $N(\varepsilon)$ the minimal number of ε-balls covering Y. The *upper* (respectively *lower*) *capacity* of a set Y is defined by

$$\overline{C}(Y) = \overline{\lim_{\varepsilon \to 0}} \frac{\log N(\varepsilon)}{\log(1/\varepsilon)} \quad \left(\text{respectively } \underline{C}(Y) = \underline{\lim_{\varepsilon \to 0}} \frac{\log N(\varepsilon)}{\log(1/\varepsilon)}\right).$$

It is not difficult to show that $\dim_H Y \leq \underline{C}(Y) \leq \overline{C}(Y)$. Let μ be a Borel measure on X. Then

$$\dim_H \mu = \inf\{\dim_H Y: Y \subset X, \mu(Y) = 1\},$$

$$\overline{C}(\mu) = \overline{\lim_{\delta \to 0}} \inf\{\overline{C}(Y): Y \subset X, \mu(Y) \geq 1 - \delta\},$$

$$\underline{C}(\mu) = \underline{\lim_{\delta \to 0}} \inf\{\underline{C}(Y): Y \subset X, \mu(Y) \geq 1 - \delta\}$$

are called respectively the *dimension of a measure*, the *upper* and the *lower capacity of a measure*.

For given $\varepsilon, \delta > 0$ we use $N(\varepsilon, \delta)$ to denote the minimal number of ε-balls needed to cover some set of μ-measure $\geq 1 - \delta$. Then

$$\underline{C}_L(\mu) = \overline{\lim_{\delta \to 0}} \underline{\lim_{\varepsilon \to 0}} \frac{\log N(\varepsilon, \delta)}{\log(1/\varepsilon)},$$

$$\bar{C}_L(\mu) = \varlimsup_{\delta \to 0} \varlimsup_{\varepsilon \to 0} \frac{\log N(\varepsilon, \delta)}{\log(1/\varepsilon)}$$

are called respectively the *lower* and the *upper Ledrappier capacity* of μ. It is easy to see that $\dim_H \mu \leq \underline{C}(\mu) \leq \bar{C}(\mu)$, $\underline{C}_L(\mu) \leq \underline{C}(\mu)$, $\bar{C}_L(\mu) \leq \bar{C}(\mu)$, $\underline{C}_L(\mu) \leq \bar{C}_L(\mu)$. It can also be proved that $\dim_H \mu \leq \underline{C}_L(\mu)$. Let ξ be a measurable partition of X. Fix $\varepsilon > 0$ and set

$$H_\mu(\varepsilon) = \inf\{H_\mu(\xi) : \operatorname{diam} \xi \leq \varepsilon\}$$

($H_\mu(\xi)$ is the entropy of ξ see Chap. 3, Sect. 1). Then

$$\bar{R}(\mu) = \varlimsup_{\varepsilon \to 0} \frac{H_\mu(\varepsilon)}{\log(1/\varepsilon)}, \qquad \underline{R}(\mu) = \varliminf_{\varepsilon \to 0} \frac{H_\mu(\varepsilon)}{\log(1/\varepsilon)}$$

are called respectively the *upper* and the *lower information dimension* (or the *upper* and the *lower Renyi dimension*).

Theorem 6.4 (cf [Y2]). *Let μ be a Borel probability measure on the compact metric space X. Suppose that for μ-almost every x*

$$\lim_{\rho \to 0} \frac{\log \mu(B(x, \rho))}{\log \rho} = \alpha$$

($B(x, \rho)$ is a ρ-ball centred in x). Then

$$\dim_H \mu = \bar{C}(\mu) = \underline{C}(\mu) = \bar{C}_L(\mu) = \underline{C}_L(\mu) = \bar{R}(\mu) = \underline{R}(\mu) = \alpha.$$

If Λ is an S^t-invariant set, then the dimension characteristics introduced above may be used for some kind of description of the geometrical structure of Λ. A general approach to the characteristics of this type is given in [Pe5]. Here we formulate the results which sometimes allow to calculate these characteristics.

Theorem 6.5 (cf [Y1]). *Let Λ be an invariant set of a $C^{1+\alpha}$ diffeomorphism S of a smooth compact manifold M, μ be a Sinai measure (see 3.9), and $\chi_\mu^1 \geq \cdots \geq \chi_\mu^k > 0 > \chi_\mu^{k+1} \geq \cdots \geq \chi_\mu^k$ the corresponding Lyapunov exponents. Then*

$$\bar{C}(\operatorname{supp} \mu) \geq k + (\chi_\mu^1 + \cdots + \chi_\mu^k)/|\chi_\mu^{k+1}| = A.$$

Corollary (cf [Y1]). *If Λ is a hyperbolic attractor of a $C^{1+\alpha}$-diffeomorphism S, $S|\Lambda$ is topologically transitive and μ is a Sinai-Ruelle-Bowen measure on Λ (see 3.6), then $\bar{C}(\Lambda) \geq A$.*

Theorem 6.6 (cf [Y2]). *Let S be a $C^{1+\alpha}$-diffeomorphism of a smooth compact two-dimensional Riemannian manifold M, μ be a Borel probability measure with characteristic exponents $\chi_\mu^1 > 0 > \chi_\mu^2$ (see Section 3).*
Then

$$\lim_{\rho \to 0} \frac{\log \mu(B(x, \rho))}{\log \rho} = \alpha = h_\mu(S)\left(\frac{1}{\chi_\mu^1} - \frac{1}{\chi_\mu^2}\right)$$

and consequently

$$\dim_H \mu = \bar{C}(\mu) = \underline{C}(\mu) = \bar{C}_L(\mu) = \underline{C}_L(\mu) = \bar{R}(\mu) = \underline{R}(\mu).$$

Consider a hyperbolic periodic point p of a C^2-diffeomorphism S of a smooth surface M, and let x be a transversal homoclinic point corresponding to p (see Sect. 2). Fix some small $\varepsilon > 0$ and consider a LMHS Λ_ε contained in the ε-neighbourhood of the trajectory $\{S^n(x)\}$ (see Theorem 2.4). Let λ, γ be the eigenvalues of $DS(p)$, $0 < \lambda < 1 < \gamma$.

Theorem 6.7 (cf [AP]). *There exist constants $C_1 > 0$, $C_2 > 0$ such that*

$$\beta \leqslant \dim_H(\Lambda_\varepsilon \cap V^u(x)) \leqslant \alpha,$$

where $\alpha = \alpha(\varepsilon)$, $\beta = \beta(\varepsilon)$ coincide with the unique roots of the equations

$$\alpha \ln \frac{C_1 \gamma}{\varepsilon} = -\ln(\gamma^\alpha - 1), \qquad \beta \ln \frac{C_2 \gamma}{\varepsilon} = -\ln(\gamma^\beta - 1).$$

Corollary. *When $\varepsilon \to 0$ the function $\varphi(\varepsilon) = \dim_H(\Lambda_\varepsilon \cap V^u(x))$ behaves asymptotically as $\varphi(\varepsilon) = \ln\ln\frac{1}{\varepsilon} \Big/ \ln\frac{1}{\varepsilon} - \ln\ln\ln\frac{1}{\varepsilon} \cdot \ln\gamma / \ln\frac{1}{\varepsilon}$.*

The analogous result (with γ replaced by λ^{-1}) holds for $\dim_H(\Lambda_\varepsilon \cap V^s(x))$.

Chapter 8
Dynamical Systems of Hyperbolic Type with Singularities

L.A. Bunimovich

There are many important problems in physics where dynamical systems of hyperbolic type with singularities have arisen. Moreover, a Poincaré map for a smooth (and even analytic) flow often possesses singularities. One should also mention that the representation of a flow as a special flow (see Chap. 1, Sect. 4) and the passage to the corresponding induced (Poincaré) map is one of the most effective tools to study ergodic properties of dynamical systems with continuous time.

§ 1. Billiards

Dynamical systems with elastic reflections (or billiards) form the most important class of dynamical systems with singularities both in the general theory and in applications. By billiard we mean a dynamical system corresponding to the motion by inertia of a point mass within a region which has a piecewise smooth boundary. The reflections from the boundary are elastic.

1.1. The General Definition of a Billiard. Let Q be a compact Riemannian manifold with piecewise smooth boundary, $\dim Q = d$. The boundary ∂Q consists of a finite number of smooth compact submanifolds $\Gamma_1, \ldots, \Gamma_r$ of H1 codimension. A point $q \in \bigcup_{i=1}^{r}(\Gamma_i \backslash \bigcup_{i \neq j}\Gamma_j)$ and a set $\tilde{\Gamma}_i = \Gamma_i \backslash \bigcup_{i \neq j}\Gamma_j$ will be respectively called a *regular point* and a *regular component of the boundary*. A point which is not regular will be called *singular*.

Let $\mathcal{T}_q$ be the tangent space to the submanifold Γ_i at the point q, and $n(q)$ the unit normal vector at q directed toward the interior of Q. If q is a regular point then $n(q)$ is uniquely defined. At a singular point there can be several vectors $n(q)$.

We denote by M a unit tangent bundle over Q, $M = \{x: x = (q, v), q \in Q, v \in S^{d-1}\}$ where $S^{d-1} \subset \mathbb{R}^{d-1}$ is a unit sphere. A point $x = (q, v) \in M$ is called a line element with a footpoint q. Let $\pi: M \to Q$ be the natural projection of M onto Q, i.e. $\pi(q, v) = q$. If Q is a compact smooth manifold with piecewise smooth boundary, then M is also a manifold with the boundary $\partial M = \pi^{-1}(\partial Q)$, consisting of a finite number of regular components $\partial M_i = \pi^{-1}(\Gamma_i)$. Obviously, $\dim M = 2d - 1$.

We introduce a measure μ in M by setting $d\mu = d\rho(q) d\omega_q$, where $d\rho(q)$ is an element of the Riemannian volume on Q and ω_q is the Lebesgue measure on the sphere $S^{d-1} = \pi^{-1}(q)$.

We consider now a geodesic flow (see Chap. 1, Sect. 1) on the space M. It is determined by the vector field $X = \{X(x), x \in M\}$ where $X(x)$ is a tangent vector to M at the point x. X therefore characterizes the motion of points along geodesic lines with unit speed. Let N_{ij} be a set of inner points $x = (q, v) \in M$ such that the segment of the geodesic drawn along the direction[1] X intersects ∂Q in a point of the set $\Gamma_i \cap \Gamma_j$. It is clear that N_{ij} is a closed submanifold of codimension 1 and therefore $\mu(\bigcup_{i \neq j} N_{ij}) = 0$.

We shall assume that for almost every point $x \in \text{Int } M$ (with respect to μ) the geodesic drawn along the direction X intersects with the boundary. This property is fulfilled for all known interesting examples of billiards. Let s be the smallest positive number such that the geodesical segment of length s drawn along the direction x ends in a regular point of a boundary ∂Q. (It can be shown that the set of points x, such that the corresponding geodesics fall into singular points of the boundary, has measure 0). Let y be the tangent vector to Q which arose from x by parallel displacement along the geodesic of length s. We shall now construct the new tangent vector $y' = y - 2(n(q), y)n(q)$ at the point $q = \pi(y)$. This means that the vector y is reflected from the boundary according to the law of elastic reflections, "the angle of incidence equals the angle of reflection". Then we take

---

[1] By definition the geodesic passing through q along the direction of the vector v will be called the geodesic drawn along the direction $x = (q, v)$.

a segment of the geodesic drawn along the direction y' up to the next intersection with the boundary and so on. It can be shown that the set of trajectories which have an infinite number of reflections from the boundary within a finite time has measure 0. We shall also assume that for almost all points x, all the geodesical segments that are generated by the process just described have finite lengths.

Therefore one can define a one-parameter group of transformations $\{T^t\}$ on a subset $M' \subset M$ of full measure by associating to any $x \in M'$ and t, $-\infty < t < \infty$, the tangent vector $T^t x$, obtained by the parallel displacement of x along the direction defined by its geodesic for a distance t. If t is a moment of reflection at the boundary, we set $T^t x = \lim_{t' \to t+0} T^{t'} x$. It is convenient to identify a point y of the boundary ∂M with a point $y' = y - 2(n(q), y)n(q)$. The resulting set will also be denoted M'.

Definition. The group of transformations $\{T^t\}$ is called a *billiard* in Q.

It can be proved that the group $\{T^t\}$ preserves the measure μ (cf [KSF]). Therefore billiard is a flow in the sense of ergodic theory. Usually $M(Q)$ is called a phase (configuration) space of a billiard.

It can be easily seen from this definition that billiards are geodesic flows on nonclosed manifolds with reflection from the boundary according to the law "the angle of incidence equals the angle of reflection". A billiard in a region Q can be also defined as a Hamiltonian system with a potential $V(q) = 0$ if $q \in Q \setminus \partial Q$ and $V(q) = \infty$ if $q \in \partial Q$.

A billiard $\{T^t\}$ has the following natural special representation (see Chap. 1, Sect. 4). Consider the following transformation T_1 of the set $M_1 = \{x \in \partial M: (n(q), x) \geq 0, q = \pi(x)\}$ into itself. Draw the geodesic along the direction x up to its first intersection with the boundary. A vector y, equal to the reflection of the tangent vector at that intersection point, will be denoted $T_1 x$. We thus obtain the special representation of the flow $\{T^t\}$ generated by the transformation T_1 and the function $\tau(x)$, where $\tau(x)$ is the length of the geodesic segment under consideration. T_1 preserves the measure ν which is the projection of μ on ∂M.

1.2. Billiards in Polygons and Polyhedrons. Let $Q \subset \mathbb{R}^d$ be a convex polyhedron, i.e. a closed bounded set $Q = \{q \in \mathbb{R}^d: f_i(q) \geq 0, i = 1, \ldots, r\}$ where the functions $f_1, \ldots, f_r$ are linear. In this case regular components of ∂Q are faces $\Gamma_i, i = 1, \ldots, r$, of a polyhedron Q. We denote by n_i a unit vector orthogonal to Γ_i and directed inside Q.

It follows from definition that trajectories of billiards in domains of Euclidean space are broken lines. Let us consider the isometric mapping $\sigma_j: S^{d-1} \to S^{d-1}$ acting in every point $x = (q, v)$, $q \in \Gamma_i$, according to the formula $\sigma_i(v) = v - 2(n_i, v)n_i$. We assume that some trajectory of a billiard in Q has vertices in faces with numbers $i_1, i_2, \ldots$. Then by means of successive reflections of Q we can obtain a straight line instead of the broken one. This straight line intersects with the polyhedrons $Q, Q_{i_1}, Q_{i_1, i_2}, \ldots$ where $Q_{i_1, \ldots, i_k}$ is the result of successive reflections of Q, relative to faces $\Gamma_{i_1}, \ldots, \Gamma_{i_k}$, where Γ_{i_l} is a face of $Q_{i_1, \ldots, i_{l-1}}$.

Consider a point $x_0 = (q_0, v_0) \in M$. The vector $v_0 \in S^{d-1}$ defines the initial velocity of the billiard trajectory originating from the point $q_0 \in Q$. The velocity vector becomes $v_k = (\sigma_{i_k} \sigma_{i_{k-1}} \ldots \sigma_1) v_0$ between the kth and $(k+1)$th reflections.

Let G_Q be a subgroup of the group of all isometries of S^{d-1} generated by $\sigma_1, \ldots, \sigma_r$.

Theorem 1.1. *If G_Q is a finite group, then the billiard in the polyhedron Q is nonergodic. Moreover, to every orbit $\Omega = \Omega(v_0) = \{g v_0 \in S^{d-1} : g \in G_Q, v_0 \in S^{d-1}\}$ of G_Q acting on S^{d-1}, there corresponds an invariant set A_Ω with respect to the action of the flow $\{T^t\}$ consisting of all points $x = (q, v) \in M$, such that $v \in \Omega$.*

The geometric meaning of this theorem is as follows: if the group G_Q is finite, then only finite sets of directions can arise from the initial direction during the motion along trajectories of a billiard. For $d = 2$ the finiteness of the group is equivalent to the commensurability of all angles of a polygon Q.

The simplest example in the class under consideration is a billiard in a rectangle. In this case G_Q contains four elements: **Id**, σ_1, σ_2, $\sigma_2 \sigma_3$. The phase space of this billiard can be decomposed into $\{T^t\}$-invariant sets A_Ω, where $\Omega = \Omega(v)$. It is easy to see that if $\sigma_1 v \neq v$ and $\sigma_2 v \neq v$, then A_Ω is a two-dimensional torus. The flow $\{T_\Omega^t\}$ induced on A_Ω by $\{T^t\}$ is a one-parameter group of shifts on a torus. Hence (see Chap. 1, Sect. 3), the flow $\{T_\Omega^t\}$ is ergodic if the number $v_2 a_1 / v_1 a_2$ is irrational where v_1, v_2 are projections of the velocity in the directions of the sides of the rectangle, and where a_1, a_2 are the lengths of these sides. So the decomposition of the phase space into invariant tori A_Ω coincides with its decomposition into ergodic components. Analogously, ergodic components of a billiard in a rectangular parallelepiped are (mod 0) one-parameter groups of shifts acting on d-dimensional invariant tori.

The general problem of investigating ergodic properties of billiards in arbitrary polyhedra (or even polygons) is still open. There are two main results on this subject (cf [BKM], [KSF]).

Theorem 1.2. *An entropy of a billiard in an arbitrary (not necessarily convex) polyhedron is equal to zero.*

Theorem 1.3. *If angles of a polygon Q are commensurable, then almost all trajectories of the corresponding billiard are dense in Q.*

It was conjectured that every trajectory of a billiard in a convex polygon must be periodic or everwhere dense. G.A. Galperin has proved recently that the conjecture is wrong (cf [Ga]). In his example there is a trajectory which is everywhere dense in some subdomain of Q. Apparently this situation often occurs.

Some problems in classical mechanics can be reduced to the investigation of billiards in polygons and polyhedrons. Consider $n \geqslant 2$ point particles moving along a straight segment, colliding elastically between them and also with the ends of the segment. The order of particles on this segment is preserved. Therefore the configuration space of this dynamical system is a simplex. It is easy to show that the elastic collisions of the particles correspond to the reflections of trajectories from the boundary of this simplex according the law "the angle of incidence equals the angle of reflection" (cf [KSF]).

1.3. Billiards in Domains with Smooth Convex Boundary. Let $Q \subset \mathbb{R}^2$ be a domain bounded by a smooth convex curve $\Gamma = \partial Q$. In the simplest case Γ is a circle. It is easy to see that every link of a broken line which corresponds to an arbitrary trajectory in a configuration space of a billiard is tangent to some concentrix circle to Γ but with a smaller radius. A billiard in a circle is therefore a completely integrable Hamiltonian system. Let Γ now be an ellipse $\Gamma = \Gamma_c = \{q \in \mathbb{R}^2 : \text{dist}(q, A_1) + \text{dist}(q, A_2) = c\}$ with foci A_1 and A_2. On can verify that all links of any trajectory of a billiard in Q are tangent to one and the same ellipse Γ_{c_1}, $c_1 < c$, confocal with Γ_c or to a hyperbola $H_{c_1} = \{q \in \mathbb{R}^2 : \text{dist}(q, A_1) - \text{dist}(q, A_2) = c_1\}$ confocal with Γ_c. (In the latter case, points of tangency may belong not only to links of a broken line but also to their continuations). Therefore a billiard in an ellipse is nonergodic.

The investigation of ergodic properties of billiards is important as well in partial differential equations.

Consider the following problem in the domain Q

$$(\Delta + \lambda)u(x, y) = 0$$
$$u|\partial Q = 0 \qquad (8.1)$$

where $(x, y) \in Q$, Δ is Laplace operator and λ is a spectral parameter. This problem arises in the study of small oscillations of a membrane with fixed boundary. But one may also consider it as a problem of quantum mechanics, concerning the determination of eigenvalues and eigenfunctions of a Schrödinger equation, with a potential equal to zero inside Q and equal to infinity on the boundary. An important problem is the study of the limiting behavior of the eigenvalues and eigenfunctions of (8.1) when $\lambda \to \infty$. The limit corresponds to the quasiclassical approximation ($\hbar \to 0$). It is thus natural to suggest that solutions of the quantum problem are connected with solutions of the corresponding problem in classical mechanics, i.e. with the billiard in Q.

Definition. A *caustic* for a billiard in Q is a smooth closed curve $\gamma \subset Q \subset \mathbb{R}^2$, such that if one link of the billiard trajectory is tangent to γ, then any other segment of this trajectory is also tangent to γ.

There is a unique family of caustics for a circle (concentric circles), but for ellipses there are two such families (confocal ellipses and hyperbolas).

V.F. Lazutkin has proved that there exists an uncountable set of caustics which has a positive measure in Q if a boundary $\Gamma = \partial Q$ is convex and sufficiently smooth (cf [La1], [La2]). Moreover Γ is a limit point of this set and the measure of the set of all line elements tangent to caustics is positive in M. Therefore a

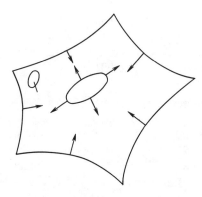

Fig. 5

billiard inside a sufficiently smooth convex curve on the plane is non-ergodic. It was shown in [La2] that with the help of invariant sets of a billiard, one can construct "quasi"—eigenfunctions (quasimodes) and corresponding "quasi"—eigenvalues for the Dirichlet problem in Q. A carrier of any such eigenfunction is localized in a neighborhood of one of the invariant sets of a billiard defined by caustics.

A.I. Shnirelman has shown that for ergodic billiards, carriers of eigenfunctions of the Dirichlet problem in Q fill in some sense the whole domain when $\lambda \to \infty$ (cf [Sh]).

1.4. Dispersing (Sinai) Billiards. Consider the billiard in the domain Q illustrated in Fig. 5. We shall assume that all regular components of the boundary ∂Q are smooth curves (of class C^3) which are convex inside Q and have a strictly positive curvature with respect to a framing of ∂Q in each point $q \in \partial Q$ by inner unit normal vectors $n(q)$. The ergodic properties of such billiards are quite different from those described above.

It is easy to see that the difference between this example and the billiards considered in 1.2 and 1.3 stems from the structure of the boundary ∂Q. In 1.2, ∂Q consisted of pieces of hyperplanes (straight lines), in 1.3, it was convex outside Q, and in this example it is convex inside Q.

Ergodic properties of billiards in domains in Euclidean spaces ($Q \subset \mathbb{R}^d$) or on a torus with Euclidean metric are completely defined by the structure of the boundary ∂Q. In particular, a billiard in a domain Q with a boundary convex inside Q is a hyperbolic dynamical system.

It was Ya.G. Sinai who began the rigorous mathematical investigating of billiards satisfying a condition of hyperbolicity (cf [Si4], [Si8]). In [Si2] he introduces the important class of dispersing billiards (which are now called Sinai billiards).

Let us recall that a boundary ∂Q is framed by unit normal vectors $n(q)$ directed inside Q. Therefore for any regular point $q \in \partial Q$ one can define a linear self-

Chapter 8. Dynamical Systems of Hyperbolic Type with Singularities

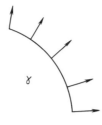

Fig. 6

adjoint operator of a second quadratic form $K(q)$ which acts in the tangent space $\mathcal{T}_q$ to the boundary ∂Q at q.

Definition. A billiard is called Sinai (*dispersing*) if $K(q) > 0$ for any regular point $q \in \partial Q$.

In the case $d = 2$, this condition is equivalent to positivity of the curvature of ∂Q. We shall now explain why dispersing billiards are the natural analog of smooth dynamical systems of hyperbolic type. Consider again the billiard in the domain represented in Fig. 5. We will assume for simplicity that regular components of the boundary ∂Q intersect in their endpoints transversally.

Take a smooth curve $\tilde{\gamma} \subset Q$ and a continuous set γ of unit vectors normal to $\tilde{\gamma}$. Then γ is a smooth curve in M. It is evident that two curves γ correspond to every curve $\tilde{\gamma}$ according to the choice of a field of normal vectors. After fixing the curve γ we can speak about the curvature of the curve $\tilde{\gamma}$. Therefore it is more correct, in general, to speak about the curvature of the curve γ. We shall say that the curve γ is *convex* when its curvature is positive everywhere. Fig. 6 shows a convex curve γ.

We denote by $\kappa(x_0)$ the curvature of γ at the point x_0. Let t be so small that no point of the curve γ could reach the boundary ∂Q during the time interval $[0, t]$. It is easy to calculate that the curvature of the curve $\gamma_t = T^t \gamma$ at the point $x_t = T^t x_0$ is given by the equality $\kappa(x_t) = \kappa(x_0)(1 + t\kappa(x_0))^{-1}$. Therefore as the motion proceeds inside Q, the length of γ_t increases locally linearly with respect to t and its curvature decreases according to a $1/t$ law. From the equality given above it follows that if $\kappa(x_0) > 0$, then $\kappa(x_t) > 0$, that is, if γ_0 is a convex curve, then γ_t is a convex curve too.

We shall now consider what happens upon reflections from the boundary ∂Q. Let $\tau(x_0) > 0$ be a moment of the first reflection from ∂Q of a point x_0. It can be shown, using elementary geometrical arguments, that $\kappa(x_{\tau+0}) = \kappa(x_{\tau-0}) + \dfrac{2k(q_\tau)}{\cos \varphi_\tau}$, where $k(q_\tau)$ is the curvature of the boundary at the point of reflection and $\varphi(x_\tau)$ is the angle between the incident ray and the normal vector with $\cos \varphi(x_\tau) \geq 0$. Thus, reflection from a boundary which is convex inside Q turns every convex beam that approaches ∂Q with a decreased curvature into a beam with a curvature not less than $2\min_{q \in \partial Q} k(g)$. From this fact one can already derive the local exponential instability of dispersing billiards. In fact, from the equalities given above it follows that the length of a convex curve γ increases locally between its ith and $(i+1)$th reflections from the boundary with a coefficient $1 + \tau_{i+1}(x_0)k_{i+1}(x_0)$. Here, $x_0 \in \gamma$, $\tau_{i+1}(x_0)$ is the time between the ith and $(i+1)$th reflections from ∂Q and $k_{i+1}(x_0)$ is the curvature of ∂Q at the point of the $(i+1)$th reflection. Finally it is easy to see that for a "generic" trajectory, the number of reflections from the boundary increases linearly with time.

Thus dispersing billiards are very similar to geodesic flows in Riemannian manifolds of negative curvature (see Chap. 7, Sect. 4). The role of a negative curvature is played here by a boundary convex inside Q. Nevertheless there is an essential difference between these classes of hyperbolic dynamical systems. It will be shown below that dispersing billiards are not uniformly, but nonuniformly hyperbolic dynamical systems (see Chap. 7, Sect. 1 for a definition).

In view of the local exponential instability one could try to construct local stable and unstable manifolds (LSM and LUM) for dispersing billiards (see Chapter 7, Section 1). The corresponding theorem for such billiards was proved in [Si4] (see [Si8] also).

Consider again the above given formulae which describe the evolution of a smooth curve in M under the action of billiard dynamics. These formulae allow to write at once differential equations for the vector fields tangent to LSM and LUM. Let $0 < t_1 < t_2 < \cdots < t_n < \cdots$ be the moments of the successive reflections of the semitrajectory of the point $x \in M$ from the boundary, $t_n \to \infty$ as $n \to \infty$, let $\tau_i = t_i - t_{i-1}$, $t_0 = 0$, $q_i \in \partial Q$, be the point of the boundary where the i-th reflection occurs, let v_i^- and v_i^+ be the velocities directly before and after the i-th reflection, and let $\cos \varphi_i = -(v_i^+, n(q_i))$, K_i be the operator of the second fundamental form of the boundary ∂Q at the point q_i, U_i be the isometric operator which maps in the direction parallel to the normal vector $n(q_i)$ the hyperplane $A_i^- \subset \mathbb{R}^d$ which contains the point q_i and is orthogonal to the vector v_i^- onto the hyperplane A_i^+, which contains q_i also and is orthogonal to v_i^+, V_i be the operator which maps A_i^- in the direction parallel to the vector v_i^- onto the hyperplane $A_i \subset \mathbb{R}^d$, which is tangent to the boundary ∂Q at the point q_i and let V_i^* be the operator adjoint to it. Consider the following infinite operator-valued continuous fraction

$$B(x) = \cfrac{I}{\tau_i I + \cfrac{I}{2\cos\varphi_1 V_1^* K_1 V_1 + U_1^{-1} \cfrac{I}{\tau_2 I + \cfrac{I}{2\cos\varphi_2 V_2^* K_2 V_2 + \ldots}} U_1}} \quad (8.2)$$

where I is the identity operator. If $d = 2$ (see Fig. 5), then the operator K_i in (8.2) must be replaced by the curvature of ∂Q at the point q_i and the operators U_i,

U_i^{-1}, V_i, V_i^* must be replaced by unities. It could be shown that the operator $B(x)$ defines the tangent plane to the local stable manifold of the point x in the phase space of the billiards under consideration (cf [Si8]).

Definition. A smooth $(d-1)$-dimensional submanifold $\gamma \subset \mathbb{R}^d$ is called *convex* or *concave* if the operator of the second fundamental form is positively or negatively defined correspondingly at every point $q \in \gamma$.

Theorem 1.4. *For almost every point $x \in M$ there exist convex and concave $(d-1)$-dimensional submanifolds $\gamma^{(u)}(x) \ni x$ and $\gamma^{(s)}(x) \ni x$ such that*

$$\lim_{t \to \infty} \left[-\frac{\ln \operatorname{diam} T^{-t}(\gamma^{(u)}(x))}{t} \right] > 0$$

and

$$\lim_{t \to \infty} \left[-\frac{\ln \operatorname{diam} T^t(\gamma^{(s)}(x))}{t} \right] > 0.$$

It was mentioned above that there exist essential differences between the dispersing billiards and the geodesic flows in manifolds of negative curvature. The main difference is that for dispersing billiards the flow $\{T^t\}$ is defined in M only almost everywhere and is nonsmooth. The corresponding singularities arise in trajectories falling into singular points of the boundary ∂Q and in trajectories tangent to ∂Q (see Fig. 7). This implies the non-existence of LUM and LSM in this set of measure zero. Besides, LUM and LSM do not exist in points such that their trajectories come too close to singular points of ∂Q or to trajectories tangent to ∂Q.

Thus the dispersing billiards are nonuniformly completely hyperbolic dynamical systems (see Chap. 7, Sect. 1). It could be shown that these local manifolds satisfy the property of absolute continuity (see Chap. 7, Sect. 3). Thus the general theory of NCH-systems implies that ergodic components of the

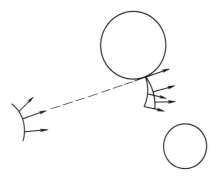

Fig. 7

Fig. 8

dispersing billiards have positive measure, its entropy is positive and a flow $\{T^t\}$ is the K-flow in almost every ergodic component.

Due to the existence of the singular set, global stable and unstable manifolds of dispersing billiards are submanifolds consisting of a countable number of smooth components. Fig. 8 shows a typical picture of part of such a global unstable manifold in case $d = 2$. The cusps on this curve correspond to trajectories tangent to the boundary, and break-points to trajectories falling into singular points of the boundary ∂Q.

The investigation of ergodic properties of dispersing billiards is therefore much more complicated in comparison with the smooth uniformly completely hyperbolic dynamical systems (see Chap. 7, Sect. 3). Really, for UCH-systems one can prove ergodicity at once. The corresponding proof is based on the approach which was first applied by E. Hopf to the proof of ergodicity of the geodesic flows on the surfaces of constant negative curvature. According to this approach, one has to construct for almost any pair x_1 and x_2 of the phase space of the dynamical system under consideration the so-called Hopf chain, i.e. a finite set W_1^s, W_1^u, $W_2^s, \ldots, W_n^s$ which consists of LSM and LUM such that $W_i^s \cap W_j^u \neq \emptyset$, where $j = i \pm 1$. Then one can derive easily from the Birkhoff-Khinchin ergodic theorem (see Chap. 1, Sect. 2) that both points x_1 and x_2 belong to one and the same ergodic component. However for the dispersing billiards, such a chain cannot be constructed in an analogous manner because a global manifold changes its direction in a singular point "almost" to the opposite direction. Therefore in order to construct the Hopf chain, one has to prove that regular components of LSM and LUM which pass through the majority of points of M are sufficiently large. The corresponding statement was formulated and proved in [Si4]. From that time it was generalized several times and the proof was modified (cf [BS1], [Si8], [Si7]). A proof of an analogous assertion is necessary before the proof of ergodicity for all billiards which are completely hyperbolic systems (see for instance [Bu1], [Si8], [Bu2]). Thus it is natural to call it the main theorem of ergodic theory of billiards. We shall formulate this theorem in the simplest variant in order to demonstrate clearly its meaning and the idea of the proof.

First we introduce some notation. We shall consider throughout this paragraph a dynamical system with a discrete time generated by the transformation T_1 (see Sect. 1.1). We shall assume for simplicity that Q_0 is a two-dimensional torus and its boundary ∂Q consists of one connected component only. We define on the phase space M_1 a system of coordinates (l, φ), where the parameter l denotes the arc length on ∂Q and φ is an angle between a line element $x = (q, v)$ and a normal vector $n(q)$, $-\frac{\pi}{2} \leqslant \varphi \leqslant \frac{\pi}{2}$. It is easy to see that, with respect to these coordinates, M_1 is a cylinder. Let S_0 be the set of all line elements tangent to the boundary. The transformation T_1 is a discontinuous one. It has discontinuities on the set $S_{-1} = T_1^{-1} S_0$ (see Fig. 7). This set consists of a denumerable number of smooth curves on the cylinder M_1. Each such curve in the coordinates (l, φ) can be defined by a monotonically increasing function $\varphi(l)$ (see Fig. 9) which satisfies the differential equation $\frac{d\varphi}{dl} = k(l) + \frac{\cos \varphi}{\tau(l, \varphi)}$, where $k(l)$ is a curvature of the boundary at the point l and $\tau(l, \varphi)$ is the time up to the next reflection from the boundary of the trajectory of the line element (l, φ). The limit points of the set S_{-1} are such points which have neighborhoods where the function $\tau(l, \varphi)$ is unbounded. To these points correspond such periodic trajectories of the billiard which are always tangent to the boundary ∂Q where, at the points of tangency, the boundary lies on one side of the trajectory (see Fig. 10). There are not more than a finite number of such trajectories. The analogous structure has the set of

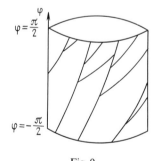

Fig. 9

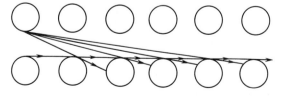

Fig. 10

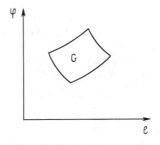

Fig. 11

discontinuities of the transformation T_1^{-1}; the only difference is that in the coordinates (l, φ) their regular components are decreasing curves. We shall call a piecewisely smooth curve $\varphi = \varphi(l)$ *increasing* (respectively *decreasing*) if
$$a_i \leq \frac{d\varphi}{dl} \leq a_2 \left(\text{respectively } -b_1 \leq \frac{d\varphi}{dl} \leq -b_2\right), \text{ where } 0 < a_1 < a_2, 0 < b_2 < b_1$$
are some constants which depend on geometrical properties of the domain Q. It can be demonstrated that LSM (respectively LUM) are increasing (respectively decreasing) curves on the cylinder M_1. By a *quadrilateral* we mean such a domain $G \subset M_1$ that its boundary consists of four piecewise continuously differentiable curves and one pair of opposite curves consists of increasing curves while the curves of the other pair are decreasing (Fig. 11). We shall denote by $l(\gamma)$ the length of the piecewise smooth curve $\gamma \subset M_1$. We can formulate now for the example under consideration the main theorem of ergodic theory of billiards.

Theorem 1.5. *Let $x_0 \in M_1$ be a point such that its positive semitrajectory $T^i x_0$, $i = 0, 1, \ldots$, never coincide with the singular points of the boundary. Then for every $\alpha (0 < \alpha \leq 1)$ and an arbitrary number C $(0 < C < \infty)$ there exists $\varepsilon = \varepsilon(x_0, \alpha, C)$, such that the ε-neighborhood U_ε of the point x_0 satisfies the following condition: for any increasing curve $\gamma_0 \subset U_\varepsilon$, $l(\gamma_0) = \delta_0$, there exists a quadrilateral $G_1(G_2)$ for which γ_0 is the upper (lower) side such that, if $G_i' = \{x : x \in G_i, \text{through } x \text{ passes a regular segment of a LSM joining the upper and lower sides of } G_i, l(\gamma^{(s)}(x)) > C\delta_0\}, i = 1, 2$, then $v(G_i') \geq (1 - \alpha)v(G_i), i = 1, 2$.*

The idea of the proof of this theorem is the following. Consider a quadrilateral G for which the increasing curves joining the upper and the lower sides of G have lengths greater than $C\delta_0$. If ε is sufficiently small, one can take a large number k_0 such that $T^{k_0}|U_\varepsilon$ is a smooth transformation. Then the image $T^{k_0}G$ becomes a very narrow and elongated quadrilateral, which is split up into connected components by curves which belong to S_{-1}. It is easy to see that the preimages of points lying in a small neighborhood of these curves do not belong to $G_1'(G_2')$. To every connected component obtained by intersection $T_1^{k_0}$ with S_{-1} we apply T_1 and consider the intersection of its image with S_{-1}. Then $G_1'(G_2')$ does not contain the preimages of points which belong to smaller neighborhoods of the curves of discontinuity. We apply again the transformation T_1 to the connected components arising from the intersection $T_1^{k_0+1} G \cap S_{-1}$ and repeat the same arguments, etc. Since in the direction of decreasing curves contraction occurs at each successive step, one can throw out more and more narrow neighborhoods of curves which belong to S_{-1}. Thus the total area of the sets thrown out is relatively small. The analogous theorem holds if one changes LSM onto LUM and the upper and the lower sides of G onto its left and right sides.

In order to prove the following statement, we can now use a Hopf chain whose links are formed by sets of type G'_1 and G'_2 constructed in the theorem 1.5, instead of LSM and LUM.

Theorem 1.6. *Sinai (dispersing) billiard is ergodic and a K-system.*

It can also be shown (cf [G]) that a dispersing billiard is isomorphic to a Bernoullian shift.

1.5. The Lorentz Gas and the Hard Spheres Gas. We shall consider in this section two examples of billiards which belong to the set of the most popular models of statistical mechanics.

In connection with the problem of description of electronic motion in metals, H. Lorentz introduced in 1905 the dynamical system which is now called the Lorentz gas. We shall consider here its simplest version. The general definition of the Lorentz gas will be given in Sect. 1.10.

Let D be a compact domain with a piecewise smooth boundary in the Euclidean space $\mathbb{R}^d$, $d \geq 1$, $B_1, \ldots, B_r$ be a collection of nonintersecting d-dimensional balls in D which are called scatterers. By definition, the Lorentz gas is a billiard in the domain $Q = D \setminus \bigcup_{i=1}^r B_i$.

We shall now consider the gas of hard, or absolutely elastic spheres. Assume that r hard spheres with radius and mass 1 are moving inside a compact domain $D \subset \mathbb{R}^d$, $d \geq 1$, and suffer elastic collisions between themselves and with the boundary ∂D.

The position of the ith sphere is determined uniquely by coordinates $q_j^{(i)}$, $1 \leq i \leq r$, $1 \leq j \leq d$, of its center. Let $D^- \subset D$ consist of all points which are situated not closer to the boundary ∂D than at the distance ρ. Consider a direct product $D^{(r)} = \underbrace{D^- \times D^- \times \ldots \times D^-}_{r \text{ times}} \subset \mathbb{R}^{dr}$ and exclude from $D^{(r)}$ all inner points of the sets

$$C_{i_1, i_2} = \left\{ q \in D^{(r)} : \sum_{j=1}^d (q_j^{(i_1)} - q_j^{(i_2)})^2 \leq (2\rho)^2 \right\}, \quad 1 \leq i_1, i_2 \leq r, i_1 \neq i_2.$$

It is easy to see that the set C_{i_1, i_2} is a product of the $(d-1)$-dimensional sphere and the Euclidean space $\mathbb{R}^{d(r-2)}$, i.e. it is a cylinder. The resulting set $Q \subset \mathbb{R}^{dr}$ is the domain with a piecewise smooth boundary. To each configuration of r spheres of radius ρ in D corresponds some point $q \in Q$. Thus the motion of spheres described above induces a group of transformations of the set Q. It can be easily checked that the laws of elastic collisions of spheres between themselves and with the boundary ∂Q correspond to elastic collisions of a moving point q from the boundary ∂Q, i.e. a billiard arises in Q.

1.6. Semidispersing Billiards. It was shown in Sect. 1.5 that in a phase space of a dispersing billiard, the condition of complete hyperbolicity holds in a set of full measure. It is known from the general theory of hyperbolic dynamical systems (see Chap. 7, Sect. 3) that systems which satisfy some weaker conditions of hyperbolicity could possess strong statistical properties as well. It happens that the situation for billiards is in some sense analogous. The corresponding classes of billiards will be considered in this and in the following sections.

Definition. A billiard is called *semidispersing* if, for any regular point $q \in \partial Q$, the second fundamental form $K(q) \geq 0$.

In other words, a billiard is semidispersing if a boundary ∂Q is not strictly convex inside Q but its curvature equals zero in some directions. Semidispersing billiards can serve as analogs of partially hyperbolic dynamical systems. The most important example of semidispersing billiards is the gas of hard spheres. One could expect that a semidispersing billiard possesses sufficiently good ergodic properties only in the case when plane directions at different points of the boundary are nonparallel. For instance, the system of two disks elastically interacting in the torus $\mathbb{T}^2$ has a full momentum as the additional integral of motion, and thus is non-ergodic, while the same system in a square is a K-system (cf [Si4]). It is easy to see that in the first case the directions with zero curvature are parallel, while in the last one they are not.

As usual, the first step in the studying of hyperbolic billiard systems is connected with the investigation of the corresponding operator $B(x)$ (see (8.2)). It was proved in [Si8] that a continuous fraction converges almost everywhere in a phase space M and it defines a symmetrical non-negatively defined operator which acts in a hyperplane $J(x)$, $x = (q, v)$ which is orthogonal to a vector v. Therefore $J(x)$ can be decomposed into a direct sum of two $B(x)$-invariant, zero and positive subspaces, correspondingly $J_0(x)$ and $J_+(x)$.

In the case of dispersing billiards, dimensions of both these foliations equal $d - 1$, where d is the dimension of the phase space. In the case of semidispersing billiards, a dimension of foliations depends on geometrical properties of a boundary ∂Q. It was shown in [Ch] that local transversal fibers exist for all points of a set $\hat{M} \subset M$ containing an open everywhere dense subset of M. Besides, a function which equals the dimension of an expanding (contracting) fiber is equal to a constant on the negative (positive) semitrajectory of the point x. From this result it follows in particular that an entropy of a semidispersing billiard is positive.

We should like to mention in the conclusion of this section that it was N.S. Krylov who first pointed out the exponential instability of the hard spheres gas (cf [Kr]).

1.7. Billiards in Domains with Boundary Possessing Focusing Components. In this section we shall consider billiards in domains Q on the plane or on the two-dimensional torus, such that a boundary ∂Q contains convex-outwards Q components. We shall call these focusing, in analogy as convex-inwards Q components of ∂Q will be called dispersing.

In accordance with the traditional ideas, stochastic properties of billiards are generated by the scattering of trajectories which results from collisions with the convex-inwards boundary of a domain Q. Therefore the first intuitive idea was that if a billiard is dispersing and the angles of ∂Q will be "smoothed", then a generic trajectory would spend the most time on the dispersing part of the boundary and occur on the focusing part very rarely, hence the resulting billiard could obey good statistical properties.

However the situation turns out to be essentially different. In order to understand what really occurs, let us consider the evolution of a curvature of a smooth curve $\gamma \subset M$ which suffers a series of consecutive reflections from some focusing component of ∂Q.

The main difficulty of the corresponding analysis is to study properties of the continuous fraction (8.2). In the case under consideration ($d = 2$) elements of the continuous fraction are not operators but numbers, and in the difference with the dispersing billiards these elements have different signs, thus it is harder to study the problem of its convergence. The following proposition is fundamental for studying reflection from the focusing part of the boundary (cf [Bu2], [Bu1]).

Theorem 1.7. *Suppose that on an arbitrary focusing component $\Gamma \subset \partial Q$ with constant curvature there is a bundle of parallel trajectories corresponding to a curve $\gamma \subset M_1$ of zero curvature (Fig. 12) which has n successive reflections from Γ. Then for any line element $x \in \gamma$ and for any number m, $1 \leqslant m \leqslant n$, the following inequalities hold: $\kappa_-(T_1^m x_0) < 0$, $\kappa_+(T_1^{m-1} x_0) > 0$, $\kappa_+(T_1^{m-1} x_0) > |\kappa_-(T_1^m x_0)|$ while*

$$\lim_{\substack{m \to \infty \\ n \to \infty}} \frac{\kappa_+(T_1^{m-1} x_0)}{|\kappa_-(T_1^m x_0)|} = 1$$

where $x_0 = T^{\tau(x)+0} x$, $\tau(x)$ is the nearest positive moment of reflection from the boundary by the trajectory of the point x, $\kappa_-(T_1^m x_0)$ is the curvature of the image of γ in the point $T_1^m x_0$ before-and $\kappa_+(T_1^m x_0)$ after the mth reflection from the boundary of the given trajectory.

This theorem has the following meaning qualitatively. A bundle of parallel trajectories with a plane front (i.e. with zero curvature) after reflection from a focusing component $\Gamma \subset \partial Q$ becomes contracting (i.e. has negative curvature)

Fig. 12

but on its way between any two successive reflections from Γ it passes through a conjugate point and comes to the boundary before the following reflection as an expanding bundle. Moreover, the time during which the bundle under consideration has positive curvature (expands) is more than half of the whole time interval between any two successive reflections from Γ in the considered series (see Fig. 12). During this series of reflections a curvature of a bundle under consideration stays bounded from above in its absolute value. So, by successive reflections from any focusing component of a constant curvature, the length of a curve γ in the phase space increases linearly.

The main classes of ergodic billiards in domains with focusing components of the boundary were studied in [Bu1] and [Bu2].

In [Bu1] it was shown that a billiard in a domain whose boundary consists of focusing as well as dispersing components is a K-system if the following conditions hold:

1) the curvature of each focussing component $\Gamma \subset \partial Q$ is constant;

2) there is no pair of focusing components which are arcs of one and the same circle;

3) a complement of each focusing component up to the whole circle is contained entirely inside the domain Q.

We would like to mention that these conditions do not force the focusing part of the boundary ∂Q to be small but instead its length could be much greater than the length of the dispersing part of the boundary (Fig. 13).

In [Bu2] the K-property was proved for some classes of billiards in the case when the boundary ∂Q has no dispersing components. Besides, ∂Q has to have at least one focusing component and each focusing component has to have a constant curvature. The last condition can be essentially relaxed (cf [Bu6]). In particular, if for any point $x \in \pi^{-1}(\Gamma)$ such that $Tx \in \pi^{-1}(\Gamma)$, where Γ is a focusing component $4\tau(x) > (k(\pi(x)))^{-1} + (k(\pi(Tx)))^{-1}$ then such a component could be part of a boundary of a domain which generates a billiard possessing the K-property. Therefore K-property is preserved under small perturbations of such focusing components. More severe conditions on a curvature of focusing

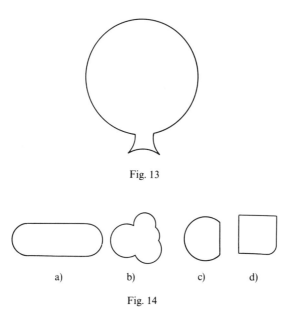

Fig. 13

Fig. 14

components was considered in [Wo] where it was proved by the other approach that some examples of billiards have nonvanishing Lyapunov exponents. The general approach which is based on the analysis of the continuous fraction (8.2) gives, for these billiards, K-property as well (cf [Bu2]).

Some examples of domains which generated billiards satisfying the K-property are shown in Fig. 14. The most popular example is a stadium (Fig. 14a), i.e. a convex domain whose boundary consists of two identical half-circles and two parallel segments. From the other side a billiard in a domain bounded by a sufficiently smooth convex curve is non-ergodic (see Sect. 1.2). The boundary of a stadium is of class C^1. Therefore one of the most interesting problems in this subject is to study how one could smooth out a boundary of a stadium in order to preserve the K-property. The hypothesis is that the smoothness of class C^2 is critical.

We should like to mention here the interesting work [BeS] where the computer simulation of the passage which arises under continuous deformation of a boundary was made from the billiard inside a circle to the billiard inside a stadium. In particular, it was shown in this work that an entropy increases during this passage but not monotonically, because at some moment several ergodic components arise in the dynamical system under consideration.

1.8. Hyperbolic Dynamical Systems with Singularities (a General Approach). The classes of billiards considered above form examples of hyperbolic dynamical systems, which act on a manifold with boundary and have singularities (for instance discontinuities) on some set of Lebesgue measure zero.

The proofs of the theorems obtained for the corresponding billiard systems are based on a detailed analysis of the structure of the set of singularity. It is natural to study what type of results could be obtained, having on the structure of this set the information of a general character only. This problem was studied by A. Katok and J.-M. Strelcyn [KS].

Let M be a smooth compact Riemannian manifold with boundary ∂M, ρ be the metric in M induced by the Riemannian metric, $N \subset M$ a union of a finite number of compact submanifolds, $f: M\setminus N \to f(M\setminus N)$ a diffeomorphism of class C^2, and μ an invariant Borel measure. The pair (f, N) is said to be a discontinuous dynamical system if the following conditions are satisfied:

1) $\int_M \ln^+ \|df_x\| \, d\mu < +\infty$, $\int_M \ln^+ \|df_x^{-1}\| \, d\mu < +\infty$ where df_x is the differential of f at a point x, $\|\cdot\|$ is a norm in the tangent space, $\ln^+ x = \max(\ln x, 0)$;

2) for every $\varepsilon > 0$ there exists a $C_1 > 0$ and a_1, $0 < a_1 \leqslant 1$, such that $\mu(U_\varepsilon(N)) \leqslant C_1 \varepsilon^{a_1}$, where $U_\varepsilon(N)$ denotes a ε-neighborhood of the set N;

3) there exist $C_2 > 0$, $C_3 > 0$, a_2, a_3, $0 < a_2$, $a_3 \leqslant 1$, such that for every $x \in M\setminus N$

$$\|df_x\| < C_2 \rho(x, N)^{-a_2},$$

$$\|d^2 f_x\| < C_3 \rho(x, N)^{-a_3}.$$

It follows from condition 1) and theorem 2.7 (see Chapter 1, Section 2) that μ-almost every point $x \in M\setminus N$ will be regular in the sense of Lyapunov (see Chap. 7, Sect. 2). From condition 2) it follows that $\mu(N) = 0$. The conditions 1)–3) enable one to construct LSM and LUM at every regular point $x \in M\setminus N$ at which the characteristic Lyapunov exponents are different from zero, to prove the property of absolute continuity and to obtain an exact upper estimation of the entropy of the dynamical system under consideration.

The proofs repeat in great detail the proofs of the corresponding statement considered in Chap. 7. We shall formulate the main results obtained in this way.

Theorem 1.8. *Let (f, N) be a discontinuous dynamical system, preserving the measure μ, equivalent to the Riemannian volume, the set $\Lambda \subset M\setminus N$ consists of points with nonzero characteristic Lyapunov exponents and $\mu(\Lambda) > 0$. Then*

1) the ergodic components of the automorphism $f|\Lambda$ have positive measure;

2) if the automorphism $f|\Lambda$ has continuous spectrum, then it is isomorphic to a Bernoulli shift;

3) the corresponding formula of Chap. 7 is valid for the entropy of the mapping f.

Apparently, the only known example of conservative hyperbolic dynamical systems with singularities, besides the billiard systems, is formed by toral linked twist mappings. Let $\mathbb{T}^2 = \mathbb{R}^2/\mathbb{Z}^2$ be the standard torus and let P, Q be closed annuli in $\mathbb{T}^2$ defined by

$$P = \{(x, y) \in \mathbb{T}^2 : y_0 \leqslant y \leqslant y_1, |y_1 - y_0| \leqslant 1\}$$

and

$$Q = \{(x, y) \in \mathbb{T}^2 : x_0 \leq x \leq x_1, |x_1 - x_0| \leq 1\}.$$

Let $f: [y_0, y_1] \to \mathbb{R}$, $g: [x_0, x_1] \to \mathbb{R}$ be C^2-functions such that $f(y_0) = g(x_0) = 0$, $f(y_1) = k$, $g(x_1) = l$ for some integers k and l.

Define mappings F_f and G_g of the set $P \cup Q$ onto itself by $F_f(x, y) = (x + f(y), y)$, $G_g(x, y) = (x, y + g(x))$ on P and Q respectively and $F_f|Q\setminus P = \text{Id}$, $G_g|P\setminus Q = \text{Id}$. We shall be interested in the composition of these mappings $H = H_{f,g} = G_g \circ F_f$. It is easy to see that H preserves the Lebesgue measure on $P \cup Q$ and has singularities on the set $\partial P \cup \partial Q$.

Suppose that $\dfrac{df}{dy} \neq 0$ and $\dfrac{dg}{dx} \neq 0$ for all $y \in [y_0, y_1]$, $x \in [x_0, x_1]$ and denote $\alpha = \inf\left\{\dfrac{\partial f}{\partial y} : y \in [y_0, y_1]\right\}$, $\beta = \inf\left\{\dfrac{dg}{dx} : x \in [x_0, x_1]\right\}$. We shall then call $F|P$ a (k, α)-twist, $G|Q$ a (l, β)-twist and H a toral linked twist mapping.

The results concerning the properties of this class of mappings belong to F. Przytycki [Pr1].

Theorem 1.9. *Let H be a toral linked twist mapping composed from (k, α)- and (l, β)-twists. Then*

If $\alpha\beta > 0$, then H is isomorphic to a Bernoulli shift

If $\alpha\beta < -4$, then H is almost hyperbolic, i.e. all Lyapunov exponents are almost everywhere (with respect to Lebesgue measure) different from zero and the set $P \cup Q$ decomposes into a countable family of positive measure invariant pairwise disjoint sets Λ_i such that for every i $(i = 1, 2, \ldots)$ $H|\Lambda_i$ is ergodic and $\Lambda_i = \bigcup_{j=1}^{j(i)} \Lambda_i^j$, where $\Lambda_i^{j'} \cap \Lambda_i^{j''} = \emptyset$ for $j' \neq j''$, $H|\Lambda_i$ permutes Λ_i^j and for each j a mapping $H^{j(i)}|\Lambda_i^j$ is isomorphic to a Bernoulli shift.

1.9. The Markov Partition and Symbolic Dynamics for Dispersing Billiards. For dispersing billiards a Markov partition can be constructed (see the definition in Chap. 7, Sect. 3). Nevertheless if for smooth hyperbolic systems (for instance A-systems) there exists a Markov partition with a finite number of elements, one could not expect to construct such a partition for billiards. In fact as the transformation T_1 is discontinuous, then regular components of global stable and unstable manifolds can be arbitrary small. Hence a Markov partition has to contain elements of arbitrary small sizes.

A Markov partition was constructed in [BS2] for two-dimensional dispersing billiards, such that transformations T_1 and T_1^{-1} have a finite number of curves of discontinuity and satisfy some additional technical conditions.

The existence of a Markov partition with a countable number of elements does not allow us to obtain immediately the same consequences as holds for a smooth hyperbolic dynamical system which has a finite Markov partition. In the case of an infinite partition, an induced symbolic dynamics has to satisfy some additional conditions, i.e. a union of its elements with small sizes must have a small measure and images of small elements have to transform under the action of $\{T_1^n\}$ mainly into elements with larger sizes.

Images of every LUM under the action of a dynamical system will have large (exponential in time) sizes and therefore, after a sufficiently large time, images of all LUMs would fill the whole phase space almost uniformly. Dispersing billiards are discontinuous systems. Hence, simultaneously with the expansion of LUM it is also decomposed when its images intersect the manifolds of discontinuity of the transformation T_1. Using some additional properties of the Markov partition constructed in [BS2] it was proved that in dispersing billiards, expansion is more effective than the decomposition due to discontinuities and that there exists a subset of a large measure (but not of full measure) as in the case of smooth hyperbolic systems, which consists of LUMs, such that its images under the action of large iterates of T_1 sufficiently dense and uniform fill the whole phase space of a billiard under consideration.

1.10. Statistical Properties of Dispersing Billiards and of the Lorentz Gas. The properties of symbolic dynamics stated in [BS2] allow us to prove for dispersing billiards in two-dimensional domains some assertions which are analogous to limit theorems of the probability theory.

Let $h(\omega)$ be a function defined on the space of sequences Ω such that $|h(\omega)| < C_1$, where C_1 is a constant and let there also exist a number λ_5, $0 < \lambda_5 < 1$, satisfying the property that for all n sufficiently large one can find functions $h_n(\omega) = h_n(\omega_{-n} \ldots \omega_n)$, $\int_\Omega h_n \, d\mu_0 = 0$, depending only on coordinates ω_i with $|i| \leq n$, such that $\sup_\omega |h(\omega) - h_n(\omega)| < \lambda_5^n$. We define by T_0 the shift in the space Ω which corresponds to the transformation T_1 by the symbolic representation $\varphi: M \to \Omega$ of the dynamical system under consideration. For those dispersing billiards (in particular, with a finite number of curves of discontinuity) for which a Markov partition was constructed in [BS2] the following theorem was proved in [BS4].

Theorem 1.10. *Let μ_0 be a measure which satisfies conditions 1–3, stated in [BS2] (see lemmas 6.2, 6.5, 6.6). If $\int_\Omega h \, d\mu_0 = 0$, then there exists a number α, $0 < \alpha < 1$, such that $|\int_\Omega h(T_0^n \omega) h(\omega) \, d\mu_0| < \exp(-n^\alpha)$ for all n sufficiently large.*

One can prove (cf [BS4]) that for functions satisfying conditions of Theorem 1.10 the central limit theorem of the probability theory holds.

The most important example of a dispersing billiard is the *Lorentz gas*. We shall give now the general definition of this dynamical system. Let an infinite number of balls (scatterers) be randomly distributed in the d-dimensional Euclidean space. In the complement of the set of scatterers an infinite number of point particles are distributed according to the Poisson law. Every particle moves with a constant velocity and reflects from scatterers according to the law "the angle of incidence equals the angle of reflection". The dynamical system which corresponds to the motion of this infinite ensemble of particles is called the Lorentz gas.

In view of the absence of interactions between moving particles we can consider a dynamical system generated by the motion of a single particle. The dynamical system generated by the motion of an infinite ensemble of particles will be considered in Chap. 10.

The Lorentz gas is one of the most popular models in the non-equilibrium statistical mechanics, for which, for instance, it is useful to discuss the problem of existence of transport coefficients and the related problem of slow decay of correlation functions. Among transport coefficients there are coefficients of diffusion, viscosity, heat conductivity, electroconductivity and so on. In view of the character of interactions in the Lorentz gas a momentum does not conserve, and hence the diffusion coefficient D is a single transport coefficient for this model. According to the Einstein formula

$$D = \frac{2}{d} \int_0^\infty \int_M (v(x(0)) \cdot v(x(t))) \, d\mu(x(0)) \, dt$$

where $v(x(0))$ is the velocity of the particle at the moment $t = 0$.

We shall consider a periodic configuration of scatterers. One can then consider the motion of the point particle in a torus $\mathbb{T}^d$ with Euclidean metric. We shall say that the Lorentz gas has a *finite horizon* if a free path length (i.e. a path between two successive reflections from scatterers) of the point particle is uniformly bounded from above by some constant. In that case, the number of curves of discontinuity of the transformation T_1 is finite.

It was proved in [BS4] that for the two-dimensional periodic Lorentz gas with a finite horizon, the diffusion coefficient exists and is positive.

Let a rectangle $\Pi = \{q = (q_1, q_2): 0 \leqslant q_1 < B_1, 0 \leqslant q_2 < B_2\}$ be a fundamental domain which corresponds to the configuration of scatterers in the plane. We shall consider the set $M \cap (\Pi \times S^1) = \hat{M}$ where M is the phase space of the Lorentz gas. Let μ be a probability measure concentrated on $\hat{M}$ which is absolutely continuous with respect to the Lebesgue measure on M and its density $f(x) \in C^1$. Then, a point $x \in M$ can be considered as a random variable, distributed according to the measure μ. If $T^t x = x(t) = (q(t), v(t))$, then $q(t), v(t)$ are also random variables. For every t we put $q_t(s) = \frac{1}{\sqrt{t}} q(st), 0 \leqslant s \leqslant 1$. The measure μ induces the probability distribution μ_t on the set of all possible trajectories $q_t(s), 0 \leqslant s \leqslant 1$, which are points of the space $C_{[0,1]}(\mathbb{R}^2)$ of continuous functions defined on the unit segment with values in $\mathbb{R}^2$.

The following statement which is an analog of Donsker's invariance principle in the theory of random processes was proved in [BS4] for the dynamical system under consideration.

Theorem 1.11. *The measures μ_t converge weakly to a Wiener measure.*

We want to stress that this is the first rigorous result on convergence of a purely deterministic dynamical system without any random mechanism to the random process of Brownian motion. Usually the motion of a heavy particle under collisions with a gas of light particles that do not interact between themselves is considered as a "physical" image of the random walk. Nevertheless the imagination of the random walk as of the motion of a light particle in the periodic field of heavy immobile scatterers with elastic reflections from their boundaries seems to be even more natural.

If the length of a free path of the particle is not bounded from above, then the rate of correlation decay in the case of continuous time is power, but in the case of discrete time it is of the same type as for bounded free path (see the Theorem 1.10) (cf [Bu5]). It is generated by the existence of singular periodic trajectories which are tangent to the boundary ∂Q at every point of reflection (Fig. 10).

We would like to mention in conclusion that the most interesting and important problem in this subject is the study of ergodic properties of the Lorentz gas with random configuration of scatterers. This problem is closely connected with the yet unsolved classical problem of the random walk in a random environment. A natural hypothesis is that in this system, the velocity autocorrelation function decays as $\text{const } t^{-(d/2+1)}$.

§2. Strange Attractors

2.1. Definition of a Strange Attractor. The notion of a "strange attractor" was introduced in the work by D. Ruelle and F. Takens [RT]. They proposed to use this term for invariant attracting sets of dynamical systems which are not manifolds and have in some cross-sections the structure of the perfect Cantor set. According to their general idea there are sets namely with such a structure which arise in the dynamical system which corresponds to the Navier-Stokes equations. Although this idea is not realized yet this term has become very popular, especially among physicists since such strange attractors had been found, usually on the basis of computer simulations, in various physical systems and one has begun to connect the existence of a strange attractor with the chaotic behavior of trajectories of a dynamical system. Besides, it is a general belief that trajectories on a strange attractor are (in some sense) unstable which is the cause of stochasticity of the corresponding dynamical system. Now there are only a few rigorous mathematical results which support this point of view. Nevertheless it became in some sense a fashion to find a strange attractor. The role of a mathematician is to formulate and to prove rigorous results on the behavior of the studied dynamical system on the basis of available results of computer simulations.

At present many works devoted to calculations of different dimension-like characteristics of invariant sets of dynamical systems have appeared. The corresponding basic idea is that among these sets only strange attractors can have a fractional dimension. The general theorems concerning this subject can be found in Chap. 7, Sect. 6.

In accordance with the general idea of the given volume, we shall consider only the rigorous mathematical results relating to ergodic theory of strange

Chapter 8. Dynamical Systems of Hyperbolic Type with Singularities

attractors. First of all we would like to change slightly the corresponding terminology. Let the flow $\{S^t\}$ be induced by a smooth vector field on a smooth compact manifold M.

Definition. An invariant closed set A is called an *attractor* if there exists a neighborhood $U_0 \supset A$ such that $U_t = S^t U_0 \subset U_0$ as $t > 0$ and $\bigcap_t U_t = A$.

The simplest examples of attractors are a stable stationary point and a stable periodic trajectory. In Chap. 7, Sect. 2, hyperbolic attractors of smooth dynamical systems and, in particular, the Smale-Williams attractor, which has a more complex structure, were considered. We shall introduce the notion of the stochastic attractor taking as a basis the existence of an invariant measure which is analogous to the u-Gibbs measure $\mu_{\varphi^{(u)}}$ for the smooth hyperbolic systems.

Definition. An attractor A will be called *stochastic* if the following conditions hold:
1) for any absolutely continuous measure μ_0 with the support in U_0, its shifts μ_t converge weakly to an invariant measure λ which does not depend on μ_0;
2) the dynamical system $(A, \lambda, \{S^t\})$ is mixing.

The choice of absolutely continuous measures as a class of initial measures μ_0 has, besides analytical advantages, the following ones: this class of measures is stable with respect to the passage to a discrete version of the dynamical system under consideration, to its numerical simulations on an electronic machine and to its small random perturbations.

The hyperbolic attractors are stochastic ones (see Chap. 7, Sect. 2.5). We shall now consider a structure and the properties of the stochastic attractor arising in the famous Lorenz model.

2.2. The Lorenz Attractor. Let us consider the following system of three ordinary differential equations

$$\begin{cases} \dfrac{dx}{dt} = -\sigma x + \sigma y \\ \dfrac{dy}{dt} = rx - y - xz \\ \dfrac{dz}{dt} = -bz + xy \end{cases} \quad (8.3)$$

where σ, b, r are some parameters with positive values. The system (8.3) has been obtained with the help of the Galerkin procedure in the Rayleigh-Benard problem on the convection in a layer of fluid of uniform depth between two parallel smooth planes, with a constant temperature difference between the upper and the lower surfaces (cf [Lor]). Quite the same system as (8.3) arises in the theory of lasers and in geophysics as the simplest model of a geomagnetic "dynamo".

It was E. Lorenz who first studied the system (8.3) with the help of computer simulation and discovered the rather irregular behavior of its trajectories when $r = 28, \sigma = 10$ and $b = 8/3$.

We shall now describe the simplest properties of the system (8.3):

i) The divergence of the right-hand side of (8.3) is equal to $-(\sigma + 1 + b) < 0$ and therefore any volume in the phase space of the Lorenz system shrinks exponentially with time.

ii) Infinity is an unstable point. Any trajectory of the system will reach a compact subset of the phase space and will remain there at all subsequent times.

iii) The system (8.3) is invariant with respect to the transformation $x \to -x$, $y \to -y, z \to z$.

iv) If $\sigma = 10, b = 8/3, r > \sigma(\sigma + b + 3)(\sigma - b - 1)^{-1} = 24,73 \ldots$, then there are only three stationary points in the phase space of the system $O = (0, 0, 0)$, $O_1 = (\sqrt{b(r-1)}, \sqrt{b(r-1)}, r-1)$ and $O_{-1} = (-\sqrt{b(r-1)}, -\sqrt{b(r-1)}, r-1)$ which are all hyperbolic; the stable separatrix $W^{(s)}$ of the point O is two-dimensional and the stable separatrices of the points O_1 and O_2 are the both one-dimensional. The branches of the unstable separatrix at point O will be denoted by Γ_1 and Γ_{-1} (see Fig. 15).

The sequence of bifurcations leading to the arising of the stochastic attractor in the Lorenz model when $r = 28, b = 8/3$ and σ varies between 0 and 10 was considered by V.S. Afraimovich, V.V. Bykov and L.P. Shilnikov (cf [ABS1]) (see also paper [ABS2] which contains the detailed proofs for a more general class of dynamical systems). The results of [ABS1] were obtained by the combination of the rigorous mathematical analysis of qualitative behavior of the system (8.3) with the quantitative estimations obtained by computer simulations.

It was shown in [ABS1] that Γ_1 and Γ_{-1} become doubly asymptotic when $r = 28, b = 8/3$ and $\sigma = \sigma_1 \approx 3.42$. This means that Γ_1 and Γ_{-1} belong locally to the stable separatrix of the point 0. From the results of [ABS1] it follows that two periodic hyperbolic trajectories (L_1 and L_{-1}) appear from the loops Γ_1 and Γ_{-1} when σ crosses the value σ_1. Moreover the branch $\Gamma_1(\Gamma_{-1})$ is attracted to the

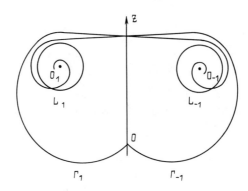

Fig. 15

point $O_{-1}(O_1)$ when $\sigma_1 < \sigma < 5.87$. When $\sigma = \sigma_2 \approx 5.87$ another bifurcation takes place. As a result the branch Γ_1 is attracted to a periodic motion L_{-1} and symmetrically Γ_{-1} to L_1 (Fig. 15). A limiting set which is called the Lorenz attractor, appears in the phase space of the system (8.3) when σ crosses the value σ_2.

Let us consider the Poincaré mapping of the plane $P = \{z = r - 1\}$ which contains the stable points O_1 and O_{-1}. We denote by z_i^*, $i = 1, -1$, the points of the first intersection of the branches Γ_i, $i = 1, -1$, with the plane P and it is assumed that Γ_i, $i = 1, -1$, intersects P at the point z_i^* from above to below; by z_i^0, $i = 1, -1$, the points of intersection of the hyperbolic cycles L_i, $i = 1, -1$, with P and by S the set $W^{(s)} \cap P$. (The set S is, generally speaking, not locally closed since the separatrix $W^{(s)}$ is strongly bending and does not divide the space locally. Thus the unstable branch $\Gamma_1(\Gamma_{-1})$ can come close to the point $O_{-1}(O_1)$ without intersection with $W^{(s)}$). The Poincaré mapping T associates to any point $x \in P$ such point $y = Tx \in P$ where the integral curve with origin at x at first intersects the plane P from above to below. It is obvious that T is discontinuous on the set S. Actually points which belong to P and lie on opposite sides of $W^{(s)}$ move along different branches (Γ_1 and Γ_{-1}) of the unstable separatrix. In [ABS1] the properties of the mapping T were formulated axiomatically on the basis of results of the computer simulations and of the qualitative analysis of the system (8.3) in the following manner.

Let $\Pi = \{|x| \leq 1, |y| \leq 1\}$ be the rectangle in the plane. We denote $\Pi_1 = \{|x| \leq 1, 0 < y \leq 1\}$, $\Pi_{-1} = \{|x| \leq 1, -1 \leq y < 0\}$. A mapping T is defined by the following conditions:

1) $T_i(x, y) = (f_i(x, y), g_i(x, y))$, $i = 1, -1$, where f_i, g_i are functions of class C^2.
2) A mapping T_i has a stationary point $z_i^0 \in K_i = \{|x| \leq 1, y = i\}$ of saddle type.
3) The segment K_i belongs to a stable separatrix of a point z_i^0, $i = 1, -1$.
4) The functions f_i and g_i can be continued to the segment S so that $\lim_{y \to 0} f_i(x, y) = x_i^*$, $\lim_{y \to 0} g_i(x, y) = y_i^*$ and $\operatorname{sgn} x_i^* = \operatorname{sgn} i$, $\operatorname{sgn} y_i^* = -\operatorname{sgn} i$, $i = \pm 1$. We denote $z_i^* = (x_i^*, y_i^*)$.
5) The mappings T_1 and T_{-1} are uniformly completely hyperbolic (see Chap. 7, Sect. 1) on the set $\Pi \setminus (\bigcup_{k=0}^{\infty} T_1^{-k} S \cup \bigcup_{k=0}^{\infty} T_{-1}^{-k} S)$, they are expanding in the vertical direction and contracting in the horizontal one.
6) $f_1(x, y) = -f_{-1}(-x, -y)$, $g_1(x, y) = -g_{-1}(-x, -y)^2$

Let us define the mapping T by letting $T = T_i | \Pi_i \cup S$, $i = 1, -1$, (see Fig. 16).

It is easy to see that the image $T\Pi$ consists of two curvilinear triangles (Fig. 16). In [SV] it was shown with the help of computer simulations that when $\sigma = 6$, $b = 8/3$, $r = 28$ there exist two narrow triangles with vertices z_1^* and z_{-1}^* situated on P in the opposite sides of S such that the Poincaré mapping of the plane P in the phase space of the system (8.3) satisfies on these triangles the property of

---

[2] Condition 6) is not necessary, in the sense that all results which will be formulated below for the mapping T also hold without it.

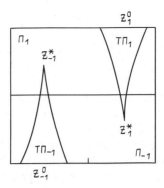

Fig. 16

hyperbolicity. We should like to note that P.V. Gachok had considered a more general class of systems which contains (8.3) and had shown that in this class there exists a strange attractor (cf [Gac]).

The following assertion was proved in [ABS1][3]

Theorem 2.1. *If the mapping T is topologically transitive (see Chap. 7, Sect. 2.2), then there is a single stable limiting set of T in Π which has the form $\Lambda = \bigcap_{n \geq 0} T^n \Pi$ and is a one-dimensional hyperbolic set with the following properties*
  i) *Λ consists of two connection components,*
  ii) *Λ is the closure of the set of periodic points.*

The attractor Λ contains continuum of smooth curves which can be projected uniquely on the y-axis, and any such curve is a regular segment of an expanding fibre. We shall denote by ξ the partition of Λ into such maximal expanding curves and by $C_\xi(x)$ the element of ξ which contains the point x.

We therefore have the discontinuous hyperbolic mapping of the square Π into itself. Thus the situation is in some respects similar to the case of Sinai billiards considered in the preceding paragraph. However the Lorenz system has some essential differences with respect to Sinai billiards. First of all, it has no natural invariant measure.

Theorem 2.2. *Let v be a measure on Π which is absolutely continuous with respect to the Lebesgue measure and the corresponding density is continuously differentiable. Then the sequence of measures $T_*^n v$ converges weakly to an invariant with respect to T measure μ. Moreover the conditional measure $\mu(\cdot | C_\xi)$ induced by μ on regular expanding fibres is absolutely continuous with respect to the Lebesgue measure on C_ξ.*

---

[3] In [ABS1] some other results on the properties of the mapping T were obtained too, but we shall not use them in what follows.

One could prove a stronger assertion which states, analogously to the corresponding one in the Chap. 7, Sect. 3, the strong convergence almost everywhere.

Nevertheless it occurs that under the above formulated conditions on the mapping T, the attractor Λ does not have to be stochastic, i.e. the restriction of T onto Λ does not have to be mixing. In this case any element of ξ contains lacunae[4]. Therefore there is a cyclic component in a spectrum of the dynamical system under consideration. In order for T to be mixing it is necessary that in condition 5) a stronger expansion holds than in the usual condition of hyperbolicity. Under this condition the following theorem is true (cf ([BS3]).

Theorem 2.3. *Let $f(x)$ be a continuous function defined in some neighborhood U, $\Lambda \subset U \subset \Pi$, and let v be an absolutely continuous measure on Π whose support is contained in U. Then, for almost every point $x \in \Pi$ with respect to v*

$$\lim_{n \to \infty} \int_\Pi f(T^k x) \, dv = \int_\Lambda f(x) \, d\mu.$$

In [Bu4] the central limit theorem (see Chap. 6, Sect. 1) was proved for a wide class of functions on Π, and for sufficiently large n and for some positive number $\gamma < 1$ the following estimation of the rate of mixing was obtained

$$\left| \int_\Lambda f(x) g(T^n x) \, d\mu - \int_\Lambda f(x) \, d\mu \int_\Lambda g(x) \, d\mu \right| < \text{const} \exp(-n^\gamma) \qquad (8.4)$$

We should like to mention that the mapping T has a global stable foliation which is constructed analogously to the partition of the square Π into segments $y = \text{const}$ (cf [ABS1]). It allows to represent T as a skew product (see Chap. 1, Sect. 4) over a monotonic transformation of the unit segment which has a single point of discontinuity, and to use the theory of one-dimensional mappings (see Chap. 9, Sect. 1). However, in order to prove mixing, one has to use a smoothness of the stable foliation which apparently does not hold for the dynamical system under consideration.

Let us discuss some available generalizations of the considered example. First of all it is clear that all results hold in case when the mapping T has discontinuities of the first class not on one but on a finite number of curves (compare with dispersing billiards). The phase space may not be two-dimensional but can have any finite dimension. In the latter case, manifolds of discontinuity have to have codimension one. In the phase space of the corresponding dynamical system on the n-dimensional cube, a one-dimensional stochastic attractor exists as well. Hence one could use a factorization along contracting fibres in order to pass to a one-dimensional mapping.

---

[4] This phenomenon is analogous to the arising of lacunae in piecewise monotonic one-dimensional mappings (see Chap. 9).

2.3. Some Other Examples of Hyperbolic Strange Attractors.

V.N. Belykh considered the two-dimensional mapping which has a hyperbolic attractor and cannot be reduced to a one-dimensional one (cf [Be]). This example also arose in the study of concrete dynamical systems in physics, which are called the discrete systems of the phase synchronization. We shall consider its simplest version when the corresponding mapping is piecewise-linear and property 6) holds.

Let T be the following mapping of the square Π into itself

$$\begin{cases} x_{n+1} \mp 1 = \lambda_1(x_n \mp 1) \\ y_{n+1} \mp 1 = \lambda_2(y_n \mp 1) \end{cases} \text{ if } y \gtrless kx$$

where $0 < \lambda_1 < 1/2$, $1 < \lambda_2 < 2$, $k \neq 0$, $|k| < 1$. The main difference between this mapping and the mapping which arises in the Lorenz system is that there is no global contracting foliation now, since the curve of discontinuity is not parallel to contracting fibers and one cannot perform the factorization.

R. Lozi considered the following class of linear mappings of the plane into itself $(x, y) \to (1 + y - a|x|, bx)$ and by computer simulations found that there is a strange attractor in the system when $a = 1.7$ and $b = 0.58$ (cf [Loz] [69]). The introduction of this class of mappings was stimulated by numerous works devoted to the study of Henon's attractor for the following class of mappings of the plane into itself $(x, y) \to (1 + y - ax^2, bx)$ (cf [H]). However in these works there is no proof of existence of such an attractor. The experience accumulated by studying ergodic and topological properties of one-dimensional mappings (see Chap. 9, Sect. 2) shows that it is easier to investigate a Lozi mapping than a Henon one. In fact the existence and hyperbolicity of a Lozi attractor for some region in the space of parameters a and b was soon proved (cf [Mi1]). The metric properties of a Lozi attractor were investigated in [Le], where an invariant measure was constructed and its uniqueness and stochasticity were proved.

Ya.B. Pesin recently introduced a class of generalized hyperbolic attractors, which includes Lorenz, Belykh and Lozi attractors, and studied some of their topological and ergodic properties (cf [Pe6]). In [Pe6] u-Gibbsian measures were constructed as well for these attractors.

The appearance of hyperbolic strange attractors in model dynamical systems is not a rule but rather an exception. Numerous computer simulations and rigorous mathematical investigations have shown that, in a generic dissipative dynamical system with a chaotic behavior, there are usually stable periodic trajectories besides nontrivial hyperbolic subsets. Thus a generic situation in dissipative systems is apparently the same as in Hamiltonian systems, and the dynamics on a generic attractor is similar to the dynamics of generic conservative systems where there are stochastic layers and invariant KAM-tori as well (see Chap. 6, Sect. 2).

Chapter 9
Ergodic Theory of One-Dimensional Mappings

M.V. Jakobson

In this chapter we shall consider one-dimensional mappings which have been intensely studied during the last few years, from the point of view of both dynamical systems and ergodic theory. Phase spaces of these systems are intervals $I \subset \mathbb{R}^1$ and transformations are real-valued functions determined on I and taking their values in I. We shall study invariant measures of one-dimensional maps, especially absolutely continuous invariant measures. The topological aspect of the problem will be discussed in one of the further volumes.

§1. Expanding Maps

1.1. Definitions, Examples, the Entropy Formula. We start the study of one-dimensional dynamics by considering the map $T_2: [0,1] \to [0,1]$ given by

$$T_2(x) = \begin{cases} 2x, & x \in [0, \frac{1}{2}] \\ 2x - 1, & x \in (\frac{1}{2}, 1]. \end{cases}$$

If we denote by $T_2^{-1}(x)$ the preimage of x, then for any measurable set C we have $l(T_2^{-1}C) = l(C)$ which means that the Lebesgue measure $l(dx) = dx$ is invariant under T_2.

Let us denote by $(\Omega_2, S, \mu(\frac{1}{2}, \frac{1}{2}))$ the *Bernoulli shift* $S: (\omega_0, \omega_1, \omega_2, \ldots) \mapsto (\omega_1, \omega_2, \ldots)$ on the set Ω_2 of one-sided sequences with the measure μ on Ω_2 given by $\mu(K_n) = \frac{1}{2^n}$ for any cylindric set $K_n = \{\omega: \omega_{i_1} = \omega_{i_1}^0, \ldots, \omega_{i_n} = \omega_{i_n}^0\}$. Consider the measurable partition $\xi = \{A_0 = [0, \frac{1}{2}], A_1 = (\frac{1}{2}, 1]\}$ and let the point $\pi(\omega) = \bigcap_{k=0}^{\infty} T_2^{-k} A_{\omega_k}$ correspond to a sequence $\omega = (\omega_0, \omega_1, \ldots, \omega_n, \ldots) \in \Omega_2$. Then the equality $l(\bigcap_{k=0}^{n-1} T_2^{-k} A_{\omega_k}) = \frac{1}{2^n}$ shows that π is an isomorphism (mod 0) between noninvertible transformations of measure spaces (endomorphisms) $([0,1], T_2, dl) \stackrel{\pi}{\approx} (\Omega_2, S, \mu(\frac{1}{2}, \frac{1}{2}))$. Similarly the map $T_n: x \mapsto nx \pmod 1$, $2 \leq n \in \mathbb{N}$ is isomorphic (mod 0) to the one-sided shift on the alphabet of n symbols.

Now let $T: [0,1] \to [0,1]$ be a map, such that the derivative $|dT/dx|$ is not constant (Fig. 17) but satisfies the *expanding property*

$$|dT/dx| \geq C_1 > 1.$$

For any expanding map the distance between the trajectories of near-by initial points grows exponentially. This provides the analogy between such maps and

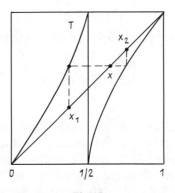

Fig. 17

hyperbolic systems considered in Chap. 7, 8. In particular expanding maps generally admit absolutely continuous invariant measures.

In this section we shall consider piecewise monotone transformations T with a finite or countable number of maximal intervals (laps) on which T is monotone. The following argument due to Ya.G. Sinai [Si5] shows that an expanding map $T: [0, 1] \to [0, 1]$ with a finite number of laps which is C^2 on each lap and maps it onto $[0, 1]$ admits an invariant measure $\mu(dx) = h(x)\, dx$ where $h(x)$ is continuous and strictly positive.

The map T induces in L^1_{dl} the so-called *Perron-Frobenius operator*

$$\mathscr{L}\varphi(x) = \sum_{y \in T^{-1}(x)} \varphi(y)/|T'(y)|.$$

Suppose for definiteness that T is monotone increasing from 0 to 1 on the intervals $[0, \tfrac{1}{2}]$ and $[\tfrac{1}{2}, 1]$. Then

$$\mathscr{L}\varphi(x) = \frac{\varphi(x_1)}{T'(x_1)} + \frac{\varphi(x_2)}{T'(x_2)}$$

where $\{x_1 \in [0, \tfrac{1}{2}], x_2 \in [\tfrac{1}{2}, 1]\} = T^{-1}(x)$ (see Fig. 17). For the density of an invariant measure we have $\mathscr{L}\varphi = \varphi$. We shall find some φ satisfying this equation in the set $\mathbb{A}_c = \{\varphi(x)\}$ such that, for any $x, y \in [0, 1]$

$$\exp(-cd(x, y)) \leq \varphi(x)/\varphi(y) \leq \exp(cd(x, y))$$

where $c > 0$ is some constant. We have

$$\frac{\mathscr{L}\varphi(x)}{\mathscr{L}\varphi(y)} \leq \max\left\{\frac{\varphi(x_1)}{\varphi(y_1)} \cdot \frac{T'(y_1)}{T'(x_1)}; \frac{\varphi(x_2)}{\varphi(y_2)} \cdot \frac{T'(y_2)}{T'(x_2)}\right\}$$

Since $\varphi \in \mathbb{A}_c$, we obtain

$$\frac{\varphi(x_i)}{\varphi(y_i)} \cdot \frac{T'(y_i)}{T'(x_i)} \leq \exp[c \cdot d(x_i, y_i) + \ln T'(y_i) - \ln T'(x_i)]$$

Now we use $|\ln T'(y_i) - \ln T'(x_i)| = \left|\dfrac{T''(\theta)}{T'(\theta)}\right| \cdot d(x_i, y_i)$ where $\theta \in [x_i, y_i]$ and the expanding property $d(x_i, y_i) \leq c_1^{-1} d(x, y)$. If we denote by c_2 the quantity $\max_{z \in [0,1]} |T''(z)/T'(z)|$, then we have

$$\frac{\mathscr{L}\varphi(x)}{\mathscr{L}\varphi(y)} < \exp[d(x,y)(c_1^{-1} \cdot c + c_1^{-1} \cdot c_2)].$$

If c is sufficiently large for $c_1^{-1}(c + c_2) < c$, then

$$\mathscr{L}\mathbb{A}_c \subset \mathbb{A}_c$$

and the Shauder-Tychonoff theorem implies the existence of a fixed point $h = \mathscr{L}h \in \mathbb{A}_c$ which is the density of an absolutely continuous invariant measure. $\square$

A traditional object of ergodic theory is the distribution of fractional parts for various functions. Namely, for a monotone function $g(x)$ defined on $(0,1)$, we consider the transformation

$$Tx = \{g(x)\}.$$

The question is whether or not T admits an invariant measure μ equivalent to the Lebesgue measure and what are the ergodic properties of μ which define ergodic properties of the corresponding fractional parts.

A classical example due to Gauss is $g(x) = 1/x$. In this case the invariant measure is given by the explicit formula

$$\mu(X) = \frac{1}{\ln 2} \int_X \frac{dx}{1+x}.$$

A. Rényi and V.A. Rokhlin considered a wide class of examples (cf [RO]). Their results concern the mappings T satisfying the following conditions. Suppose that T has a finite or a countable set of points of discontinuity $\{a_i\}$ (these are points where $g(x)$ takes integer values). The points a_i partition I into intervals $\Delta_i = (a_{i-1}, a_i)$ which are referred to as intervals of the first rank. Let this partition be denoted by $\xi^{(1)}$ and let $\xi^{(n)} = \bigvee_{i=0}^{n-1} T^{-i} \xi^{(1)}$ be the partition of $[0,1]$ into the intervals $\Delta_{i_1 i_2 \ldots i_n} = \Delta_{i_1} \cap T^{-1}\Delta_{i_2} \cap \cdots \cap T^{-(n-1)}\Delta_{i_n}$ of rank n. Then T^n is uniquely defined on $\Delta_{i_1 i_2 \ldots i_n}$. We assume that:

(i) There exists $k_0 \in \mathbb{N}$ such that T^{k_0} is an expanding map.

(ii) There exists $c_1 > 0$ such that for any n and for any y, z belonging to $\Delta_{i_1 i_2 \ldots i_n}$, $c_1^{-1} \leq \dfrac{dT^n}{dx}(y) \Big/ \dfrac{dT^n}{dx}(z) \leq c_1$ holds.

(iii) $T\Delta_i \supset (0,1)$ for any interval of rank 1.

Notice that (iii) is a Markov condition of special type. If (iii) holds, then we have $T^n \Delta_{i_1 i_2 \ldots i_n} \supset (0,1)$ for intervals of any rank. Condition (ii) which concerns intervals of any rank is much more cumbersome.

If for any i $T|\Delta_i$ is a C^2-map, then we may substitute (ii) by another condition which concerns only the intervals of the first rank

(ii)'
$$\sup_i \sup_{z \in \Delta_i} \left|\frac{d^2 T(z)}{dT(z)}\right| |\Delta_i| \leq c_2.$$

If T satisfies (i), (ii)' and (iii) then (ii)' holds for $T^n|\Delta_{i_1 i_2 \ldots i_n}$ on any interval of any rank (cf [J1]). Thus (i), (ii)' and (iii) imply (ii). For T satisfying these conditions, the above arguments prove the existence of an absolutely continuous measure μ invariant under T^{k_0}. Then $\mu' = \sum_{i=0}^{k_0-1} T^{i*}\mu = \rho(x)\,dx$ is T-invariant and the density $\rho(x)$ is continuous and strictly positive.

V.A. Rokhlin proved in [Ro] that (T, μ') is an exact endomorphism (see Chapter 1, Section 4 for definition) and the following *entropy formula* holds:

$$h_{\mu'}(T) = \int_0^1 \log|dT/dx|\mu'(dx).$$

We shall now show that the entropy formula follows from (i)–(iii). Condition (i) implies that $\xi^{(1)}$ is a one-sided generator. Therefore using Shannon-McMillan-Breiman theorem we have $-1/n \log \mu'(\Delta_{i_1 \ldots i_n}(x)) \to h_{\mu'}(T)$ for μ' almost all x. By the intermediate value theorem we have

$$l(\Delta_{i_1 \ldots i_n}(x)) = 1 \bigg/ \left|\frac{dT^n(\bar{x})}{dx}\right|, \quad \bar{x} \in \Delta_{i_1 \ldots i_n}(x).$$

Then (ii) and the properties of μ' give for some $c_2, c_3 > 0$

$$c_3 \leq \mu'(\Delta_{i_1 \ldots i_n}(x)) \bigg/ \left|\frac{dT^n(x)}{dx}\right|^{-1} \leq c_2.$$

Since by the Birkhoff theorem $\dfrac{1}{n}\log\left|\dfrac{dT^n}{dx}\right| = \dfrac{1}{n}\sum_{i=0}^{n-1}\log\left|\dfrac{dT}{dx}(T^i(x))\right| \to \int_0^1 \log\left|\dfrac{dT}{dx}\right| \cdot \mu'(dx)$ for μ' almost all x, the entropy formula is proved. □

In particular for $Tx = \{1/x\}$ we obtain

$$h_\mu(T) = \frac{\pi^2}{6} \frac{1}{(\ln 2)^2}.$$

1.2. Walters Theorem. In [W2], Walters studied ergodic properties of mappings which expand distances. The examples considered above satisfy the conditions of the Walters theorem.

Let $\bar{X}$ be a compact metric space with the metric denoted by d, $X \subset \bar{X}$ an open dense subset of X, X_0—an open dense subset of X and let $T: X_0 \to X$ be a continuous surjective map with the following properties.

I. There exists an $\varepsilon_0 > 0$ such that $\forall x \in X$ the set $T^{-1}(B_{\varepsilon_0}(x) \cap X)$ is a disjoint union of open subsets (components) $A_i(x) \subset X_0$, such that $T: A_i(x) \to B_{\varepsilon_0}(x) \cap X$ is a homeomorphism satisfying $d(Ty, T'y) \geq d(y, y')$ for each i.

II. For any $\varepsilon > 0$ there exists $m \in \mathbb{N}$ such that for any $x \in X$ $\{T^{-m}x\}$ is ε-dense in X.

Let v be a probability Borel measure on $\bar{X}$ non-singular with respect to T (i.e. $v(E) = 0$ implies $v(TE) = 0$, $v(T^{-1}E) = 0$). Suppose that Radon-Nikodym derivative dvT/dv is continuous on X_0 and the following conditions hold.

(a) $$\frac{dvT^{-1}}{dv}(x) = \sum_{y \in T_x^{-1}} 1 \bigg/ \frac{dvT}{dv}(y) \leq K_1 \quad \forall x \in X$$

(b) $$\sup_{n \geq 1} \sup_{y \in T_x^{-1}} \frac{dvT^n}{dv}(y') \bigg/ \frac{dvT^n}{dv}(y) \leq R_2$$

when $d(x, x') < \varepsilon_0$ and this expression converges to 1, as $d(x, x') \to 0$.

Here we use the fact that for x, $x' \in X$, $d(x, x') < \varepsilon_0$ the property I gives rise to a natural bijection: $y \in T^{-n}x \leftrightarrow y' \in T^{-n}x'$ for y, y' lying in the same component.

The Perron-Frobenius operator is defined for $f \in C(X)$ by

$$\mathcal{L}f(x) = \sum_{y \in T_x^{-1}} \left(1 \bigg/ \frac{dvT}{dv}(y)\right) f(y).$$

It follows from a) and b) that $\mathcal{L}$ extends to a continuous linear operator $z: C(\bar{X}) \to C(\bar{X})$.

Theorem 1.1 (cf [W2]).
(1) *There exists $h(x) \in C(\bar{X})$, $h(x) > 0$ such that for any $f \in C(\bar{X})$*

$$\mathcal{L}^n f \rightrightarrows h(x) \cdot v(f);$$

(2) *The measure $\mu(dx) = h(x) \cdot v(dx)$ is T-invariant and (T, μ) is an exact endomorphism;*
(3) $v \circ T^{-n} \to \mu$ *in weak topology;*
(4) *The Variational principle. For any T-invariant Borel probability measure m*

$$0 = H_\mu(\mathcal{B}/T^{-1}\mathcal{B}) - \mu\left(\log\left(\frac{dvT}{dv}\right)\right) \geq H_m(\mathcal{B}/T^{-1}\mathcal{B}) - m\left(\log\frac{dvT}{dv}\right)$$

and μ is the unique measure with this property.

If there exists a finite or a countable partition ξ which is a one-side generator, then $H_\mu(\mathcal{B}/T^{-1}\mathcal{B}) = h_\mu(T)$ and the left part of (4) gives the entropy formula. Suppose further that any element of ξ is contained in some set $T^{-1}B_{\varepsilon_0}(x)$ for some x and that ξ is a Markov partition satisfying $\mu(\partial \xi) = 0$. Then we have
(5) *The natural extension of (T, μ) is isomorphic to a Bernoulli shift.*

Apart from one-dimensional examples considered above, Theorem 1 gives the existence and describes the ergodic properties of absolutely continuous invariant measures for expanding maps of compact manifolds (cf [W2]).

According to the property 1 of the map T we have for any $x \in X$ $T^{-1}B_{\varepsilon_0}(x) = \bigcup_i A_i(x)$ where $T_i A_i(x) \to B_{\varepsilon_0}(x)$ are homeomorphisms for all i. This is a restrictive condition: for example β-transformations $x \mapsto \beta x$ (mod 1) with irrational β do not satisfy it.

Hofbauer and Keller obtained a result similar to Theorem 1 for a wide class of piecewise-monotone transformations of the interval including β-transformations and studied ergodic properties of the corresponding invariant measures (cf [HK]).

§2. Absolutely Continuous Invariant Measures for Nonexpanding Maps

2.1. Some Examples. A classical example of a one-dimensional nonexpanding map admitting absolutely continuous invariant measure is

$$F: x \mapsto 4x(1-x).$$

It was studied by Ulam and von Neiman (cf [CE]). It is interesting to point out that Fatou is his famous work (1920) on the iteration of rational maps of the Riemann sphere considered the maps of that type and studied their relation with expanding maps (cf [MO]). For $F: x \to 4x(1-x)$ there is a change of variable $y = \varphi(x) = \frac{2}{\pi} \arcsin \sqrt{x}$ transforming F into the piecewise linear map

$$T(x) = \begin{cases} 2x, & x \in [0, \tfrac{1}{2}] \\ 2x - 1, & x \in (\tfrac{1}{2}, 1]. \end{cases}$$

Since the Lebesgue measure dl is T-invariant, its image $\varphi^{-1*} dl = \mu(dx) = \dfrac{dx}{\pi\sqrt{x(1-x)}}$ is F-invariant. Let $\xi = \{A_0 = [0, \tfrac{1}{2}], A_1 = (\tfrac{1}{2}, 1]\}$ be the partition of $[0,1]$ into intervals of monotonicity. The correspondence $\omega = (\omega_0 \omega_1 \ldots) \leftrightarrow \pi(\omega) = \bigcap_{k=0}^\infty T^{-k} A_{\omega_k}$ generates an isomorphism between the system $([0,1], T, dl)$ and the one-sided Bernoulli shift. Thus the system $\left([0,1], F, \dfrac{dx}{\pi\sqrt{x(1-x)}}\right)$ is also Bernoullian.

Generally the invariant measure is not given by an explicit formula, and it is a more difficult task to study ergodic properties of invariant measures for nonexpanding maps than for expanding ones. The first results in this direction were obtained for unimodal maps, satisfying the following property: some iteration of the critical point coincides with an unstable periodic point (see references in [CE], [J2]). For such mappings it is possible to overcome the difficulties related to the critical point by considering the induced map (see Chapter 1, Section 4).

Let $F \in C^2([0,1], [0,1])$ be a unimodal map with a nondegenerate critical point c satisfying $F(0) = F(1) = 0$, $F'(0) = A > 1$, $F(c) = 1$. Let $t \in (c, 1)$ be the

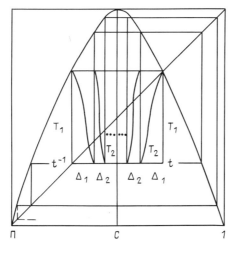

Fig. 18

fixed point of F, $t^{-1} \in (0, c)$, the preimage of t. We consider the induced map T on the interval $[t^{-1}, t] = I$ (Fig. 18).

This map has a countable number of laps Δ_i with $T|\Delta_i = T_i = F^{i+1}$, $i = 1, 2, 3, \ldots$, $T_i: \Delta_i \to I$. Using the nondegeneracy of c it is straightforward to check that for some $c_1, c_2, c_3 > 0$ independent of i, the following inequalities hold:

$$c_1 A^{i/2} < \left|\frac{dT_i(x)}{dx}\right| < c_2 A^{i/2} \left|\frac{d^2 T_i(x)}{dx^2}\right| < c_3 A^i.$$

If i is sufficiently large then we obtain

$$\left|\frac{dT_i}{dx}\right| > c > 1.$$

Besides for any i and for any $x, y, z \in \Delta_i$, we have

$$\left|\frac{d^2 T_i(x)}{dx^2}\right| \bigg/ \left|\frac{dT_i}{dx}(y)\right| \cdot \left|\frac{dT_i}{dx}(z)\right| < c'.$$

If some iteration of T is expanding, then T satisfies the conditions of Theorem 1.1 and we conclude that an absolutely continuous T-invariant measure ν exists and obtain the ergodic properties of ν. The measure ν uniquely defines an F-invariant ergodic measure μ absolutely continuous with respect to the Lebesgue measure on $[0, 1]$. The density of μ is separated from zero and is continuous everywhere except the orbit of the critical point (which in our case consists of two points 1 and 0), where it has singularities of the type $1/\sqrt{x}$. The above conditions are satisfied for unimodal *maps with negative Schwarzian derivative*

$$SF = \frac{F'''}{F'} - \frac{3}{2}\left(\frac{F''}{F'}\right)^2 < 0.$$

The topological structure of these maps is well known (cf [CE]). The induced map method alllows us to prove the existence of an absolutely continuous invariant measures for different kinds of maps with negative Schwarzian derivatives. Namely this is the case when some iterate of the critical point falls into an unstable orbit or when it falls into an invariant unstable Cantor set (cf [J1], [Mi2]). K. Ziemian generalized the results of [HK] to this case (cf [Z]).

2.2. Intermittency of Stochastic and Stable Systems. When considering several one-parameter families of one-dimensional maps $x \to f_\lambda(x)$ we see different types of behavior. For the maps of the first type the trajectories of almost every point (with respect to dx) converge to a stable periodic orbit, and the nonwandering set of f_λ consists of a finite number of such orbits and of an invariant unstable Cantor set. These maps are structurally stable.

The maps of the second type display stochastic behavior on a set of positive Lebesgue measure. As pointed out above this happens for a unimodal map with negative Schwarzian derivative when some iterate of the critical point falls into a periodic unstable orbit or into an invariant unstable Cantor set. In such cases the critical point is not recurrent: $c \notin \omega(c)$. These maps do not present all the possibilities for stochastic dynamics, it seems quite probable that the corresponding parameter values form a set of measure zero, although the whole set of parameter values corresponding to the stochastic dynamics has positive measure (cf [J2]).

Let $F: [0, 1] \to [0, 1]$, $F(0) = F(1) = 0$, $F'(0) \neq 0$ be a unimodal C^2-map with a nondegenerate critical point. Consider a one parameter family of piecewise smooth mappings

$$T_\lambda: x \to \lambda \cdot F(x) \pmod{1}, \quad \lambda > 1.$$

T_λ has a finite number of monotone branches depending on λ and a middle parabolic branch (Fig. 19). If $\lambda \cdot F(c) \in \mathbb{N}$, then the middle branch bifurcates into

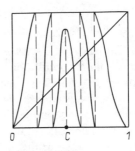

Fig. 19

two monotone branches and a new middle branch is born. We shall use λ_n to denote the corresponding parameter values and we shall use mes to denote the Lebesgue measure on the axis of parameter.

Theorem 2.1 (cf [J2]). *For any $\varepsilon > 0$ there exists $\Lambda(\varepsilon)$ such that for $\lambda_n > \Lambda(\varepsilon)$ there is a set of positive measure $M = \{\lambda \in [\lambda_n, \lambda_{n+1}]: T_\lambda$ admits an ergodic absolutely continuous invariant measure with positive entropy$\}$ and mes $M > (\lambda_{n+1} - \lambda_n)(1 - \varepsilon)$.*

We construct the set M as a complement to an open set $V = \bigcup_{n=0}^\infty V_n$, where V_n consists of λ such that $T_\lambda^{k_n}(c)$ falls into an open interval containing c. The presence of a large parameter implies that the lengths of intervals removed at the n-th step decrease very quickly and mes $V \to 0$, as $\lambda \to \infty$.

Theorem 2.1 may be generalized to the mappings with several extrema. For a smooth one-parameter family of maps such as $x \mapsto ax(1 - x)$, $a \in [0, 4]$ there is no large parameter. In order to reduce the situation to the preceding one we construct the induced map as described above. Then for a close to 4 we obtain a piecewise smooth map T_a similar to the one depicted in Fig. 19. The limit map T (Fig. 18) has a countable number of laps and is expanding. This is essential in proving the following results (cf [J2]).

Theorem 2.2. *Let F be a map of $[0, 1]$, $F(0) = F(1) = 0$, C^2-close to $x \to x(1 - x)$. Define λ_0 by $\lambda_0 \cdot F(c) = 1$. Then the Lebesgue measure of the set $M = \{\lambda \in (0, \lambda_0] | F_\lambda: x \to \lambda \cdot F(x)$ has an ergodic absolutely continuous invariant measure of positive entropy$\}$ is positive, and λ_0 is a density point of M.*

A similar theorem holds for the family of smooth unimodal maps $x \to a \cdot F(x)$, where $F(x)$ is a map with negative Schwarzian derivative.

Several results of that kind were obtained for different families of unimodal maps by J. Guckenheimer [Gu], M. Benedicks and L. Carleson [BC], P. Collet and J.P. Eckmann (cf [CE]), M. Rychlic ([Ry]). For generalizations see [J3].

The following questions concern the problem of intermittency between structurally stable periodic systems and stochastic systems.

1. Is the set of structurally stable systems dense in C^r-topology for $r \geq 2$? (The last results in that direction are due to R. Mañé [M]).

2. Is it true that any unimodal map with negative Schwarzian derivative which is topologically conjugate to a map with a constant slope admits an absolutely continuous invariant measure?

3. Does the union of structurally stable and stochastic systems form a set of full measure in the parameter space?

2.3. Ergodic Properties of Absolutely Continuous Invariant Measures. If μ is an absolutely continuous f-invariant measure of positive entropy, then f exhibits strong stochastic properties with respect to μ. The most general results in this direction, due to F. Ledrappier, are summarized in Theorems 2.3–2.6 below.

We consider a map $f: [0,1] \to [0,1]$ with a finite number of points of discontinuity $0 = b_0 < b_1 < \cdots < b_m < b_{m+1} = 1$ and we suggest that for $j \in [0, m]$ the restriction $f|[b_j, b_{j+1}]$ satisfies the following conditions.

C_1. f is a $C^{1+\varepsilon}$ map, i.e. f' satisfies the Hölder condition of order $\varepsilon > 0$.
C_2. f has a finite number of critical points.
C_3. There are positive numbers k_i^-, k_i^+, such that

$$\log \frac{|f'(x)|}{|x - a_i|^{k_i-(+)}}$$

is bounded in a left (right) neighborhood of a_i.

Theorem 2.3 (cf [L1]). *Let f satisfy $C_1 - C_3$ and let μ be an ergodic absolutely continuous f-invariant measure of positive entropy. If μ is ergodic for f^k for all $k > 0$, then the natural extension of f is isomorphic to a Bernoulli shift. In any case there exists $k_0 \geq 1$ such that the natural extension of f^{k_0} is Bernoulli on every ergodic component. The Rokhlin formula holds*

$$h_\mu(f) = \int \log |f'| \, d\mu.$$

In order to prove Theorem 2.3 the local unstable manifolds are constructed, and the absolute continuity of the unstable foliation is proved similarly to the case of hyperbolic attractors considered in Chaps. 7, 8.

According to Chap. 7, we classify f as a nonuniformly completely hyperbolic map. The size of a local unstable manifold at some point of the natural extension may become small: it is related to the iterates of critical points and of points of discontinuity (similar to the systems considered in Chap. 8).

The similarity becomes quite clear if we establish an isomorphism between the natural extension of a one-dimensional map and the action of some multi-dimensional map on an attractor with one dimensional unstable manifolds. A wellknown example of that type corresponding to a one-dimensional map with discontinuities is the Lorenz attractor (cf [R] and Chap. 8). An example corresponding to a one-dimensional map with critical points is the so-called twisted horseshoe (cf [J3]).

Let us denote by Q a partition of I into intervals whose end points are the critical ones and the points of discontinuity, and let $Q_i = [\alpha_i, \alpha_{i+1}]$ be the elements of Q. Then the points of the natural extension $(Y, \tilde{f}, \mu)$ of the endomorphism (I, f, μ) may be identified with $y = (x; z) = (x; z_1, z_2, \ldots)$, where $x \in I$ and z_k is the index of the element Q_{z_k} containing x_k (see Chap. 1, Sect. 4.6 for definition). We denote by Z the set of sequences z and by $\Pi: y = (x; z) \mapsto x$ the projection on the first coordinate.

Theorem 2.4 (cf [L1]). *Suppose that f is a map satisfying $C_1 - C_3$ and μ is an absolutely continuous f-invariant measure, and let χ be a number satisfying $0 < \chi < \int \log|f'| d\mu$. Then there exist measurable functions $\alpha, \beta, \gamma, \tilde{\gamma}$ on Y and a*

constant D such that
  (i) μ-almost everywhere on Y holds $\alpha > 0$, $1 < \beta < \infty$, $0 < \gamma < \tilde{\gamma} < \infty$
  (ii) If $y = (x; z)$ and $t < \alpha(y)$ then $y_t = ((x + t); z) \in Y$
  (iii) $|\Pi(\tilde{f}^{-n}y) - \Pi(\tilde{f}^{-n}y_t)| \leq |t| \cdot \beta(y) \exp(-\chi \cdot n)$
  (iv) for any n

$$1/D \leq \frac{f'(\Pi(\tilde{f}^{-n}y))}{f'(\Pi(\tilde{f}^{-n}y_t))} \leq D$$

  (v) $\gamma(y) \leq \prod_{n=1}^{\infty} \frac{f'(\Pi(\tilde{f}^{-n}y))}{f'(\Pi(\tilde{f}^{-n}y_t))} = \Delta(y, y_t) \leq \tilde{\gamma}(y)$.

Theorem 2.4 asserts in particular that classes of equivalence relation on Y-defined by the projection on Z are open subsets of the intervals. These classes define a measurable partition ξ. Conditional measures of $\tilde{\mu}$ with respect to ξ are described in the following theorem.

Theorem 2.5 (cf [L1]). *Let f, μ be as in Theorem 2.4, and let $q(y, \cdot)$ denote the conditional measures of μ with respect to ξ. Then for $\tilde{\mu}$-almost every y, the measure $q(y, \Pi^{-1}, \cdot)$ on I is absolutely continuous with respect to the Lebesgue measure. There exists a partition $\eta > \xi$ such that $f^{-1}\eta > \eta$, $\bigvee_{k=0}^{\infty} f^{-k}\eta = \varepsilon$ and*

$$q(y, B) = \frac{\int_{B \cap \eta(y)} \Delta(y, y') \, dy'}{\int_{\eta(y)} \Delta(y, y') \, dy'},$$

where $\Delta(y, y')$ is defined in (v) of Theorem 2.4.

When considering individual trajectories it is natural to call a point satisfying the following conditions *regular*:

R_1) the sequence of measures $\frac{1}{n}\sum_{i=0}^{n-1} \delta_{f^i_{(x)}}$ weakly converges towards an ergodic measure μ_x

R_2) the sequence of numbers $\frac{1}{n}\log|df^n/dx|$ converges to

$$\lambda_x = \int \log|f'| \, d\mu_x.$$

A regular point is called *positive regular* if $\lambda_x > 0$.

Theorem 2.6 (cf [L1]). *Let $f: I \to I$ be a smooth map satisfying $C_1 - C_3$. Then the set of positive regular points has positive Lebesgue measure if and only if there is an absolutely continuous invariant measure with positive entropy.*

The "if" of this theorem may be considered as a one-dimensional version of the Sinai-Ruelle-Bowen theorem about limit measures on attractors (see Chap. 7).

§3. Feigenbaum Universality Law

3.1. The Phenomenon of Universality. Several one-parameter families of differential equations depending on some external parameter μ (the Lorenz system, nonlinear electrical circuits, Galerkin approximations for Navier-Stokes equations, etc) demonstrate a cascade of successive *period doubling bifurcations* of stable periodic orbits when μ is varied.

Let γ be some stable periodic orbit continuously varying with μ. For some μ_0 an eigenvalue $\lambda(\mu)$ of the Poincaré map along γ takes the value $\lambda(\mu_0) = -1$. It is a generic property that for $\mu > \mu_0$ the orbit γ becomes unstable and a new stable periodic orbit γ' is born which coincides for $\mu = \mu_0$ with γ passed twice. If we denote by $\lambda'(\mu)$ the eigenvalue of the Poincaré map along γ', then $\lambda'(\mu_0) = (\lambda(\mu_0))^2 = 1$. The orbit γ' varies with μ and in turn for some $\mu_i > \mu_0$ becomes unstable, so that $\lambda'(\mu_1) = -1$ and a new periodic orbit is born whose period is twice the period of γ', and so on. The parameter values μ_i corresponding to successive bifurcations converge to some limit μ_∞. When $\mu_i \to \mu_\infty$, the form of the bifurcating orbits becomes more and more complicated and they converge to an invariant set whose structure does not depend on the family under consideration.

Several one-parameter families of discrete transformations, in particular of one-dimensional maps exhibit a similar phenomenon. Moreover, it follows from the theory of kneading, by Milnor and Thurston, that for any one-parameter family of smooth unimodal maps $F_t: [0, 1] \to [0, 1]$, $t \in [0, 1]$ continuously depending on t in C^1-topology and satisfying $F_0(x) \equiv 0$, $\max_{x \in [0, 1]} F_1(x) = 1$ there is an infinite number of different period doubling series.

Let us consider for example the family of quadratic polynomials $G_\mu: x \mapsto 1 - \mu x^2$, $x \in [-1, 1]$, $\mu \in [0, 2]$. The value $\mu_0 = 0.75$ corresponds to the first period doubling: the fixed point $x_0(0.75) = 2/3$ bifurcates into a cycle of period 2 $(x_1(\mu), x_2(\mu))$. The subsequent bifurcation values μ_n corresponding to the birth of 2^n-cycles are $\mu_1 = 1{,}25$, $\mu_2 = 1{,}3681$, $\mu_3 = 1{,}3940, \ldots$. The sequence μ_n converges to $\mu_\infty = 1{,}40155\ldots$. G_{μ_∞} has orbits of period 2^n for any n and no other periodic orbits. The subsequent quotients $\dfrac{\mu_n - \mu_{n-1}}{\mu_{n+1} - \mu_n} = \delta_n$ take values: $\delta_1 = 4.23$, $\delta_2 = 4.55$, $\delta_3 = 4.65$, $\delta_4 = 4.664$, $\delta_5 = 4.668$, $\delta_6 = 4.669, \ldots$. The values δ_n tend to the limit $\delta = 4.669\ldots$. The universality law provides that δ_n computed for different families converge to δ which does not depend either on the dimension of the phase space or on the family. In other words

$$|\mu_n - \mu_\infty| \sim c \cdot \delta^{-n}$$

where c depends on the family and δ is universal. The sizes of some domains in the phase space are also characterized by universal scaling. For example if we consider a bifurcation parameter value μ_n such that for a 2^n-cycle

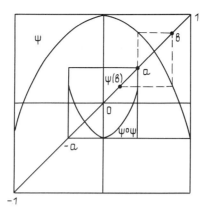

Fig. 20

$\{x_1,\ldots,x_{2^n}\}$ $\dfrac{dF^{2^n}}{dx} = \prod\limits_{i=1}^{2^n}\dfrac{dF}{dx}(x_i) = -1$ holds, and if we let $x^{(n)}$ denote the point of this cycle the closest to zero, then we have $x^{(n)} \sim c' \cdot \lambda^n$, where $\lambda = -0.3995\ldots$ is universal i.e. independent of the family.

3.2. Doubling Transformation. The universal behavior in nonlinear systems was discovered by M. Feigenbaum in 1978 and stimulated a number of theoretical and experimental works (see [CE] for references). In order to explain the Feigenbaum universality law let us consider the so-called *doubling transformation* which consists in taking the second iterate of the map with simultaneous rescaling. Namely, let $\mathfrak{U}_1$ be the set of even unimodal C^1-mappings $\psi: [-1,1] \to [-1,1]$ satisfying the following conditions (see Fig. 20):
1) $\psi'(0) = 0$; $\psi(0) = 1$;
2) $\psi(1) = -a < 0$;
3) $b = \psi(a) > a$; $\psi(b) = \psi^2(a) < a$.

For $\psi \in \mathfrak{U}_1$ we define

$$\mathcal{T}\psi(x) = -\frac{1}{a}\psi \circ \psi(-ax).$$

The nonlinear operator $\mathcal{T}$ is said to be the doubling transformation. The following theorems state the main properties of $\mathcal{T}$.

Theorem 3.1. *The doubling transformation has an isolated fixed point which is an even analytic function*

$\Phi(x) = 1 - 1.52763x^2 + 0.104815x^4 - 0.0267057x^6 + \cdots$. *The universal constant* $\lambda = -0.3993\ldots$ *coincides with* $\Phi(1)$.

We denote by $\mathfrak{H}$ the Banach space of even analytic functions $\psi(z)$ limited in some neighborhood of the interval $[-1,1]$, and real on the real axis. Let $\mathfrak{H}_0$ be a subspace of $\mathfrak{H}$ given by $\mathfrak{H}_0 = \{\psi(z): \psi(0) = \psi'(0) = 0\}$ and let $\mathfrak{H}_1 = \mathfrak{H}_0 + 1$.

Theorem 3.2. *There exists a neighborhood $\mathcal{U}_\Phi$ of Φ in $\mathfrak{H}_1$ such that the doubling transformation $\mathcal{T}$ is a C^∞-map from $\mathcal{U}_\Phi$ into $\mathfrak{H}_1$. $D\mathcal{T}_\Phi$ is a hyperbolic operator with a single unstable eigenvalue $\delta = 4.6692...$ and a stable subspace of codimension one.*

We denote by $\Sigma_0 \subset \mathfrak{U}_1$ the bifurcation surface consisting of ψ satisfying the following conditions: $\psi'(x_0) = -1$, $(\psi \circ \psi)'''(x_0) < 0$, where $x_0 = x_0(\psi)$ is the single fixed point of ψ lying in $[0, 1]$. *In consequence of Theorem 3.2 a local unstable one-dimensional manifold $W^u_{\text{loc}}(\Phi)$ and a local stable manifold $W^s_{\text{loc}}(\Phi)$ of codimension 1 are defined.*

Theorem 3.3. *A local unstable manifold may be extended to the global one $W^u(\Phi)$ which has a point of transversal intersection with Σ_0. $W^u(\Phi)$ consists of maps with the negative Schwarzian derivative.*

The proofs of Theorems 3.1–3.3 known to date use computer estimates (see [VSK] for references)[1]. Similar analytic results were obtained by Collet, Eckmann and Lanford for a class of maps of the form $f(|x|^{1+\varepsilon})$, where f is analytic and ε is sufficiently small (cf [CE]).

In a neighborhood U_Φ of the fixed point we obtain the following picture (Fig. 21). The hyperbolic structure of $\mathcal{T}$ implies that codimension one surfaces $\mathcal{T}^{-n}\Sigma_0$ are near to parallel to the stable manifold $W^s(\Phi)$. Consider a curve F_μ corresponding to a one parameter family of maps. If this curve has a point of transversal intersection with $W^s(\Phi)$, then it is the same for $\mathcal{T}^{-n}\Sigma_0$ with sufficiently large n. The parameter values μ_n corresponding to $F_{\mu_n} \in \mathcal{T}^{-n}\Sigma_0$ are the values of doubling bifurcations $2^n \to 2^{n+1}$, and the map $F_{\mu_\infty} \in W^s(\Phi)$ corresponds to the value $\mu_\infty = \lim_{n \to \infty} \mu_n$. Since the unstable eigenvalue of $D\mathcal{T}_\Phi$ equals δ, the distances between $\mathcal{T}^{-n}\Sigma_0$ and $W^s(\Phi)$ decreases as δ^{-n}. Thus

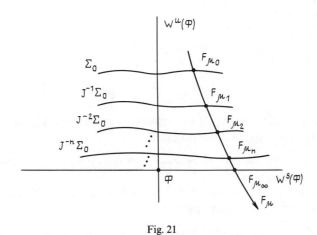

Fig. 21

---

[1] A new approach using complex variable methods has been recently suggested by Sullivan (cf [Su]$_2$).

$$|\mu_\infty - \mu_n| \sim c \cdot \delta^{-n}$$

which implies the universal scaling of parameter.

Now let $x_{(n)}$ be the closest to zero 2^n-periodic point of F_{μ_n}. In consequence of the hyperbolicity of $\mathcal{T}$, the map $\mathcal{T}^n F_{\mu_n}$ which belongs to the surface Σ_0, lies in a small neighborhood of $W^u \Phi \cap \Sigma_0 = \Sigma_*$. Let us denote by x_0 the fixed point of the map $\mathcal{T}^n F_{\mu_n}$. We have $x_{(n)} = x_0 \prod_{i=1}^n (-a_i)$ where $(-a_i)$ are the rescaling constants for $\mathcal{T}^i F_{\mu_n}$. Almost all points $\mathcal{T}^k F_{\mu_n}$, $k \in [1, n]$ lie in a small neighborhood of the fixed point Φ and the corresponding a_i are exponentially close to $a(\Phi) = 0.3995\ldots$. This explains the universal scaling in the phase space.

If we want to prove that a given family F_μ admits universal scaling and if an experimental accuracy of the relation $|\mu_n - \mu_\infty| \sim c \cdot \delta^{-n}$ does not satisfy us, then we must verify the F_μ is in the domain of the definition of $\mathcal{T}$ and that it intersects $W^s(\Phi)$ transversely. This is the case for the family of quadratic polynomials.

E.B. Vul and K.M. Khanin proposed the following method of constructing $W^u(\Phi)$ (cf [VSK]). One considers the space of one-parameter families F_μ and defines an operator T acting like the doubling transformation on F_μ and like some rescaling on the parameter. If a family F_μ lies in a small neighborhood of $W^u(\Phi)$ (which is also considered as a one-parameter family), then the iterations $T^n F_\mu$ converge, with an exponential rate, to $W^u(\Phi)$ which is an attracting fixed point of T.

3.3. Neighborhood of the Fixed Point. The following description of a C^3-neighborhood of the fixed point Φ seems to be quite probable. The stable manifold $W^s(\Phi)$ divides this neighborhood in two parts (Fig. 22). There are two fundamental domains of $\mathcal{T}^{-1}$ in $\mathcal{U}_\Phi$: the first one denoted by $\mathcal{U}$ lies above $W^s(\Phi)$, and the second denoted by $\mathcal{U}'$ lies below $W^s(\Phi)$. The boundaries of $\mathcal{U}$ are: $\Sigma = \{f : f^2(0) = 0\}$ and $\mathcal{T}^{-1}\Sigma$ (Fig. 23).

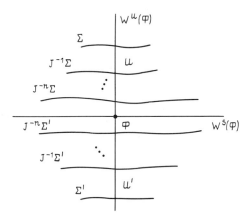

Fig. 22

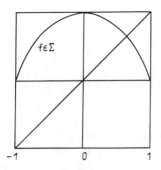
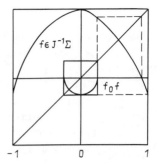

Fig. 23

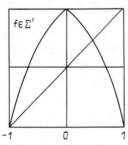
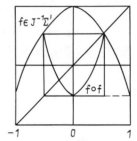

Fig. 24

The boundaries of $\mathcal{U}'$ are: $\Sigma' = \{f: f^2(0) = -1\}$ and $\mathcal{T}^{-1}\Sigma'$ (Fig. 24).

$W^s(\Phi)$ separates the mappings with a simple structure (finite non-wandering set and zero topological entropy) from the mappings with complicated structure (infinite non-wandering set and positive topological entropy).

If $\psi \in \mathcal{T}^{-n}\mathcal{U}$, then $\Omega(f)$ is a union of a stable (or neutral if $\psi \in \mathcal{T}^{-(n+1)}\Sigma_0$) orbit of period 2^{n+1} (or 2^{n+2}) and unstable periodic orbits of periods 2^k, $k < n+1$ (or $k < n+2$).

If $\psi \leqslant \mathcal{T}^{-n}\mathcal{U}'$, $n \geqslant 1$, then the nonwandering set $\Omega(\psi)$ is contained in the union of 2^n disjoint intervals (with the exception of a finite number of 2^k-cycles, $k < n$). Any map $\psi \in \mathcal{U}'$ has some periodic orbit of odd period, while for $\psi \in \mathcal{T}^{-n}\mathcal{U}'$ the period of any orbit divides 2^n (apart from 2^k-cycles mentioned above).

There are stochastic maps below and infinitely close to $W^s(\Phi)$. In order to see this, consider a curve $F_\mu \subset \mathcal{U}_\Phi$ which transversely intersects $W^s(\Phi)$ and let $\gamma^{(n)} = \{F_\mu, \mu \in [\mu_1^{(n)}, \mu_2^{(n)}]\}$ denote $F_\mu \cap \mathcal{T}^{-n}\mathcal{U}'$. The maps $G_\mu = \mathcal{T}^n F_\mu \in \mathcal{T}^n \gamma^{(n)} \subset \mathcal{U}'$ have negative Schwarzian derivatives since they are close to $W^u(\Phi)$ (recall that $W^u(\Phi)$ consists of maps satisfying $S\Phi < 0$). Since the curve $\mathcal{T}^n \gamma^{(n)}$ is transversal to Σ', the results of Section 2 are applicable. They imply that mes$\{\mu: G_\mu$ admits an absolutely continuous invariant measure$\}$ is positive. Coming back to F_μ, we obtain a positive measure set $M_n = \{\mu: F_\mu \in \gamma^{(n)}$ has absolutely continuous invariant measure with the support inside 2^n disjoint cyclically permutted intervals}. For this measure, the restriction of $F_\mu^{2^n}$ on the ergodic component is Bernoulli.

Moreover the relative measure of stochastic maps on any arc $\gamma^{(n)}$ is bounded from below. There exists $c > 0$ depending only on $\mathcal{U}_\Phi$ such that

$$\text{mes } M_n > c|\mu_1^{(n)} - \mu_2^{(n)}|$$

(cf [J5]).

Similar arguments are valid for any family of maps (not necessary close to Φ) which transversely intersects $W^s(\Phi)$. In particular this is the case for the family of quadratic polynomials.

3.4. Properties of Maps Belonging to the Stable Manifold of Φ. The stable manifold $W^s(\Phi)$ separates mappings with simple dynamics from those with complex one. Any map $\psi \in W^s(\Phi) \cap \mathcal{U}_\Phi$ satisfies two specific conditions.

(i) ψ has 2^n—periodic points for any n, and has no points of other periods.
(ii) $S\psi < 0$.

The following theorem proved independently by M. Misiurewicz and by Yu.S. Barkowski and G.M. Levin describes the properties of unimodal maps ψ satisfying (i), (ii) (cf [Mi3], [VSK]).

Theorem 3.4. $\Omega(\psi) = \sigma \cup \text{Per } \psi$, where $\text{Per } \psi$ is the set of periodic points, and σ is a closed ψ-invariant Cantor set with the following properties:

a) $\psi: \sigma \to \sigma$ is a minimal homeomorphism;
b) $\lim_{n \to \infty} \text{dist}(\psi^n x, \sigma) = 0 \ \forall x \notin \bigcup_{k=0}^\infty \psi^{-k}(\text{Per } \psi)$;
c) $\forall x \in [-1, 1] \exists t(x) = \lim_{n \to \infty} \psi^{2^n}(x)$, and $t(x) = x$ for $x \in \Omega(\psi)$;
d) The unique ψ-invariant non atomic measure μ is supported by σ; the system (σ, ψ, μ) is isomorphic to the unit translation on the group of 2-adic integers which is called the "adding machine".

This isomorphism is a result of the following construction. There exists a system of intervals $\Delta_i^{(n)}$, $n \geq 1$, $i = 0, 1, \ldots, 2^n - 1$, satisfying: $\psi \Delta_i^{(n)} = \Delta_{i+1}^{(n)}$ for $i < 2^n - 1$ and $\psi \Delta_{2^n - 1}^{(n)} \subset \Delta_0^{(n)}$; $\Delta_i^{(n)} \cap \Delta_j^{(n)} = \emptyset$ for $i \neq j$; $\Delta_k^{(n-1)} \supset \Delta_k^{(n)} \cup \Delta_{k+2^{n-1}}^{(n)}$; $\text{diam}_{n \to \infty} \Delta_k^{(n)} \to 0$. Then

$$\sigma = \bigcap_{n \geq 1} \bigcup_{i=0}^{2^n - 1} \Delta_i^{(n)}.$$

The spectrum of the dynamical system (σ, ψ, μ) consists of the binary rationals. The corresponding eigenfunctions are

$$e^{(n)}(x) = \exp(2\pi i l 2^{-n}) \text{ for } x \in \Delta_l^{(n)}$$

and

$$e_r^{(n)}(x) = (e^{(n)}(x))^{2r+1}, \quad r = 0, 1, \ldots, 2^{n-1} - 1.$$

The fixed point equation $\Phi(x) = \lambda^{-1} \Phi \circ \Phi(\lambda x)$ implies, besides a)–d) of Theorem 3.4, some additional properties of $\Phi(x)$. The following results were obtained in [VSK].

Theorem 3.5. Let $\varphi(x)$ be a C^1-function on $[-1,1]$. Consider the spectral measure of φ: $\rho_\varphi(\omega) = \sum_{n=1}^{\infty} \sum_{r=0}^{2^{n-1}-1} \rho_r^{(n)} \delta\left(\omega - \frac{2r+1}{2^n}\right)$. Then

$$\rho_r^{(n)} \leq \frac{1}{2} \max_{x \in [-1,1]} |\varphi'(x)| \cdot \max_{0 \leq l \leq 2^{n-1}-1} |\Delta_l^{(n-1)}|.$$

Theorem 3.6. There exist $\gamma < 0$, $\beta_0 > 0$ such that for x constituting some subset of σ of full μ-measure

$$\lim_{n\to\infty} \frac{1}{n} \ln |\Delta_i^{(n)}(x)| = \gamma$$

$$\lim_{n\to\infty} \frac{1}{n} \ln \left(\sum_{i=0}^{2^n-1} |\Delta_i^{(n)}|^{\beta_0} \right) = 0 \qquad \text{holds.}$$

Here β_0 is the Hausdorff dimension of σ and

$$\beta_0 > -\frac{\ln 2}{\gamma}$$

which signifies that the Hausdorff dimension is reached on the set of zero μ-measure. A similar result has been obtained by F. Ledrappier and M. Misiurewicz ([LM]).

§4. Rational Endomorphisms of the Riemann Sphere

4.1. The Julia Set and its Complement.
The dynamics of rational endomorphisms $z \mapsto R(z) = \dfrac{a_0 + a_1 z + \cdots + a_n z^n}{b_0 + b_1 + \cdots + b_m z^m}$ of the Riemann sphere was studied in the fundamental memoirs of Julia and Fatou at the beginning of the century (see [Mo] and a recent surveys of P. Blanchard ([Bl]) and M.Yu. Ljubič ([Lju4]). Similarly to one-dimensional real maps, rational endomorphisms may have a stable or a stochastic asymptotic behavior. In order to describe possible types of dynamics we shall use the following classification of periodic orbits (cycles) of $R(z)$. A cycle $\alpha = (\alpha_0, \alpha_1, \ldots, \alpha_{m-1})$ is *attracting* if $|(R^m)'(\alpha)| = |\prod_{i=0}^{m-1} R'(\alpha_i)| < 1$, *rational neutral* if $(R^m)'(\alpha) = \exp 2\pi i r$, $r \in \mathbb{Q}$, *irrational neutral* if $(R^m)'(\alpha) = \exp 2\pi i \theta$, θ-irrational and *repelling* if $|(R^m)'(\alpha)| > 1$. The points with irregular aperiodic behavior are inside the so-called *Julia set* $J(R)$ which coincides with the closure of repelling periodic orbits. $J(R)$ is a perfect set invariant under R and under R^{-1}. There exist endomorphisms R satisfying $J(R) = \overline{\mathbb{C}}$ (cf [Mo]) and moreover the measure of such R in the parameter space is positive (see below). Nevertheless a generally accepted hypothesis is that the set of such endomorphisms is not generic, i.e. it is a countable union of nowhere dense sets.

Any component of the open set $\Delta(R) = \bar{\mathbb{C}} \setminus J(R)$ consists of points with the same asymptotic behavior. Different types of dynamics for $z \in \Delta(R)$ were studied by Fatou and Julia. A final result was recently obtained by D. Sullivan (cf [Su1]).

Theorem 4.1. *Let $\mathscr{D}$ be a component of $\Delta(R)$ Then*

(i) *there exists k_0 such that the domain $\mathscr{D}_1 = R^{k_0}\mathscr{D}$ is periodic, i.e. $\mathscr{D}_1 = R^m\mathscr{D}_1$ for some m and for $|k - l| < m$ $R^k\mathscr{D}_1 \cap R^l\mathscr{D}_1 = \emptyset$,*

(ii) *the number of periodic components is finite,*

(iii) *the only possible types of dynamics on any periodic component $\mathscr{D}$ are the following:*

a) *for any $z \in \mathscr{D}$ the trajectory $\{R^n(z)\}$ converges to an attracting cycle $\alpha = (\alpha_0, \alpha_1, \ldots, \alpha_{m-1})$ where $\alpha_i \in R^i\mathscr{D}$;*

b) *for any $z \in \mathscr{D}$ the trajectory $\{R^n(z)\}$ converges to a rational neutral cycle $\alpha = (\alpha_0, \alpha_1, \ldots, \alpha_{m-1})$ where α_i lies on the boundary of $R^i\mathscr{D}$;*

c) *$\mathscr{D}$ contains a point of some irrational neutral cycle $\alpha = (\alpha_0, \alpha_1, \ldots, \alpha_{m-1})$ and $R^m|\mathscr{D}$ is topologically conjugate to an irrational rotation of a disc;*

d) *$R^m|\mathscr{D}$ is topologically conjugate to an irrational rotation of a ring.*

Let us consider for example the map $z \to R(z) = \frac{1}{2}(z + z^3)$ studied in detail by Julia. The structure of $J(R)$ and of components of $\Delta(R)$ is illustrated in Fig. 25. $\Delta(R)$ is composed of the basin of attraction to infinity, denoted by $\mathscr{D}_\infty$, of two basins of attraction to the fixed points 1 and -1 denoted respectively by $\mathscr{D}_1$ and by $\mathscr{D}_{-1}$ and of an infinite number of preimages $\bigcup_{n=1}^\infty R^{-n}(\mathscr{D}_{\pm 1})$. For R as well as for any other polynomial, the set $J(R)$ coincides with the boundary of $\mathscr{D}_\infty$, which is connected and invariant under $R(z)$ and under $R^{-1}(z)$.

4.2. The Stability Properties of Rational Endomorphisms. We have a sufficiently good understanding of the dynamics of rational maps satisfying the following hyperbolicity condition:

there exist $c > 0$ and $l_0 \in \mathbb{N}$ such that for any $z \in J(R)$

$$|(R^{l_0})'(z)| > c.$$

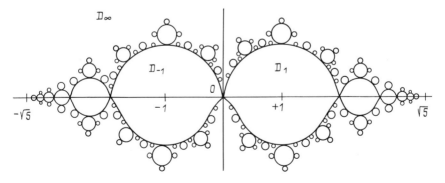

Fig. 25

In this case $\Delta(R)$ is non empty and any component of $\Delta(R)$ eventually falls in some periodic component of type a). The following condition is necessary and sufficient for hyperbolicity: the iterates of any critical point of the map R converge to some attracting cycle.

Rational endomorphisms with the hyperbolic Julia set are structurally stable. They admits Markov partitions any may be studied using the method of symbolic dynamics (cf [J5]).

The following problem formulated by Fatou is still unsolved: is the set of hyperbolic rational (polynomial) endomorphisms dense in the space of all rational (polynomial) endomorphisms? In particular the question remains unsolved for the family of quadratic polynomials $z \mapsto z^2 + a$ (as well as for the real family $x \mapsto x^2 + a$, $a \in \mathbb{R}$).

For polynomials this conjecture is implied by the following: for a dense set of polynomials the Lebesgue measure of Julia set is zero (cf [MSS]).

Although the density of hyperbolic maps is unknown, the following J-stability theorem holds as proved by Mañé, Sad, Sullivan, and in a weaker form by Ljubič (cf [Lju3]).

Theorem 4.2. *For any family $R_\omega(z)$ of rational endomorphisms holomorphicly depending on a parameter $\omega \in U \subset \mathbb{C}^k$ the set $S = \{\omega: R_\omega(z) \text{ is } J\text{-stable}\}$ is open and dense in U. If R_1 and R_2 are in the same component of S then $R_1 | J(R_1)$ is quasi-conformally conjugate to $R_2 | J(R_2)$.*

4.3. Ergodic and Dimensional Properties of Julia Sets.

The investigation of ergodic properties of rational endomorphisms i.e. of invariant measures supported by Julia sets originates from an article of Brolin who proved that for a polynomial map $z \mapsto P(z)$ the preimages of any $z \{P^{-n}(z)\}_{n=0}^\infty$ (with the possible exception of one exclusive point z_0) are asymptotically equidistributed with respect to some measure μ on $J(P)$.

Ljubič in [Lju2] generalized the result of Brolin for rational maps. The corresponding measure μ turns out to be the unique measure of maximal entropy $h\mu(R) = h(R) = \log \deg R$.

D. Ruelle in [Ru2] studied Gibbs measures for endomorphisms satisfying the hyperbolicity condition and their relation to the Hausdorff dimension. He proved that $H_{\dim}(J(R))$ is a real analytic function of R. In particular, if $R(z) = z^d + \varepsilon$ where $|\varepsilon|$ is small, then $H_{\dim}(J(R)) = 1 + (|\varepsilon|^2/4)\log d +$ higher order terms in ε.

A. Manning proved in [Ma2] that for an open set of polynomials of degree d (including in particular $z^d + \varepsilon$) the Hausdorff dimension of the measure μ of maximal entropy (which is defined as $\inf\{H_{\dim}(Y): \mu(Y) = 1\}$) equals 1 (see [Pr2] for generalization of this result).

Ljubič's result states that if R satisfies $J(R) \neq \mathbb{C}$, then any invariant ergodic measure with positive entropy is singular with respect to the Lebesgue measure (cf [Lju]$_1$). Meanwhile M. Rees proved that a set of positive measure in the parameter space consists of R such that $J(R) = \overline{\mathbb{C}}$ and for any set A of positive Lebesgue measure $\overline{\bigcup_{n \geq 0} R^n A} = \overline{\mathbb{C}} \pmod 0$.

Thus we see that in the parameter space of rational endomorphisms of the Riemann sphere as well as for the maps of the interval there exist two sets with opposite properties. The first one is an open and presumably dense set consisting of structurally stable endomorphisms with hyperbolic nonwandering sets. Under the action of such a map, the iterates of almost every (with respect to the Lebesgue measure) point z converge toward a finite number of attracting cycles. The second one is the set of positive measure consisting of endomorphisms which are ergodic with respect to the Lebesgue measure. It is natural to ask whether or not the union of structurally stable maps and of stochastic ones is a set of full measure in the parameter space.

Bibliography

The main results and methods of KAM theory are exposed in the books [Ar1], [Ar2], [Mos]. The last achievements in this area are related to the works of Aubry [AD], Parseval [P], Mather [Mat2].

The main results concerning topological and measure theoretical aspects of hyperbolic theory in different periods of its development are exposed in the surveys [KSS], [PS1] and in the books [N], which presents topological aspects of the theory, and [CSF] where measure theoretical aspects are considered. They also contain various examples, a voluminous bibliography and different applications.

The theory of Anosov systems preserving Liouville measure is exposed in the monograph [An1] which is the first systematic and fundamental research in hyperbolic theory. There is also an exposition of general results concerning Anosov systems in the book [An2] and in the review article [AS]. A theory of hyperbolic sets (topological properties, various examples) and some related problems (the A-system and so on) are treated in the book [N] (see also [K1] where a full proof of the theorem about the families of ε-trajectories is given). Symbolic dynamics for Anosov systems (Markov partitions, equilibrium states, measures of maximal entropy) is constructed in [Si6] (see also [Si2], [Si3]); a generalization to hyperbolic sets is exposed in several articles of Bowen (c.f. [Bo2]); some further generalizations may be found in [AJ], which also contains a short review of topological Markov chains theory. The foundations of the theory of UPH-systems are exposed in [BP]. NCH-systems are introduced in [Pe2] where their local properties and ergodic properties with respect to Liouville measure are considered (see also [K3]). A generalization to Sinai measures is given in [L3].

The main results concerning topological and ergodic properties of geodesic flows with hyperbolic behavior of trajectories (on manifolds without conjugate points, or without focal points, or on manifolds with negative curvature) and their relation to Riemannian geometry and classical mechanics are exposed in the survey [Pe3]. In this survey some other dynamical systems of geometric origin are considered (frame flows and horocycle flows) and a voluminous bibliography is given.

Proofs of general results about billiard systems may be found in the monograph [CSF], which also contains a detailed explanation of the relation between some models of classical mechanics and these systems. Ergodic properties of dispersed billiards are considered in [Si4] where in particular the ergodicity of a system of two disks on two-torus is proved. Some aspects of ergodic theory for dispersed and semidispersed billiards (in particularly for the gas of firm spheres) are given in [Si8]. A detailed exposition of the theory of billiards in two-dimensional convex domains and its relation to Dirichlet problem is contained in [La2]. In the survey [PS1] general stochastic properties of dynamical systems are discussed and in particular stochastic attractors. Geometrical and topological properties of attractors of the Lorenz type are considered in [ABS], their ergodic properties in [BS1] and [Bu4].

In the theory of non-invertible one-dimensional maps an important role was played by the work [Ro], where ergodic properties of an endomorphism are related to those of its natural extension. The properties of Gibbs measures for different classes of expanding maps are considered in [W2] and [HK]. Ergodic theory of one-dimensional mappings which are not expanding is treated in [J2] and [L1]. One-dimensional maps played a crucial role in the development of the renormalization group method for dynamical systems, see [CE] and [VSK]. Recently there has been an explosion of interest in ergodic theory of holomorphic dynamical systems of the Riemann sphere. Parts of results concerning this theme may be found in [Be], [Lyu]$_4$, [D].

For the convenience of the reader, references to reviews in Zentralblatt für Mathematik (Zbl.), compiled using the MATH database, and Jahrbuch über die Fortschritte der Mathematik (Jrb.) have, as far as possible, been included in this bibliography.

[AW]   Adler, R.L., Weiss, B.: Entropy be a complete metric invariant for automorphisms of the torus. Proc. Natl. Acad. Sci. USA 57, 1573–1576 (1967). Zbl. 177.80

[ABS1] Afraimovich, V.S., Bykov, V.V., Shilnikov, L.P.: On the origin and structure of the Lorenz attractor. Dokl. Akad. Nauk SSSR 234, 336–339 (1977) [Russian]. Zbl. 451.76052. English transl.: Sov. Phys., Dokl. 22, 253–255 (1977)

[ABS2] Afraimovich, V.S., Bykov, V.V., Shilnikov, L.P.: On attracting structurally unstable limit sets of Lorenz type attractor. Tr. Mosk. Mat. O.-va 44, 150–212 (1982) [Russian]. Zbl. 506.58023. English transl.: Trans. Mosc. Math. Soc. 1983, No. 2, 153–216 (1983)

[AP]   Afraimovich, V.S., Pesin, Ya.B.: An estimate of Hausdorff dimension of a basic set in a neighborhood of homoclinic trajectory. Usp. Mat. Nauk 39, No. 2, 135–136 (1984) [Russian]. Zbl. 551.58026. English transl.: Russ. Math. Surv. 39, No. 2, 137–138 (1984)

[A1]   Alekseev, V.M.: Perron sets and topological Markov chains. Usp. Mat. Nauk 24, 227–228 (1969) [Russian]. Zbl. 201.566

[A2]   Alekseev, V.M.: Quasirandom oscillations and qualitative questions in celestial mechanics. In: Ninth Summer School., Izd. Inst. Mat. Akad. Nauk Ukr. SSR, Kiev. 212–341 (1972) [Russian]. Zbl. 502.70016. English transl.: Transl., II. Ser., Am. Math. Soc. 116, 97–169 (1981)

[AJ]   Alekseev, V.M., Jakobson, M.V.: Symbolic dynamics and hyperbolic dynamical systems, Phys. Rep. 75, 287–325 (1981)

[An1] Anosov, D.V.: Geodesic flows on closed Riemann manifolds with negative curvature, Tr. Mat. Inst. Steklov 90, 210 p. (1967) [Russian]. Zbl. 163.436. English transl.: Proc. Steklov Inst. Math. 90, (1967), 235 p.

[An2] Anosov, D.V.: On a certain class of invariant sets of smooth dynamical systems. In: Proc. 5th Intern. Conf. on non-linear oscillations, pp. 39–45, 2, (1970) [Russian]. Zbl. 243.34085 English transl.: Russ. Math. Surveys 22, No. 5, 103–167 (1967)

[AS]   Anosov, D.V., Sinai, Ya.G.: Some smooth ergodic systems, Usp. Mat. Nauk 22, No. 5, 107–172 (1967) [Russian]. Zbl. 177.420

[Ar1] Arnold, V.I.: Mathematical methods of classical mechanics. Nauka, Moscow 1974 [Russian]. Zbl. 386.70001. English transl.: Springer-Verlag: New York–Heidelberg–Berlin, 462 p. (1978)

[Ar2] Arnold, V.I.: Some additional topics of the theory of ordinary differential equations. Nauka, Moscow 1978 [Russian]

[AD]   Aubry, S., Le Daeron, P.V.: The discrete Frenkel-Kontorova model and its extension. Physica D., 8, 381–422 (1983)

[Be]   Belykh, V.N.: Models of discrete systems of phase synchronization. Ch. 10 In: Systems of phase synchronization (ed. by V.V. Shakhgildyan and L.N. Belyustina), pp. 161–176. Radio and Communication, Moscow 1982 [Russian]

Bibliography

[BC] Benedicks, M., Carleson, L.: On iterations of $1 - ax^2$ on $(-1, 1)$, Ann. Math. II. Ser. *122*, 1–25 (1985). Zbl. 597.58016

[BeS] Benettin, G., Strelcyn, J.-M.: Numerical experiments on a billiard. Stochastic transition and entropy. Phys. Rev. A *17*, 773–786 (1978)

[Bil] Billingsley, P.: Ergodic theory and information. John Wiley and Sons, Inc., New York–London–Sydney; 1965. Zbl. 141.167

[Bi] Birkhoff, G.D.: Dynamical systems. New York, Am. Math. Soc., Colloq. Publ. 9 1927. Jrb. 53.732

[Bl] Blanchard, P.: Complex analytic dynamics on the Riemann sphere. Bull. Am. Math. Soc., New Ser. *11*, 85–141 (1984). Zbl. 558.58017

[BKM] Boldrighini, C., Keane, M., Marchetti, F.: Billiards in polygons. Ann. Probab. 6, 532–540 (1978). Zbl. 377.28014

[Bo1] Bowen, R.: Periodic orbits for hyperbolic flows. Am. J. Math. 94, 1–30 (1972). Zbl. 254.58005

[Bo2] Bowen R.: Methods of symbolic dynamics, collected papers, Mir, Moscow 1979 [Russian]

[BFK] Brin, M., Feldman, J., Katok, A.: Bernoulli diffeorphisms and group extensions of dynamical systems with non-zero characteristic exponents. Ann. Math., II. Ser. *113*, 159–179 (1981). Zbl. 477.58021

[BP] Brin, M.I., Pesin, Ya.B.: Partially hyperbolic dynamical systems, Izv. Akad. Nauk SSSR, Ser. Mat. *38*, 170–212 (1974) [Russian]. Zbl. 304.58017

[Bu1] Bunimovich, L.A.: On billiards close to dispersing. Mat. Sb., Nov. Ser. *94* (136), 49–73 (1974) [Russian]. Zbl. 309.58018. English transl.: Math. USSR, Sb. *95*, 49–73 (1974).

[Bu2] Bunimovich, L.A.: On the ergodic properties of nowhere dispersing billiards. Commun. Math. Phys. *65*, 295–312 (1979). Zbl. 421.58017

[Bu3] Bunimovich, L.A.: Some new advancements in the physical applications of ergodic theory. In: Ergodic theory and Related Topics (ed. by Michel, H.), pp. 27–33, Berlin, Akademie-Verlag 1982. Zbl. 523.58025

[Bu4] Bunimovich, L.A.: Statistical properties of Lorenz attractors. In: Nonlinear Dynamics and Turbulence (ed. by G.I. Barenblatt et al.), pp. 71–92, Boston-London-Melbourne: Pitman 1983. Zbl. 578.58025

[Bu5] Bunimovich, L.A.: Decay of correlations in dynamical systems with chaotic behavior. J. Exp. Theor. Phys. *89*, 1452–1471 (1985) [Russian].

[Bu6] Bunimovich, L.A.: On stochastic dynamics of rays in resonators. Radiofizika *28*, 1601–1602 (1985) [Russian].

[BS1] Bunimovich, L.A., Sinai, Ya.G.: On a fundamental theorem in the theory of dispersing billiards. Mat. Sb., Nov. Ser. *90* (132), 415–431 (1973) [Russian]. Zbl. 252.58006. English transl.: Math. USSR, Sb. *19*, 407–423 (1974)

[BS2] Bunimovich, L.A., Sinai, Ya.G.: Markov partitions for dispersed billiards. Commun. Math. Phys. *78*, 247–280 (1980), Erratum, ibid, *107*, 357–358 (1986). Zbl. 453.60098

[BS3] Bunimovich, L.A., Sinai, Ya.G.: Stochasticity of attractor in the Lorenz model. In Nonlinear waves (ed. by A.V. Gaponov-Grekhon), pp. 212–226, Moscow, Nauka 1980 [Russian].

[BS4] Bunimovich, L.A., Sinai, Ya.G.: Statistical properties of Lorentz gas with periodic configuration of scatterers. Commun. Math. Phys. *78*, 479–497 (1981). Zbl. 459.60099

[BK] Burns, K., Katok, A.: Manifolds with non positive curvature. Ergodic Theory Dyni. Syst. *5*, 307–317 (1985). Zbl. 572.58019

[C] Chentsova, N.I.: A Natural invariant measure on Smale's horseshoe, Dokl. Akad. Nauk SSSR *256*, 294–298 (1981) [Russian]. Zbl. 475.58013. English transl.: Sov. Math., Dokl. *23*, 87–91 (1981)

[Ch] Chernov, N.J.: Construction of transversal fibres in multidimensional semidispersing billiards. Funkt. Anal. Prilozh. *16*, No. 4, 35–46 (1982) [Russian]; English transl.: Funct. Anal. Appl. *16*, 270–280 (1983). Zbl. 552.28015

[Chi] Chirikov, B.V.: An universal instability of many-dimensional oscillator systems. Phys. Repts. *53*, 263–319 (1979)

Bibliography

[CE] Collet, P., Eckmann, J.-P.: Iterated maps on the interval as dynamical systems. Basel-Boston -Stuttgart Birkhäuser 1980. 248 p. Zbl. 458.58002

[CSF] Cornfeld, I.P., Sinai, Ya.G., Fomin, S.V.: Ergodic theory, Moscow, Nauka 1980, 383 p. [Russian]. English transl.: Springer 1982, 486 p. Zbl. 493.28007

[D] Douady, A.: Disques de Siegel et anneax de Herman. Sémm, Bourbaki 677 (1987)

[EW] Eckmann, J.-P., Wittwer, P.: A complete proof of the Feigenbaum conjectures, preprint (1986)

[FOY] Farmer, J.D., Ott, E., Yorke, J.A.: The dimension of chaotic attractors. Phys. $7D$, 153–180 (1983). Zbl. 561.58032

[Gac] Gachok, P.V.: Construction of models of dynamical systems with a strange attractor. Dokl. Akad. Nauk SSSR 274, 1292–1294 (1984) [Russian]. Zbl. 596.34028. English transl.: Sov. Math. Dokl. 29, 123–125 (1984)

[G] Gallavotti, G., Ornstein, D.: Billiards and Bernoulli schemes. Commun. Math. Phys. 38, 83–101 (1974). Zbl. 313.58012

[Ga] Galperin, G.A.: Non-periodic and not everywhere dense billiard trajectories in convex polygons and polyhedrons. Commun. Math. Phys. 91, 187–211 (1983). Zbl. 529.70001

[GGV] Gelfand, I.M., Graev, M.I., Vilenkin, I.Ya.: Integral geometry and some related aspects of representation theory, Moscow 1962, 656p. [Russian]. Zbl. 115.167

[GU] Guckenheimer, J.: Renormalization of one dimensional mappings and strange attractors, Contemp. Math. 58, No. 3, 143–160 (1987). Zbl. 622.58030

[H] Henon, M.: A two-dimensional mapping with a strange attractor. Commun. Math. Phys. 50, 69–76 (1976). Zbl. 576.58018

[HK] Hofbauer, F., Keller G.: Equilibrium states for piecewise monotonic transformations. Ergod. Theory and Dyn. Syst. 2, 23–43 (1982). Zbl. 499.28012

[J1] Jakobson, M.V.: Markov partitions for rational endomorphisms of the Riemann sphere, In: Multicomponent random systems, (R.L. Dobrushin, Ya.G. Sinai eds.) Advances in probability and related topics 6, 381–396 (1980). Zbl. 439.58021

[J2] Jakobson, M.V.: Absolutely continuous invariant measures for one-parameter families of one-dimensional maps. Commun. Math. Phys. 81, 39–88 (1981). Zbl. 497.58017

[J3] Jakobson, M.V.: Invariant measures of some one-dimensional attractors. Ergodic Theory Dyn., Syst. 2, 317–337 (1982). Zbl. 521.58039

[J4] Jakobson, M.V.: Families of one-dimensional maps and near-by diffeomorphisms. In: Proceedings of ICM-86, Berkeley 1987.

[J5] Jakobson, M.V.: Universal behavior and stochasticity for one-dimensional dynamical systems. In: Proceedings of the 1st world congress of the Bernoulli society: VNU Sc. Press. Utrecht 1, 85–90, 1987

[K1] Katok, A.B.: Local properties of hyperbolic sets, Appendix 1 to Nitecki Z., Differentiable dynamics, Moscow, Mir, (translated from English), pp. 214–232, 1975 [Russian]. Zbl. 246.58012

[K2] Katok, A.: Bernoulli diffeomorphisms on surfaces. Ann. Math., II. Ser. 110, 529–547 (1979). Zbl. 435.58021

[K3] Katok, A.: Lyapunov exponents, entropy and periodic orbits for diffeomorphisms, Publ. Math., Inst. Hautes Ethol. Sci. 51, 137–173 (1980). Zbl. 445.58015

[K4] Katok, A.: Some remarks on Birkhoff and Mather twist map theorems. Ergodic Theory Dyn. Syst. 2, 185–194 (1982). Zbl. 521.58048

[KSS] Katok, A.B., Sinai, Ya.G., Stepin, A.M.: Theory of dynamical systems and general transformation groups with invariant measure, Itogi Nauki Tekhn. Ser. Math. Anal. 13, 129–262, 1975 [Russian]. Zbl. 399.28011. English transl.: J. Sov. Math. 7, 974–1065 (1977)

[KS] Katok, A., Strelcyn, J.-M.: Invariant Manifolds, Entropy and Billiards; Smooth Maps with Singularities. Lect. Notes Math. 1222, Berlin-Heidelberg, Springer 1986

[Kr] Krylov, N.S.: Works on the foundations of statistical physics. Moscow: Akad. Nauk SSSR Publ., 1950 [Russian]

[L1]	Ledrappier, F.: Some properties of absolutely continuous invariant measures on an interval. Ergodic Theory Dyn. Syst. *1*, 77–93 (1981). Zbl. 487.28015
[L2]	Ledrappier, F.: Some relations between dimension and Lyapunov exponents. Commun. Math. Phys. *81*, 229–238 (1981). Zbl. 486.58021
[L3]	Ledrappier, F.: Propriétés ergodiques des mesures de Sinai, C.R. Acad. Sci., Paris, Sér. I, *294*, 593–595 (1982). Zbl. 513.58030
[LM]	Ledrappier, F., Misiurewicz, M.: Dimension of invariant measures for maps with exponent zero. Ergod. Th. Dyn. Syst. *5*, 595–610 (1985). Zbl. 608.28008
[LY]	Ledrappier, F., Young, L.-S.: The metric entropy of diffeomorphism, I, II, Ann. Math., II. Ser. *122*, 509–539, 540–574 (1985). Zbl. 605.58028
[Lanf]	Lanford III, O.E.: Computer-assisted proofs in analysis. Proceedings of ICM-86 (1987)
[La1]	Lazutkin, V.F.: The existence of caustics for a billiard problem in a convex domain. Izv. Akad. Nauk SSSR, Ser. Mat. *37*, 186–216 (1983) [Russian]. Zbl. 256.52001. English transl.: Math. USSR, Izv. *7*, 185–214 (1974)
[La2]	Lazutkin, V.F.: Convex billiards and eigenfunctions of Laplace operator. Leningrad Univ. Publ. 1981 [Russian]. Zbl. 532.58031
[Le]	Levy, Y.: Ergodic properties of the Lozi mappings. Commun. Math. Phys. *93*, 461–481 (1984). Zbl. 553.58019
[Lor]	Lorenz, E.N.: Deterministic nonperiodic flow. J. Atmos. Sci. *20*, 130–141 (1963)
[Loz]	Lozi, R.: Un attracteur etrange du type attracteur de Henon. J. Phys., Paris *39*, 9–10 (1978)
[Lyu1]	Lyubich, M.Yu: On typical behavior of trajectories of rational map of the sphere, Dokl. Akad. Nauk SSSR *268*, 29–32 (1982) [Russian]. Zbl. 595.30034. English transl.: Sov. Math., Dokl. *27*, 22–25 (1983)
[Lyu2]	Lyubich, M.Yu.: On the measure of maximal entropy for rational endomorphism of Riemann sphere, Funkt. Anal. Prilozh. *16*, No. 4, 78–79 (1982) [Russian]. Zbl. 525.28021. English transl.: Funct. Anal. Appl. *16*, 309–311 (1983)
[Lyu3]	Lyubich, M.Yu.: The study of stability for dynamics of rational maps, In: Teor. Funkts., Funkt. Anal. Prilozh. *42*, 72–91 (1984) [Russian]. Zbl. 572.30023
[Lyu4]	Lyubich, M.Yu.: Dynamics of rational maps: topological picture. Usp. Mat. Nauk *41*, No. 4 (250) 35–95 (1986) [Russian]. Zbl. 619.30033. English transl.: Russ. Math. Surv. *41*, No. 4, 43–117 (1987)
[M]	Mañé, R.: Hyperbolicity, sinks and measure in one dimensional dynamics, Commun. Math. Phys. *100*, 495–524 (1985). Zbl. 583.58016
[MSS]	Mañé, R. Sad P., Sullivan D.: On the dynamics of rational maps. Ann. Sci. E.N.S. *16*, 193–217 (1982). Zbl. 524.58025
[Ma1]	Manning, A.: A relation between Lyapunov exponents, Hausdorff dimension and entropy. Ergodic Theory Dyn. Systems *1*, 451–459 (1981). Zbl. 478.58011
[Ma2]	Manning, A.: The dimension of the maximal measure for a polynomial map., Ann. Math., II. Ser. *119*, 425–430 (1984). Zbl. 551.30021
[Mar]	Margulis, G.A.: On some measures related to C-flows on compact manifolds, Funkt. Anal. Prilozh. *4*, No. 1, 62–76 (1970) [Russian]. Zbl. 245.58003. English transl.: Funct. Anal. Appl. *4*, 55–67 (1970)
[Mat1]	Mather, J.: Characterization of Anosov diffeomorphisms. Indagaiones Math., *30*, 479–483 (1968). Zbl. 165.570
[Mat2]	Mather, J.: Concavity of the Lagrangian for quasiperiodic orbits. Comment Math. Helv. *57*, 356–376 (1982). Zbl. 508.58037
[Mi1]	Misiurewicz, M.: Strange attractors for the Lozi mappings. In: Nonlinear dynamics (ed. by R.G. Helleman), pp. 348–358, New York, Ann. N.Y., Acad. Sci, 357, 1980. Zbl. 473.58016
[Mi2]	Misiurewicz, M.: Absolutely continuous measures for certain maps of an interval. Publ. Math., Inst. Hautes Etud. Sci. *53*, 17–51 (1981). Zbl. 477.58020
[Mi3]	Misiurewicz, M.: Structure of mappings of an interval with zero entropy, Publ. Math., Inst. Hautes Etud. Sci. *53*, 5–17 (1981). Zbl. 477.58030

[Mo] Montel, P.: Leçons sur les familles normales de fonctions analytiques et leurs applications. Paris, Gauthier-Villars 1927. Jrb. 53.303

[Mor] Mori, H.: Fractal dimension of chaotic flows autonomous dissipative systems. Progr. Theor. Phys. *63*, 1044–1047 (1980)

[Mos] Moser, J.: Lectures on Hamiltonian systems. New York, Courant Inst. Math. Sci. 1968

[N] Nitecki, Z.: Differentiable dynamics. MIT Press, London, 1971, 282 p. Zbl. 246.58012

[OS] Osserman, R., Sarnak, P.: A new curvature invariant and entropy of geodesic flows, Invent. Math. *77*, 455–462 (1984). Zbl. 536.53048

[P] Parseval, I.C.: Variational principles for invariant tori and cantori. In: Symp. on Nonlinear Dyn. and Beam—Beam Interactions, Amer. Inst. Phys. Conf. Proc., pp. 310–320 (1980)

[Pe1] Pesin, Ya.B.: An example of non ergodic flow with non zero characteristic exponents., Funkt. Anal. Prilozh. *8*, 81–82 (1974) [Russian]. Zbl. 305.58012. English transl.: Funct. Anal. Appl. *8*, 263–264 (1975)

[Pe2] Pesin, Ya.B.: Characteristic Lyapunov exponents and smooth ergodic theory, Usp. Mat. Nauk *32*, No. 4 (196), 55–112 (1977) [Russian]. Zbl. 359.58010. English transl.: Russ. Math. Surv. *32*, No. 4, 55–114 (1977)

[Pe3] Pesin, Ya.B.: Geodesic flows with hyperbolic behavior of the trajectories and objects, connected with them, Usp. Mat. Nauk *36*, No. 4, 3–51 (1981) [Russian]. Zbl. 482.58002. English transl.: Russ. Math. Surv. *36*, No. 4, 1–59 (1981)

[Pe4] Pesin, Ya.B.: On the notion of the dimension with respect to a dynamical system. Ergodic Theory Dyn. Syst. *4*, 405–420 (1984). Zbl. 616.58038

[Pe5] Pesin, Ya.B.: The generalization of Carathéodory's construction for dimensional characteristics of dynamical systems. In: Statistical physics and dynamical systems, Progr. Phys. *10*, pp. 191–202, Birkhäuser 1985

[Pe6] Pesin, Ya.B.: Ergodic properties and dimensional—like characteristics of strange attractors closed to hyperbolic one. Proc. of ICM-86, Berkeley 1987

[PP] Pesin, Ya.B., Pitskel, B.S.: Topological pressure and variational principle for noncompact sets, Funkt. Anal. Prilozh. *18*, 50–63 (1984) [Russian]. Zbl. 567.54027. English transl.: Funct. Anal. Appl. *18*, 307–318 (1984)

[PS1] Pesin, Ya.B., Sinai, Ya.G.: Hyperbolicity and stochasticity of dynamical systems. In: Mathematical Physics Reviews, Gordon and Breach Press, Harwood Acad. Publ., USA, *2*, pp. 53–115, 1981. Zbl. 561.58038

[PS2] Pesin, Ya.B., Sinai, Ya.G.: Gibbs measures for partially hyperbolic attractors. Ergodic Theory Dyn. Syst. *2*, 417–438 (1982). Zbl. 519.58035

[Pl] Plykin, R.V.: About hyperbolic attractors of diffeomorphisms, Usp. Mat. Nauk *35*, No. 3, 94–104 (1980) [Russian]. Zbl. 445.58016. English transl.: Russ. Math. Surv. *35*, No. 3, 109–121 (1980)

[Pla] Plante, J.F.: Anosov flows. Am. J. Math. *94*, 729–754 (1972). Zbl. 257.58007

[Pr1] Przytycki, F.: Ergodicity of toral linked twist mappings. Ann. Sci. Ec. Norm. Super., IV. Ser. *16*, 345–354 (1983). Zbl. 531.58031

[Pr2] Przytycki, F.: Hausdorff dimension of harmonic measure on the boundary of an attractive bassin for a holomorphic map, Invent. Math. *80*, 161–179 (1985). Zbl. 569.58024

[R] Rand, D.: The topological classification of Lorenz attractors. Math. Proc. Camb. Philos. Soc. *83*, 451–460 (1978). Zbl. 375.58015

[Re1] Rees, M.: Ergodic rational maps with dense critical point forward orbit. Ergodic Theory and Dyn. Syst. *4*, 311–322 (1984). Zbl. 553.58008

[Re2] Rees, M.: Positive measure sets of ergodic rational maps. Ann. Sc. Ec. Norm. Supér. *19*, 383–407 (1986). Zbl. 611.58038

[Ro] Rokhlin, V.A.: Exact endomorphisms of Lebesgue space, Izv. Akad. Nauk SSSR, Ser. Mat. *25*, 499–530 (1961) [Russian]. Zbl. 107.330

[Ru1] Ruelle, D.: Thermodynamic formalism. Addison-Wesley, Publ. Comp., London 1978. Zbl. 401.28016

Bibliography

[Ru2] Ruelle, D.: Repellers for real analytic maps. Ergodic Theory Dyn. Syst. 2, 99–107 (1982). Zbl. 506.58024

[RT] Ruelle, D., Takens, F.: On the nature of turbulence. Commun. Math. Phys. 20, 167–192 (1971). Zbl. 223.76041

[Ry] Rychlik, M.R.: Another proof of Jakobson's theorem and related results. Preprint, Univ. of Washington (1986)

[S] Series, C.: The infinite word problem and limit sets in Fuchsian groups. Ergodic Theory Dyn. Syst. 1, 337–360 (1981). Zbl. 483.30029

[Sh] Shnirelman, A.I.: Statistical properties of eigenfunctions. In: Trans. of Dilijan Math. School. Erevan, pp. 267–278 (1974) [Russian]

[Si1] Sinai, Ya.G.: Geodesic flows on compact surfaces of negative curvature. Dokl. Akad. Nauk SSSR 136, 549–552 (1961) [Russian]. Zbl.133.110. English transl.: Sov. Math., Dokl. 2, 106–109 (1961)

[Si2] Sinai, Ya.G.: Markov partitions and C-diffeomorphisms. Funkt. Anal. Prilozh. 2, No. 1, 64–89 (1968) [Russian]. Zbl. 182.550. English transl.: Funct. Anal. Appl. 2, 61–82 (1968)

[Si3] Sinai, Ya.G.: Construction of Markov partitions. Funkt. Anal. Prilozh. 2, No. 3, 70–80 (1968) [Russian]. Zbl. 194.226. English transl.: Funct. Anal. Appl. 2, 245–253 (1968)

[Si4] Sinai, Ya.G.: Dynamical systems with elastic reflections. Ergodic properties of dispersing billiards. Usp. Mat. Nauk 25, No. 2 (152), 141–192 (1970) [Russian]. Zbl. 252.58005. English transl.: Russ. Math. Surv. 25, No. 2, 137–189 (1970)

[Si5] Sinai, Ya.G.: Some rigorous results on decay of correlations. Supplement to the book: G.M. Zaslavskij, Statistical irreversibility in nonlinear systems, pp. 124–139, Moscow, Nauka 1970 [Russian]

[Si6] Sinai, Ya.G.: Gibbs measures in ergodic theory. Usp. Mat. Nauk 27, No. 4 (166), 21–64, (1972) [Russian]. Zbl. 246.28008. English transl.: Russ. Math. Surv. 27, No. 4, 21–69 (1973)

[Si7] Sinai, Ya.G.: Ergodic properties of Lorentz gas. Funkt. Anal. Prilozh. 13, No. 3, 46–59, (1979) [Russian]. Zbl. 414.28015. English transl.: Funct. Anal. Appl. 13, 192–202 (1980)

[Si8] Sinai, Ya.G.: Development of Krylov ideas. An addendum to the book: N.S. Krylov, Works on the foundations of statistical physics, pp. 239–281, Princeton, Princeton Univ. Press 1979

[SC] Sinai, Ya.G., Chernov, N.I.: Entropy of hard spheres gas with respect to the group of space-time translations. Tr. Semin. Im. I.G. Petrovskogo 8, 218–238, 1982 [Russian]. Zbl. 575.28014

[SV] Sinai, Ya.G., Vul, E.B.: Hyperbolicity conditions for the Lorenz model. Phys. D. 2, 3–7 (1981)

[Su1] Sullivan, D.: Quasi conformal homeomorphisms and dynamics, I. Solution of the Fatou–Julia problem on wandering domains, Ann. Math., II. Ser. 122, 401–418 (1985). Zbl. 589.30022

[Su2] Sullivan, D.: Quasi-conformal conjugacy classes and the stable manifold of the Feigenbaum operator, preprint 1986

[T] Theory of solitons. Inverse scattering method. (ed. by S.P. Novikov), Moscow, Nauka 1980 [Russian]. Zbl. 598.35003

[VSK] Vul, E.B., Sinai, Ya.G., Khanin, K.M.: Feigenbaum universality and thermodynamic formalism. Usp. Mat. Nauk, 39, No. 3 (237), 3–37 (1984) [Russian]. Zbl. 561.58033. English transl.: Russ. Math. Surv. 39, No. 3, 1–40 (1984)

[W1] Walters, P.: A variational principle for the pressure of continuous transformations. Am. J. Math. 97, 937–971 (1975). Zbl. 318.28007

[W2] Walters, P.: Invariant measures and equilibrium states for some mappings which expand distances. Trans. Am. Math. Soc. 236, 127–153 (1978). Zbl. 375.28009

[Wo] Wojtkowski, M.: Principles for the design of billiards with nonvanishing Lyapunov exponent. Commun. Math. Phys. 105, 391–414 (1986). Zbl. 602.58029

[Y1]   Young, L.-S.: Capacity of attractors. Ergodic Theory Dyn. Syst. *1*, 381–388 (1981). Zbl. 501.58028
[Y2]   Young, L.-S.: Dimension, entropy and Lyapunov exponents. Ergodic Theory Dyn. Syst. *2*, 109–124 (1982). Zbl. 523.58024
[Z]   Ziemian, K.: Almost sure invariance principle for some maps of an interval, Ergodic Theory Dyn. Syst. *5*, 625–640 (1985). Zbl. 604.60031

III. Dynamical Systems of Statistical Mechanics and Kinetic Equations

Contents

Chapter 10. Dynamical Systems of Statistical Mechanics
(R.L. Dobrushin, Ya.G. Sinai, Yu.M. Sukhov) 208
§1. Introduction 208
§2. Phase Space of Systems of Statistical Mechanics and Gibbs Measures 210
    2.1. The Configuration Space 210
    2.2. Poisson Measures 212
    2.3. The Gibbs Configuration Probability Distribution 212
    2.4. Potential of the Pair Interaction. Existence and Uniqueness of a
         Gibbs Configuration Probability Distribution 214
    2.5. The Phase Space. The Gibbs Probability Distribution 217
    2.6. Gibbs Measures with a General Potential 219
    2.7. The Moment Measure and Moment Function 220
§3. Dynamics of a System of Interacting Particles 222
    3.1. Statement of the Problem 222
    3.2. Construction of the Dynamics and Time Evolution 224
    3.3. Hierarchy of the Bogolyubov Equations 226
§4. Equilibrium Dynamics 227
    4.1. Definition and Construction of Equilibrium Dynamics 227
    4.2. The Gibbs Postulate 229
    4.3. Degenerate Models 231
    4.4. Asymptotic Properties of the Measures P_t 232
§5. Ideal Gas and Related Systems 232
    5.1. The Poisson Superstructure 232
    5.2. Asymptotic Behavior of the Probability Distribution P_t as
         $t \to \infty$ 234
    5.3. The Dynamical System of One-Dimensional Hard Rods 235
§6. Kinetic Equations 236
    6.1. Statement of the Problem 236
    6.2. The Boltzmann Equation 239

6.3. The Vlasov Equation 234
　　6.4. The Landau Equation 244
　　6.5. Hydrodynamic Equations 245
Bibliography.. 247
Chapter 11. Existence and Uniqueness Theorems for the Boltzmann
Equation (*N.B. Maslova*)...................................... 254
§ 1. Formulation of Boundary Problems. Properties of Integral
　　Operators.. 254
　　1.1. The Boltzmann Equation 254
　　1.2. Formulation of Boundary Problems.................... 258
　　1.3. Properties of the Collision Integral 259
§ 2. Linear Stationary Problems 261
　　2.1. Asymptotics 261
　　2.2. Internal Problems 262
　　2.3. External Problems 263
　　2.4. Kramers' Problem.................................. 265
§ 3. Nonlinear Stationary Problems 265
§ 4. Non-Stationary Problems 267
　　4.1. Relaxation in a Homogeneous Gas..................... 267
　　4.2. The Cauchy Problem 268
　　4.3. Boundary Problems 269
§ 5. On a Connection of the Boltzmann Equation with Hydrodynamic
　　Equations.. 270
　　5.1. Statement of the Problem 270
　　5.2. Local Solutions. Reduction to Euler Equations 272
　　5.3. A Global Theorem. Reduction to Navier-Stokes Equations 274
Bibliography.. 276

Chapter 10
Dynamical Systems of Statistical Mechanics

R.L. Dobrushin, Ya.G. Sinai, Yu.M. Sukhov

§ 1. Introduction

　　The motion of a system of N particles in d dimensions is described in Statistical Mechanics by means of a Hamiltonian system of $2Nd$ differential equations, which generates the group of transformations of the phase space. The object of the investigation is the time evolution of probability measures on the phase space determined by this group of transformations. The principal feature of problems in Statistical Mechanics is the fact that one deals with systems consisting of a large number of particles of the same type (a mole of a gas contains $6 \cdot 10^{23}$ particles). Therefore, only those results in which all estimates are uniform with respect to the number of degrees of freedom are of interest here. This restriction, which is unusual from the point of view of the standard theory of dynamical systems, specifies the mathematical feature of the problems of Statistical Mechanics.

Thus, the problem of studying asymptotically the properties of a system, as $N \to \infty$, arises naturally. In our view, the fundamental mathematical approach to this problem is the explicit consideration of infinitely-dimensional dynamical systems arising as the limit, as $N \to \infty$, of the system of the equations of motion of N particles. Unlike infinite systems of equations usually considered in mathematical physics which arise in other domains of application, one is interested here in systems for which all the degrees of freedom are completely "equal in importance".

Traditionally, statistical mechanics is divided into equilibrium and non-equilibrium mechanics. In the equilibrium statistical mechanics one studies properties of a special class of measures invariant with respect to the dynamics which are determined by the well-known Gibbs postulate. This wide theme which is the subject of numerous investigations (cf [S6], [D5], [R2]) remains mainly outside our exposition, although a part of the corresponding theory which is related to applications to dynamical systems with hyperbolic properties has been touched in Sect. 6 of Chap. 3 of Part I and in Chaps. 7, 8 of Part II. In the following section we shall discuss the main facts about Gibbs random fields used in further sections of this chapter.

The mathematical investigation of problems of non-equilibrium statistical mechanics is now in its initial stage. The results obtained so far are fairly odd; the remaining gaps are filled here by mathematically formulated conjectures and sometimes even by "physical" considerations. Among earlier review texts concerning this theme we mention the articles [GO], [AGL2], [DS1], [L3] (some parts of the papers [DS2], [L4] are a kind of review as well). A series of problems related to our theme is discussed in the book [CFS].

We have chosen for detailed discussion several crucial and most elaborated topics. First, there is the problem of the existence of the "infinite-particle" dynamics to which Sect. 3 and a part of Sect. 4 are devoted. Secondly, we discuss, in particular simple cases, ergodic properties of infinite-particle dynamical systems with invariant measures (see Sect. 5). The fundamental question about asymptotical properties of time evolution, as $t \to \pm\infty$, is closely connected with the problem of describing the set of invariant measures. This subject is considered in Sect. 4. The matter of Sects. 4, 5 is immediately related to the problem of mathematical foundation of the Gibbs postulate. In Sect. 6 we discuss some results which concern deriving kinetic equations, i.e., equations describing time evolution of expectation values of the principal physical variables.

The models of dynamics which are under investigation in statistical mechanics are fairly varied. In what follows we consider only one such model. This is the Newton dynamics of a system of point particles moving in Euclidean space subject to internal interaction forces—a model which is well-known from elementary courses in mechanics. As to other types, we only give references to some books and main papers where a more complete bibliography may be found. In the literature the classical "spin" dynamics is often investigated (cf [BPT], [LL], [LLL]). In models of spin dynamics one considers the evolution of coordinates which describe internal degrees of freedom of particles fastened at points of a regular lattice. Among other dynamical models usually studied are the gradient models in which, for the sake of simplicity, one deals with a system of differential equations of first order for the positions of particles (cf [FF], [F3], [L]).

Conceptually, papers on stochastic dynamics (cf [Du], [Gri], (Li]) which form now a wide chapter of probability theory, namely, the theory of Markov processes with local interaction, are close to the range of problems discussed here, as well as papers on dynamics of quantum systems with infinitely many degrees of freedom (cf [BM], [BR1], [BR2]). These themes require special reviews.

The English translation of Chapters 10 and 11 was prepared with the help of Prof. B. Hajek. The authors express to him the deep gratitude.

§2. Phase Space of Systems of Statistical Mechanics and Gibbs Measures

In this section we introduce basic notions of equilibrium statistical mechanics which will be used later: the configuration and phase spaces of dynamical systems of statistical mechanics and probability measures defined on such spaces.

2.1. The Configuration Space. According to tradition, we shall consider systems of particles in d dimensions, where d is an arbitrary integer. Although the case $d = 3$ is, of course, the most interesting, the low dimensions admit a physical interpretation as well ($d = 2$ is the case of a gossamer pellicle or face). The principal reason for such a generalization is that the properties of systems of statistical mechanics depend essentially on the dimensionality and it is interesting to investigate this dependence from both the mathematical and physical points of view.

A *particle configuration* in Euclidean space R^d is defined as a finite or countable subset $\mathbf{q} \subset R^d$. Let

$$v_\mathcal{O}(\mathbf{q}) = |\mathbf{q}_\mathcal{O}|, \qquad \mathcal{O} \subseteq R^d. \tag{10.1}$$

Here and below $\mathbf{q}_\mathcal{O} = \mathbf{q} \cap \mathcal{O}$, and the symbol $|\cdot|$ denotes the cardinality of a given set. One imposes the following condition:

$$v_\mathcal{O}(\mathbf{q}) < \infty \text{ for any bounded } \mathcal{O} \subset R^d \tag{10.2}$$

which means that infinite particle accumulations are inadmissible. The set of all such **q**'s is denoted by Q^0.

In some models of dynamical systems one admits that sometimes several particles may be found at the same point $q \in R^d$. Thereby it is convenient to extend the above definition, and to treat a configuration as a pair $(\mathbf{q}, n_\mathbf{q})$ where $\mathbf{q} \in Q^0$ and $n_\mathbf{q}$ is an integer-valued non-negative function giving the number of particles at points $q \in \mathbf{q}$. The set of all such pairs $(\mathbf{q}, n_\mathbf{q})$ is denoted by Q. Given $\mathcal{O} \subset R^d$ and $(\mathbf{q}, n_\mathbf{q}) \in Q$, we denote by $v_\mathcal{O}(\mathbf{q}, n_\mathbf{q})$ the number of particles in the set $\mathcal{O}$:

$$v_\mathcal{O}(\mathbf{q}, n_\mathbf{q}) = \sum_{q \in \mathbf{q}_\mathcal{O}} n_\mathbf{q}(q) \tag{10.3}$$

By $Q_\mathcal{O}$ one denotes the set of configurations concentrated in $\mathcal{O}$, i.e., such that $v_{\mathcal{O}^c}(\mathbf{q}, n_\mathbf{q}) = 0$ (here and below $\mathcal{O}^c$ denotes the complement of $\mathcal{O}$). In the same way one introduces the set $Q_\mathcal{O}^0$.

The space Q is equipped with natural topology: the convergence of the sequence $(\mathbf{q}_s, n_{\mathbf{q}_s}^{(s)})$, $s = 1, 2, \ldots$, to $(\mathbf{q}, n_\mathbf{q})$, means that for any bounded open $\mathcal{O} \subset R^d$ with $v_{\partial\mathcal{O}}(\mathbf{q}, n_\mathbf{q}) = 0$, where $\partial\mathcal{O}$ is the boundary of the set $\mathcal{O}$, and for any sufficiently large s,

$$v_\mathcal{O}(\mathbf{q}_s, n_{\mathbf{q}_s}^{(s)}) = v_\mathcal{O}(\mathbf{q}, n_\mathbf{q}).$$

We define the σ-algebra $\mathcal{Q}$ of subsets of Q as the smallest one generated by the functions $v_\mathcal{O}$ where $\mathcal{O}$ is an arbitrary bounded Borel subset of R^d. It is not hard to show that $\mathcal{Q}$ is the Borel σ-algebra with respect to the topology introduced above. The configuration space of a particle system in R^d is defined as the measurable space $(Q, \mathcal{Q})$.

The configurations $(\mathbf{q}, n_\mathbf{q})$ with $n_\mathbf{q} \equiv 1$ form an everywhere dense Borel subset of Q which will be identified with Q^0. The subsets of Q^0 which belong to $\mathcal{Q}$ form the σ-algebra which is denoted by $\mathcal{Q}^0$. Almost all probability distributions which appear in the sequel will be concentrated on Q^0, and we shall often determine probability measures on $(Q^0, \mathcal{Q}^0)$ to begin with, without specifying their possible interpretation as measures on $(Q, \mathcal{Q})$.

The action of the group of space translation $\{T_y, y \in R^d\}$ is defined on $(Q, \mathcal{Q})$:

$$T_y(\mathbf{q}, n_\mathbf{q}) = (\mathbf{q} + y, n_{\mathbf{q}+y})$$

where

$$\mathbf{q} + y = \{q \in R^d : q - y \in \mathbf{q}\}, n_{\mathbf{q}+y}(q) = n_\mathbf{q}(q - y), q \in \mathbf{q} + y.$$

A probability measure P on $(Q, \mathcal{Q})$ is called *translation invariant* if $P(T_y A) = P(A)$ for any $A \in \mathcal{Q}$ and $y \in R^d$.

When we speak about the convergence of probability measures on $(Q, \mathcal{Q})$ (and $(Q^0, \mathcal{Q}^0)$) we mean the weak convergence with respect to the above topology.

A probability measure on $(Q, \mathcal{Q})$ describes a "configuration state" of a particle system in R^d (cf [R2]). According to probability theory language, it is interpreted as the probability distribution of a random point field, and many results, useful for our aims may be found in monographs and papers related to the theory of random point processes and fields (see, for instance, [K1], [MKM]).

For a given Borel $\mathcal{O} \subset R^d$ it is possible to introduce the σ-subalgebra $\mathcal{Q}_\mathcal{O} \subset \mathcal{Q}$ generated by the functions $v_{\tilde{\mathcal{O}}}$ where $\tilde{\mathcal{O}}$ is an arbitrary bounded Borel subset of $\mathcal{O}$. In the same way one defines the σ-algebra $\mathcal{Q}_\mathcal{O}^0$. For any measure P on $(Q, \mathcal{Q})$ (or on $(Q^0, \mathcal{Q}^0)$) we denote by $P_\mathcal{O}$ the restriction of P to $\mathcal{Q}_\mathcal{O}$ (or to $\mathcal{Q}_\mathcal{O}^0$). It is possible to identify the measure $P_\mathcal{O}$ with a measure concentrated on $Q_\mathcal{O}$ (or $Q_\mathcal{O}^0$).

2.2. Poisson Measures. The simplest natural example of a probability measure on $(Q^0, \mathcal{Q}^0)$ is the *Poisson measure* P_z^0 with the parameter $z > 0$. It is completely determined by the following two conditions:

1) for any bounded Borel set $\mathcal{O} \subset R^d$, the random variable $v_\mathcal{O}$ has the Poissonian distribution with the average value $zl(\mathcal{O})$ where $l(\mathcal{O})$ is the Lebesgue measure of $\mathcal{O}$, i.e.,

$$P_z^0(\{\mathbf{q} \in Q^0 : v_\mathcal{O}(\mathbf{q}) = k\}) = \frac{(zl(\mathcal{O}))^k}{k!} \exp(-zl(\mathcal{O})) \tag{10.4}$$

2) for any collection of pairwise disjoint Borel sets $\mathcal{O}_1, \ldots, \mathcal{O}_n \subset R^d$, the random variables $v_{\mathcal{O}_1}, \ldots, v_{\mathcal{O}_n}$ are mutually independent.

According to physical terminology, the Poisson measure P_z^0 determines the equilibrium configuration state of the ideal gas. The parameter z determines the density of particles in this state.

We shall consider the "unnormalized" Poisson measure as well (or, as one says sometimes, the Lebesgue-Poisson measure). More precisely, given any bounded Borel $\mathcal{O} \subset R^d$, we introduce the measure $L_\mathcal{O}$ on the σ-algebra $\mathcal{Q}_\mathcal{O}^0$ which is determined by the condition: for any finite collection of pairwise disjoint Borel sets $\mathcal{O}_1, \ldots, \mathcal{O}_n \subseteq \mathcal{O}$ with $\bigcup_{j=1}^n \mathcal{O}_j = \mathcal{O}$ and for all nonnegative integers $k_1, \ldots, k_n$,

$$L_\mathcal{O}(\{\mathbf{q} \in Q^0 : v_{\mathcal{O}_j}(\mathbf{q}) = k_j, j = 1, \ldots, n\}) = \prod_{j=1}^n \frac{(l(\mathcal{O}_j))^{k_j}}{k_j!}. \tag{10.5}$$

It is possible to identify the measure $L_\mathcal{O}$ in the natural way with a measure concentrated on the set $Q_\mathcal{O}^0$. We shall often use such a mode in the sequel without specifying it every time again. The Poisson measure P_z^0 is characterized by the fact that the restriction $(P_z^0)_\mathcal{O}$ is absolutely continuous with respect to the measure $L_\mathcal{O}$ and

$$\frac{d(P_z^0)_\mathcal{O}}{dL_\mathcal{O}}(\mathbf{q}) = z^{|\mathbf{q}|} \exp(-zl(\mathcal{O})), \mathbf{q} \in Q_\mathcal{O}^0. \tag{10.6}$$

2.3. The Gibbs Configuration Probability Distribution. In this section we shall introduce the definition of a configurational Gibbs distribution. This definition is a natural generalization of the well-known definition of the grand canonical ensemble in statistical mechanics to the case of an infinite particle system. The fundamental postulate of statistical mechanics is the possibility of using this ensemble for describing equilibrium states of a particle system (see Sect. 4 for a more detailed account).

First of all we introduce a Gibbs distribution in the simplest case of a pairwise interaction which is invariant with respect to the Euclidean group of motion, and then indicate some possible generalizations. Interaction between particles is described by the pair *interaction potential* which, in the case under consideration, is a fixed measurable function $U: [0, \infty) \to R^1 \cup \{\infty\}$. The value $U(r), 0 \leq r < \infty$, is interpreted as the potential energy of a pair of point particles at the distance r. We shall suppose as well that the numbers $z > 0$ and $\beta > 0$ are fixed. In statistical mechanics the parameter z is called the *activity* (*fugacity*) of the system and the parameter β is inversely proportional to the absolute *temperature*. The so-called chemical potential $\mu = \beta^{-1} \ln z$ is sometimes used instead of the activity.

For any bounded Borel $\mathcal{O} \subset R^d$ one defines the potential energy of a configuration $\mathbf{q} \in Q_\mathcal{O}^0$ by the equality[1]

$$V(\mathbf{q}) = \frac{1}{2} \sum_{\substack{q, q' \in \mathbf{q}: \\ q \neq q'}} U(\|q - q'\|). \tag{10.7}$$

The Gibbs configuration distribution in a volume $\mathcal{O}$ with the free boundary condition, interaction potential U and parameters (z, β), is the probability measure on $Q_\mathcal{O}^0$ which is determined by the density with respect to the measure $L_\mathcal{O}$ (regarded as a measure on $Q_\mathcal{O}^0$) given by

$$\Xi_\mathcal{O}^{-1} z^{|\mathbf{q}|} \exp(-\beta V(\mathbf{q})), \qquad \mathbf{q} \in Q_\mathcal{O}^0, \tag{10.8}$$

where $\Xi_\mathcal{O}$ is the normalization factor which is called the *partition function*

$$\Xi_\mathcal{O} = \int_{Q_\mathcal{O}^0} L_\mathcal{O}(d\mathbf{q}) z^{|\mathbf{q}|} \exp(-\beta V(\mathbf{q})). \tag{10.9}$$

For the given definition to be correct, one needs to assume, of course, that $\Xi_\mathcal{O} < \infty$.

A wider class of probability distributions on $Q_\mathcal{O}^0$ is obtained by introducing a *boundary condition* $\bar{\mathbf{q}} \in Q_{\mathcal{O}^c}^0$. We set

$$V(\mathbf{q}|\bar{\mathbf{q}}) = V(\mathbf{q}) + \sum_{q \in \mathbf{q}, q' \in \bar{\mathbf{q}}} U(\|q - q'\|), \mathbf{q} \in Q_\mathcal{O}^0. \tag{10.10}$$

The probability measure on $Q_\mathcal{O}^0$ determined by the density with respect to $L_\mathcal{O}$

$$\Xi_\mathcal{O}(\bar{\mathbf{q}})^{-1} z^{|\mathbf{q}|} \exp(-\beta V(\mathbf{q}|\bar{\mathbf{q}})), \mathbf{q} \in Q_\mathcal{O}^0, \tag{10.11}$$

where

$$\Xi_\mathcal{O}(\bar{\mathbf{q}}) = \int_{Q_\mathcal{O}^0} L_\mathcal{O}(d\mathbf{q}) z^{|\mathbf{q}|} \exp(-\beta V(\mathbf{q}|\bar{\mathbf{q}})), \tag{10.12}$$

---

[1] Here and below one assumes that for any $a \in R^1$ $a + \infty = \infty$, for any $a > 0$ $a\infty = \infty$, and $\exp(-\infty) = 0$.

will be called the Gibbs configuration distribution in the volume $\mathcal{O}$ with the boundary condition **q**, interaction potential U and parameters (z, β).

For the existence of this distribution, it is sufficient to suppose that the series in the right-hand side of (10.10) is absolutely convergent for almost every $\mathbf{q} \in Q_\mathcal{O}^0$ and the integral $\Xi_\mathcal{O}(\overline{\mathbf{q}})$ is finite.

Definition 2.1. A probability measure P on $(\mathbf{Q}^0, \mathcal{Q}^0)$ is called a Gibbs configuration distribution in "infinite volume", or shortly: a Gibbs configuration distribution, with the interaction potential U and parameters (z, β) if for any bounded Borel $\mathcal{O} \subset R^d$ the following conditions are valid:

1) for P-almost all $\overline{\mathbf{q}} \in Q^0$ there exists the Gibbs configuration distribution in the volume $\mathcal{O}$ with the condition $\mathbf{q}_{\mathcal{O}^c}$, potential U and parameters (z, β),

2) for P-almost all $\overline{\mathbf{q}} \in Q^0$ the restriction onto $\mathcal{Q}_\mathcal{O}$ of the conditional probability distribution $P(\cdot | \mathcal{Q}_{\mathcal{O}^c})$ with respect to the σ-algebra $\mathcal{Q}_{\mathcal{O}^c}$ coincides with the Gibbs configuration distribution in the volume $\mathcal{O}$ with the boundary condition $\overline{\mathbf{q}}_{\mathcal{O}^c}$, potential U and parameters (z, β).

Let us briefly explain the meaning of this definition. As already mentioned, this is a generalization of the definition (10.8) to the case of the infinite particle system. It is impossible to extend the definition (10.8) to this case directly since the total energy of the infinite particle system is infinite. However, it is not difficult to calculate from (10.8) that, given $\tilde{\mathcal{O}} \subset \mathcal{O}$ and a configuration $\mathbf{q}_{\mathcal{O} \setminus \tilde{\mathcal{O}}}$ in the complement $\mathcal{O} \setminus \tilde{\mathcal{O}}$, the conditional probability density of the distribution of a configuration $\mathbf{q}_{\tilde{\mathcal{O}}} \in Q_{\tilde{\mathcal{O}}}^0$ in the volume $\tilde{\mathcal{O}}$ has the form (10.12). So, property 2) of the configurational Gibbs distribution is an analog of such a property of the usual Gibbs density (10.8). As to property 1), this is of technical character and is necessary for the correct formulation of 2).

One of the principal arguments for naturality of the given definition is that the measure P may be obtained as a limit of Gibbs configuration distributions in the volume $\mathcal{T}$ with (in general, random) boundary conditions $\mathbf{q}_{\mathcal{T}^c}$, as $\mathcal{T} \nearrow R^d$ [2]. Therefore, the term "a limit Gibbs configuration distribution" is sometimes used. The idea of such a construction arises in the work [BK] which has become classical (see also [BPK], [M1], [M2], [R1]).

Conditions 1) and 2) figuring in Definition 2.1 are often called, in literature, the DLR (Dobrushin, Lanford, Ruelle) *conditions* (cf [D2]–[D5], [LR], [R3]).

2.4. Potential of the Pair Interaction. Existence and Uniqueness of a Gibbs Configuration Probability Distribution. We shall now discuss the conditions on the interaction potential imposed in studying the configurational Gibbs distributions as well as for investigation of dynamical systems of statistical mechanics, which will be given in following sections. The simplest examples of the interaction potentials are the potential of ideal gas

---

[2] Here and below $\mathcal{T} \nearrow R^d$ denotes the inclusion-directed set of the d-dimensional cubes $\mathcal{T} = [-a, a]^d, d > 0$.

Chapter 10. Dynamical Systems of Statistical Mechanics

$$U(r) \equiv 0, \qquad r \geq 0, \qquad (10.13)$$

and that of the gas of hard (or absolutely elastic) spheres (hard rods for $d = 1$)

$$U(r) = \begin{cases} \infty, & 0 \leq r < r_0, \\ 0, & r \geq r_0, \end{cases} \qquad (10.14)$$

where r_0 is the diameter of a sphere (the length of a rod). For these cases the configurational Gibbs distribution depends on one parameter z only; for the case of ideal gas it coincides with the Poisson measure P_z^0.

A typical example of a particle interaction potential which is used in physical computations (for $d = 3$) is the Lenard-Jones potential

$$U(r) = \frac{a_1}{r^{12}} - \frac{a_2}{r^6} \qquad (10.15)$$

where $a_1, a_2 > 0$. The interaction between atoms of noble gases is described well enough by means of this potential.

The interactions arising in various physical situations are of very different character and hence the results which hold under minimal assumptions on the interaction potential are of principal interest. However, different problems require different classes of interaction potentials and the corresponding formulations in a precise form are often cumbersome. Moreover, the assumptions which one introduces thereby, do not seem to be the final ones. Therefore, we shall not formulate the precise conditions under which a given result may be proved and shall restrict ourselves to giving qualitative descriptions and referring to the original papers [BPK], [BK], [D4]–[D6], [M1], [M2], [Su2], [GMS], [R1] [R3]. Let us list the main types of restrictions which are usually imposed in the literature:

I) a condition of boundedness from below for the interaction potential:

$$\inf_{r \geq 0} U(r) > -\infty.$$

This condition is physically natural and we shall suppose in the sequel that it is fulfilled.

II) conditions of increase as $r \to 0$. A typical example is

$$U(r) \geq cr^{-\gamma} \text{ for sufficiently small } r > 0 \qquad (10.16)$$

where $c > 0$. Usually one supposes that $\gamma > d$. One considers often the potential with a hard core for which

$$U(r) \begin{cases} = \infty, & 0 \leq r < r_0, \\ < \infty, & r > r_0, \end{cases} \qquad (10.17)$$

where $r_0 > 0$ is a diameter of the core (cf (10.14)). The advantage of the last condition is that it simplifies essentially several constructions. Sometimes totally bounded interaction potentials are considered, but in these cases one usually assumes that the values $U(r)$ for r near 0 are positive and the positive part of interaction dominates over the negative one.

The physical meaning of such conditions is that they correspond to a sufficiently strong repulsion of particles at short distances which is necessary for preventing the collapse of particles which should be excluded in an equilibrium situation.

III) conditions of decrease as $r \to \infty$. A typical example of such a condition is

$$|U(r)| \leqslant cr^{-\gamma} \text{ for sufficiently large } r \qquad (10.18)$$

where $c > 0$. As above it is supposed that $\gamma > d$. For simplicity, one sometimes introduces the condition of *finite range* for the potential U:

$$U(r) = 0 \qquad \text{for } r > r_1. \qquad (10.19)$$

The physical meaning of the conditions of III) is connected with an assumption about the locality of the interaction, i.e. its weakening at distances much longer than the average distance between particles.

IV) conditions of smoothness. It is supposed that the interaction potential has a certain number of continuous derivatives in the domain where it has finite values. One often introduces in addition some restrictions on the behaviour of these derivatives as $r \to 0$ ($r \to r_0$ in the case of potential with hard core) and $r \to \infty$.

These last conditions usually arise in the proof of existence of a time dynamics (cf Sections 3, 4).

We shall now discuss the question about the existence and uniqueness of a Gibbs configuration distribution. For simplicity assume that the potential U has a hard core of diameter $r_0 > 0$ and that the condition (10.19) with $\gamma > d$ is satisfied. For extending these results to other classes of interaction potentials see the papers cited above.

The theorem of existence of a configurational distribution is formulated as follows.

Theorem 2.1 (cf [D4]). *Given $z > 0$ and $\beta > 0$, there exists at least one translation invariant Gibbs configuration distribution with the interaction potential U and parameters (z, β). The set of Gibbs configuration distributions $\mathfrak{P}_{U,z,\beta}$ with the potential U and parameters (z, β) forms a convex compact set in the space of probability measures on $(Q, \mathcal{Q})$ (and hence coincides with the closure of the convex envelope of the set of its extreme points).*

The formulation of the uniqueness theorem is different for the multi-dimensional ($d \geqslant 2$) and one-dimensional ($d = 1$) case. Let $\mathcal{T}(y, b)$ denote the d-dimensional cube centered at a point $y \in R^d$ with the edges parallel to the coordinate axes and with the edge length $2b$.

Theorem 2.2 (cf [D4], [M1], [M2], [R1]). *Assume that $d \geqslant 1$. Then for any $\beta > 0$ one can find the value $z_0 = z_0(\beta) > 0$ such that for all $z \in (0, z_0)$ there exists* only one Gibbs configuration distribution with the potential U and parameters (z, β) (i.e., the set $\mathfrak{P}_{U,z,\beta}$ consists of one point). This distribution P is translation invariant and has the following mixing property:

$$\sup_{y \in R^d} \sup_{\substack{A_1 \in \mathscr{Q}_{\mathscr{T}(y,u)}, \\ A_2 \in \mathscr{Q}_{\mathscr{T}(y,u+s)^c}}} |P(A_1 \cap A_2) - P(A_1)P(A_2)| \leq c_1 u^{d-1} s^{-\gamma'}, \quad u, s > 0, \quad (10.20)$$

with constants $\gamma' \geq \gamma - d$ and $c_1 > 0$.

Theorem 2.3 (cf [D4], [GMS], [Su2]). *Assume that $d = 1$ and that the potential U has property (10.18) with $\gamma > 2$. Then for any $z > 0$ and $\beta > 0$ there exists only one Gibbs configuration distribution with the potential U and parameters (z, β). This distribution is translation invariant and has the Rosenblatt mixing property:*

$$\sup_{u \in R^1} \sup_{\substack{A_1 \in \mathscr{Q}_{(-\infty,u)}, \\ A_2 \in \mathscr{Q}_{(u+s, \infty)}}} |P(A_1 \cap A_2) - P(A_1)P(A_2)| \leq c_1 s^{-(\gamma-2)}, \quad s > 0, \quad (10.21)$$

where $c_1 > 0$ is a constant.

Let us discuss the condition $z \in (0, z_0)$ figuring in the formulation of Theorem 2.2. From the physical point of view, it means that we consider particle systems with low density. According to a wide-spread conjecture, for large values z, i.e. for high densities of particles, phase transitions may occur; this can be manifested, in particular, by a nonuniqueness of the Gibbs configuration distribution. On the other hand, in Theorem 2.3 there is no restriction on the value of the parameter z: this is explained by the fact that, according to physical pictures, there is no phase transitions in one-dimensional systems (under wide conditions on the interaction potential).

2.5. The Phase Space. The Gibbs Probability Distribution. To construct dynamical systems of statistical mechanics it is necessary to introduce a more detailed description of a particle system where one considers both particle positions q and momenta p (in the following we assume that the mass of every particle is equal to 1; this fact allows us to identify the momentum and the velocity of a particle). The situation where two particles have the same positions and momenta usually does not arise and we shall restrict ourselves to considering the phase space which is analogous to the configuration space Q^0.

Let M denote the collection of all finite or countable subsets $\mathbf{x} \subset R^d \times R^d$ satisfying the condition

$$v_{\mathcal{O} \times R^d}(\mathbf{x}) < \infty \text{ for any bounded } \mathcal{O} \subset R^d. \quad (10.22)$$

Here, as above,

$$v_{\mathscr{D}}(\mathbf{x}) = |\mathbf{x}_{\mathscr{D}}|, \quad \mathscr{D} \subset R^d \times R^d, \quad (10.23)$$

and $\mathbf{x}_{\mathscr{D}} = \mathbf{x} \cap \mathscr{D}$. Let us denote by $M_{\mathcal{O}}$, $\mathcal{O} \subset R^d$, the collection of those $\mathbf{x} \in M$ for which $v_{\mathcal{O}^c \times R^d}(\mathbf{x}) = 0$.

The space M is endowed with natural topology: the convergence of the sequence $\mathbf{x}_s, s = 1, 2, \ldots$, to $\mathbf{x}$ means that for any bounded open $\mathcal{O} \subset R^d$, any open $C \subseteq R^d$ such that $v_{\partial(\mathcal{O} \times C)}(\mathbf{x}) = 0$ and for all sufficiently large s,

$$v_{\mathcal{O} \times C}(\mathbf{x}_s) = v_{\mathcal{O} \times C}(\mathbf{x}).$$

We shall define the σ-algebra $\mathcal{M}$ of subsets of M as the σ-algebra generated by the functions $v_{\mathcal{O} \times C}$ where $\mathcal{O}$ is an arbitrary bounded Borel subset and C is an arbitrary Borel subset of R^d. It is easy to show that $\mathcal{M}$ is the Borel σ-algebra with respect to the topology on $\mathcal{M}$ introduced above. The measurable space $(M, \mathcal{M})$ is called the phase space of a particle system in R^d.

On the space $(M, \mathcal{M})$, as well as on $(Q, \mathcal{Q})$, the action of the group of space translations $\{T_y, y \in R^d\}$ is given:

$$T_y \mathbf{x} = \{(q, p) \in R^d \times R^d : (q - y, p) \in \mathbf{x}\}.$$

A probability measure P on $(M, \mathcal{M})$ is called translation invariant if for any $A \in \mathcal{M}$ and $y \in R^d$ $P(T_y A) = P(A)$. Convergence of probability measures on $(M, \mathcal{M})$ will be weak convergence with respect to the mentioned topology.

According to probability theory language, the measure on the space $(M, \mathcal{M})$ is interpreted as the probability distribution of a random marked point field (cf [MKM]).

For a given Borel $\mathcal{O} \subset R^d$ it is possible to introduce as above the σ-subalgebra $\mathcal{M}_{\mathcal{O}} \subset \mathcal{M}$ generated by the functions $v_{\tilde{\mathcal{O}} \times C}$, where $\tilde{\mathcal{O}}$ is an arbitrary bounded Borel subset of $\mathcal{O}$ and C is an arbitrary Borel subset of R^d. For any measure P on $(M, \mathcal{M})$, denote by $P_{\mathcal{O}}$ the restriction of P to $\mathcal{M}_{\mathcal{O}}$. It is possible to identify the measure $P_{\mathcal{O}}$ in a natural way with a measure concentrated on $\mathcal{M}_{\mathcal{O}}$.

Consider a measurable map $\Pi: \mathbf{x} \in M \mapsto (\mathbf{q}, n_\mathbf{q}) \in Q$ obtained by omitting the particle momenta in $\mathbf{x}$:

$$\mathbf{q} = \{q \in R^d : v_{\{q\} \times R^d}(\mathbf{x}) \geq 1\}, \quad n_\mathbf{q}(q) = |\{p \in R^d : (q, p) \in \mathbf{x}\}|. \quad (10.24)$$

This map allows us to associate with any probability distribution P on $(M, \mathcal{M})$ its "projection" ΠP which is a probability measure on $(Q, \mathcal{Q})$.

This mapping provides a natural method to construct a probability measure P on $(M, \mathcal{M})$ having a given projection ΠP. Namely, for constructing the measure P it is sufficient to determine, in addition to ΠP, the conditional probability distribution $P(\cdot | \mathcal{M}^Q)$ with respect to the σ-algebra $\mathcal{M}^Q \subset \mathcal{M}$ induced by the map Π. From the physical point of view, given a particle configuration, we consider the conditional momenta distribution. Assuming for simplicity that the measure ΠP is concentrated on Q^0, we may introduce this conditional probability distribution by means of the family of measures $P_\mathbf{q}$ on $(M, \mathcal{M})$ depending on $\mathbf{q} \in Q^0$. In addition, since the set of points $\mathbf{x} \in M$ with given $\mathbf{q} = \Pi \mathbf{x}$ is identified, in a natural way, with the space $R^\mathbf{q}$, we may interpret $P_\mathbf{q}$ as a probability measure on $R^\mathbf{q}$.

Fix a pair interaction potential U, numbers $z > 0$, $\beta > 0$ and a vector $p_0 \in R^d$.

Definition 2.2. A probability measure P on $(M, \mathcal{M})$ is called a Gibbs distribution with the interaction potential U and parameters (z, β, p_0) if

1) its projection ΠP is a configurational Gibbs distribution with the potential U and parameters z, β,

2) the conditional probability measures $P_\mathbf{q}$ correspond to the joint distribution of independent d-dimensional Gaussian vectors with the expectation value p_0 and covariance matrix $\beta^{-1} I$.

The conditional probability measure for particle momenta figuring in condition 2) is called the Maxwell distribution. In an analogous way, one introduces the notion of the Gibbs distribution in a bounded volume $\mathcal{O} \subset R^d$ with an extra condition $\mathbf{x} \in M_{\mathcal{O}^c}$.

Theorems of existence and uniqueness of a Gibbs distribution may be easily obtained from the corresponding theorems for the configurational Gibbs distribution contained in Sect. 2.3.

Sometimes it is useful to consider a more general class of measures which arises when the Gaussian probability distribution in the condition 2) is replaced by an arbitrary probability distribution σ on R^d. When $U \equiv 0$, i.e., the configurational Gibbs distribution is the Poisson measure on $(Q^0, \mathcal{Q}^0)$, the corresponding probability measure on $(M, \mathcal{M})$ is denoted by $P^0_{z,\sigma}$. We shall call the measure $P^0_{z,\sigma}$ the Poisson measure on $(M, \mathcal{M})$ with the parameters (z, σ). In the one-dimensional case we shall use this construction in Sect. 5 for the ideal hard rod potential (cf (10.14)). The corresponding measure on $(M, \mathcal{M})$ is denoted by $P^{ro}_{z,\sigma}$.

2.6. Gibbs Measures with a General Potential. In this section we shall discuss the notion of a Gibbs measure corresponding to a general "potential" which depends on positions and momenta of arbitrary finite collections of particles. Such a generalization is meaningful, in particular, because the probability measures arising thereby may be distinguished in the class of all measures on $(M, \mathcal{M})$ by means of certain conditions which are connected with the decay of "correlations" in some natural sense (cf [Ko]). Hence, the limits of the mathematical applications of Gibbs distributions are much wider than those dictated by "physical traditions" and some of the results formulated below which hold true for a large class of such "generalized" Gibbs distributions may be considered as rather universal ones.

Now we shall call a potential a sequence $\Phi = (\Phi^{(1)}, \Phi^{(2)}, \ldots)$, where $\Phi^{(n)}$ is a symmetric function of variables $x_1, \ldots, x_n$ on the set

$$\{(x_1, \ldots, x_n) \in (R^d \times R^d)^n : x_{j_1} \neq x_{j_2}, 1 \leq j_1 < j_2 \leq n\} \quad (10.25)$$

with values in $R^1 \cup \{\infty\}$ where $x_j = (q_j, p_j) \in R^d \times R^d$, $j = 1, \ldots, n$. The function $\Phi^{(n)}$ is called the n-particle potential (we reserve the previous term "pair potential" for $n = 2$). Sometimes it will be convenient to consider, as the argument of the function $\Phi^{(n)}$, the point $\mathbf{x} \in M$ with $|\mathbf{x}| = n$.

For any bounded Borel $\mathcal{O} \subset R^d$ and any $\mathbf{x} \in M_\mathcal{O}$ and $\bar{\mathbf{x}} \in M_{\mathcal{O}^c}$ denote

$$h^{(\Phi)}(\mathbf{x}|\bar{\mathbf{x}}) = \sum_{n \geq 1} \sum_{\substack{\mathbf{x}' \subseteq \mathbf{x} \cup \bar{\mathbf{x}}: |\mathbf{x}'|=n, \\ \mathbf{x}' \cap \mathbf{x} \neq \emptyset}} \Phi^{(n)}(\mathbf{x}'). \tag{10.26}$$

The definition of a Gibbs probability distribution on $(M, \mathcal{M})$ corresponding to the potential Φ is analogous to the definition of the configurational Gibbs distribution given in Sect. 2.3. The main difference is that one considers σ-algebras $\mathcal{M}_\mathcal{O}$ and $\mathcal{M}_{\mathcal{O}^c}$ instead of $\mathcal{Q}_\mathcal{O}$ and $\mathcal{Q}_{\mathcal{O}^c}$, respectively, and the underlying measure $L_\mathcal{O}$ on $\mathcal{Q}_\mathcal{O}$ is replaced by the measure $\bar{L}_\mathcal{O}$ on the σ-algebra $\mathcal{M}_\mathcal{O}$. The measure $\bar{L}_\mathcal{O}$ is determined by the following conditions: 1) projection $\Pi \bar{L}_\mathcal{O}$ coincides with $L_\mathcal{O}$ 2) the conditional measure with respect to the σ-algebra $\mathcal{M}^{\mathcal{Q}}$ generated by $\bar{L}_\mathcal{O}$ is the product of $|\mathbf{q}|$ copies of the Lebesgue measure l. The density (with respect to the measure $\bar{L}_\mathcal{O}$) of the restriction to $\mathcal{M}_\mathcal{O}$ of the conditional probability distribution $P(\cdot|\mathcal{M}_{\mathcal{O}^c})$ is

$$\Xi_\mathcal{O}(\bar{\mathbf{x}})^{-1} \exp(-h^\Phi(\mathbf{x}|\bar{\mathbf{x}})), \qquad \mathbf{x} \in M_\mathcal{O}, \tag{10.27}$$

where

$$\Xi_\mathcal{O}(\mathbf{x}) = \int_{M_\mathcal{O}} \bar{L}_\mathcal{O}(d\mathbf{x}) \exp(-h^{(\Phi)}(\mathbf{x}|\bar{\mathbf{x}})). \tag{10.28}$$

In order to prove the existence (and uniqueness) of a Gibbs distribution corresponding to a "general" potential $\Phi = (\Phi^{(1)}, \Phi^{(2)}, \ldots)$, one should impose on Φ some conditions. We shall not give here their explicit formulations since they are cumbersome; these conditions are discussed briefly in one of the possible cases in Sect. 4.2 below.

Let us comment on the connection between the general definition and Definition 2.2 given in Sect. 2.5. A Gibbs distribution with the potential U and parameters (z, β, p_0) in the context of the general definition, corresponds to the potential $\Phi = \Phi_{U;z,\beta,p_0}$ with $\Phi^{(n)} \equiv 0$ for $n \geq 3$ and

$$\Phi^{(1)}(x) = \gamma + \frac{\beta}{2} \|p - p_0\|^2, \qquad x = (q, p), \tag{10.29a}$$

$$\Phi^{(2)}(x_1, x_2) = \beta U(\|q_1 - q_2\|), \qquad x_j = (q_j, p_j), j = 1, 2, \tag{10.29b}$$

where $\gamma = -\ln z - d/2 \ln(2\pi\beta)$. The corresponding function h^Φ is of the form

$$h^\Phi(\mathbf{x}|\bar{\mathbf{x}}) = \gamma|\mathbf{x}| + \beta/2 \sum_{(q,p) \in \mathbf{x}} \|p - p_0\|^2 + \beta V(\Pi\mathbf{x}|\Pi\bar{\mathbf{x}}). \tag{10.30}$$

2.7. The Moment Measure and Moment Function. It is often convenient to determine the probability distributions on $(M, \mathcal{M})$ and $(Q, \mathcal{Q})$ by means of the

---

[3] In mathematical and physical literature one also frequently uses for these objects the terms "correlation measure", "correlation function" and "distribution function".

so-called *moment measure* or *moment function*[3] (for details, see [B1], [Le1], [Le2], [Z2]). This method is similar to the well-known method of describing random processes and fields by means of joint moments. For the sake of brevity we shall restrict ourselves by considering probability measures on $(M, \mathcal{M})$.

Let a probability distribution P on $(M, \mathcal{M})$ be given. For any $n = 1, 2, \ldots$, we define the n-moment measure $K_P^{(n)}$ on the σ-algebra of those Borel subsets of the set (10.25) which are invariant under the permutations of particles x_j, $j = 1, \ldots, n$, by the equality

$$K_P^{(n)}(A) = \int_M P(d\mathbf{x}) \sum_{\mathbf{x}' \subset \mathbf{x}: |\mathbf{x}'| = n} \chi_A(\mathbf{x}') \tag{10.31}$$

(as usual, χ_A denotes the indicator function of a set A). The density of the measure $K_P^{(n)}$ with respect to the Lebesgue measure $\prod_{j=1}^n dq_j \times dq_j$ (whenever it exists) is called the n-moment function and is denoted by $k_P^{(n)}$. This is a symmetric function of the variables $x_j = (q_j, p_j) \in R^d \times R^d$, $j = 1, \ldots, n$. Intuitively, $k_P^{(n)}((q_1, p_1), \ldots, (q_n, p_n)) \prod_{j=1}^n dq_j \times dp_j$ is interpreted as the probability that in the small volumes dq_j centered at the points q_j there are particles with momenta belonging to the small volumes dp_j centered at the points p_j, $j = 1, \ldots, n$. The sequence $K_P = (K_P^{(1)}, K_P^{(2)}, \ldots)$ is called the moment measure and the sequence $k_P = (k_P^{(1)}, k_P^{(2)}, \ldots)$ the moment function of the probability distribution.

For the Poisson probability distribution $P_{z,\sigma}^0$ (cf Sect. 2.5) the n-moment measure is given by the formula

$$K_{P_{z,\sigma}^0}^{(n)}\left(\prod_{j=1}^n dq_j \times dp_j\right) = Z^n \prod_{j=1}^n dq_j \sigma(dp_j). \tag{10.32}$$

The necessary and sufficient condition for the one-to-one correspondence between a probability distribution P and the moment measure K_P is given by the divergence of the series

$$\sum_{n \geq 0} [K_P^{(n)}((\mathcal{O} \times R^d)^n) 1/n!]^{-1/n} \tag{10.33}$$

for any bounded Borel $\mathcal{O} \subset R^d$. This result is a consequence of well-known results connected with the so-called moment problem (cf [Le1], [Z2]). It is possible to show that condition (10.33) holds for a wide class of general Gibbs distributions.

A specific role is played in the following by the 1-moment measure $K_P^{(1)}$ and 1-moment function $k_P^{(1)}$. Notice, in connection with this, that the measure $K_P^{(1)}$ gives the average values of random variables $v_{\mathscr{D}}$ with respect to the probability distribution P

$$K_P^{(1)}(\mathscr{D}) = \int_M P(d\mathbf{x}) v_{\mathscr{D}}(\mathbf{x}). \tag{10.34}$$

Thus, the function $k_P^{(1)}$ may be considered as the density of the particle distribution in the one-particle phase space $R^d \times R^d$.

When speaking of the convergence of a sequence of measures $K_s^{(1)}$ on $R^d \times R^d$, $s = 1, 2, \ldots$, to a measure $K^{(1)}$ we shall have in mind the fact that for any bounded continuous function $f: R^d \times R^d \to R^1$ with support in $\mathcal{O} \times R^d$ where $\mathcal{O}$ is a bounded subset of R^d, the integral $\int K_s^{(1)}(dq \times dp) f(q, p)$ converges, as $s \to \infty$, to $\int K^{(1)}(dq \times dp) f(q, p)$.

§3. Dynamics of a System of Interacting Particles

3.1. Statement of the Problem. We shall suppose that the particle interaction is described by a pair potential U depending on the distance between the particles only (cf Sects. 2.3, 2.4). The classical equations of motion for a finite system of identical particles of mass one, may be written in the form

$$\dot{q}_i(t) = p_i(t), \quad \dot{p}_i(t) = -\sum_{j: j \neq i} \operatorname{grad} U(\|q_j(t) - q_i(t)\|). \tag{10.35}$$

Here i and j run over a set of indices which label the particles, $q_i(t) \in R^d$ is the position vector, and $p_i(t) \in R^d$ the momentum vector of the i-th particle at time $t \in R^1$. The system of equations (10.35) may be written in the Hamiltonian form with the Hamiltonian

$$H = \sum_i \frac{1}{2} \|p_i\|^2 + \sum_{j \neq i} U(\|q_j - q_i\|). \tag{10.36}$$

We impose on U the conditions which were discussed in Sect. 2.4. Provided $U(\|q\|)$ is a smooth function of the variable $q \in R^d$, a solution of the Cauchy problem for system (10.35) exists and is unique on the whole time axis for all initial conditions $\{(q_i(0), p_i(0))\}$. If the potential U has a hard core of diameter r_0 and $\lim_{r \to r_0} U(r) = \infty$, then this assertion holds for the initial conditions with $\min_{i_1 \neq i_2} \|q_{i_1}(0) - q_{i_2}(0)\| > r_0$. If, otherwise, $\lim_{r \to r_0+} U(r) < \infty$, then one has to complete the system (10.35) with boundary conditions corresponding to collisions of particles when $\|q_{i_1}(t) - q_{i_2}(t)\| = r_0$ for some pair of indices $i_1 \neq i_2$. These boundary conditions usually correspond to the elastic collisions (compare with analogous conditions for the systems of the billiard type in Chap. 8, Sect. 1). The multiple collisions and other degenerations are neglected since they occur for subsets of initial data of Lebesgue measure zero.

Sometimes it is convenient to consider the motion of particles in a bounded domain $\mathcal{O} \subset R^d$ with a smooth or piecewise smooth boundary $\partial \mathcal{O}$. The motion equations are completed by boundary conditions for $q_i(t) \in \partial \mathcal{O}$. Usually one introduces the conditions of the elastic reflection of particles from the boundary $\partial \mathcal{O}$ (again compare with the billiard systems from Chap. 8, Sect. 1).

Formally, it is possible to write down equations (10.35) for an infinite particle system as well. However, the Hamiltonian (10.36) for this case will be infinite and therefore, some specific constructions are necessary. A family of smooth functions $\{(q_i(t), p_i(t)), t \in I\}$ for which the equations (10.35) are fulfilled for any i and $t \in I$

(and, in particular, the series in the right hand side of (10.35) converges absolutely) is called a solution of the infinite system (10.35) in the time interval $I \subseteq R^1$.

Provided the solution has the following property: for any bounded $\mathcal{O} \subset R^d$ and any $t \in I$ the number of indices i for which $q_i(t) \in \mathcal{O}$ is finite, it may be interpreted as a trajectory $\mathbf{x}(t)$ in the space M. These definitions are extended in a natural way to the case of motion with the elastic collisions of particles.

In the simplest case when $U \equiv 0$ (ideal gas) the trajectory $\mathbf{x}(t)$ with initial date $\mathbf{x}(0) = \mathbf{x}$ may be written in the explicit form

$$\mathbf{x}(t) = \{(q, p): (q - tp, p) \in \mathbf{x}\}. \tag{10.37}$$

However, already in this case some difficulties arise which become much more serious for a system with interaction. It is easy to indicate initial $\mathbf{x} \in M$ for which infinitely many particles collapse into a bounded domain $\mathcal{O} \subset R^d$ at a finite time t (because the velocities of distant particles may be large enough and "directed into the domain $\mathcal{O}$") and $\mathbf{x}(t)$ comes out of the space M. Therefore, the trajectory $\mathbf{x}(t)$ cannot be defined for arbitrary $\mathbf{x} \in M$.

For interacting particle systems one can construct, with the help of analogous arguments, examples of initial data $\mathbf{x} \in M$ for which solution (10.35) does not exist or, on the contrary, the trajectory belonging to M exists but it not unique. All these facts lead to the problem of finding a sufficiently "massive" measurable subset $\hat{M} \subset M$ such that for any point $\mathbf{x} \in \hat{M}$ there exists a unique trajectory $\mathbf{x}(t) \in \hat{M}$ on the whole axis R^1. Having such a subset, we define the one-parameter group of measurable transformations $S_t: \hat{M} \to \hat{M}$ by the formula $S_t \mathbf{x} = \mathbf{x}(t)$.

Definition 3.1. *The pair* $(\hat{M}, \{S_t, t \in R^1\})$ *is called the dynamics determined by the interaction potential U on the set $\hat{M}$.*

The construction of dynamics $(\hat{M}, \{S_t\})$ is the principal theme of this section. This construction will be meaningful, provided the set $\hat{M}$ is sufficiently massive, in the sense that the class of probability measures concentrated on $\hat{M}$ is large enough. If a measure P is of this class, we can introduce the family of measure

$$P_t(A) = P(S_{-t}(A \cap \hat{M})), \quad A \in \mathcal{M}, \quad t \in R^1. \tag{10.38}$$

Definition 3.2. The family of probability measures $\{P_t, t \in R^1\}$ given by equality (10.38) is called *time evolution of the initial measure* $P_0 = P$ generated by the dynamics $(\hat{M}, \{S_t\})$.

It is natural to claim that the class of measures for which the time evolution may be defined includes Gibbs distributions under certain restrictions on the potential (cf Sects. 2.3–2.6). In particular, if a Gibbs distribution with the interaction potential U and parameters (z, β, p_0) is concentrated on $\hat{M}$ and is not changed under the space translations in the direction of the vector p_0, then it is invariant with respect to transformations S_t (cf Sect. 4.1).

Observe that, unlike the finite-dimensional situation where there exists a distinguished class of measures (the measures which are absolutely continuous with respect to the Lebesgue measure), in the "infinite particle" situation the natural measure classes (for example, Gibbs measures with different pair potentials depending on the distance between particles) are mutually singular. Therefore, constructing trajectories for almost all initial data with respect to an "individual" probability measure does not give the complete solution of the problem under consideration.

3.2. Construction of the Dynamics and Time Evolution.
A natural way of constructing the trajectory $x(t)$ for $x \in M$ is to pass to the limit $\lim_{\mathcal{T} \nearrow R^d} x^{\mathcal{T}}(t)$ (in the topology of the space M) where $x^{\mathcal{T}}(t)$ is a trajectory determining the motion of the finite particle system with the initial date $x_{\mathcal{T} \times R^d}$ (i.e., the motion of the particles which are inside the cube $\mathcal{T}$ at the initial moment of time). There exist several natural possible ways to proceed here. First, it is possible "to ignore" the particles from $x_{\mathcal{T}^c \times R^d}$ defining $x^{\mathcal{T}}(t)$ by means of equations (10.35) with the initial data $x_{\mathcal{T} \times R^d}$. On the other hand, it is possible to introduce the boundary condition of elastic reflection from the boundary $\partial \mathcal{T}$ and to consider the motion of the particles $(q, p) \in x_{\mathcal{T} \times R^d}$ in the constant potential field generated by the "frozen" particles from $x_{\mathcal{T}^c \times R^d}$. Although there are no general results about independence of the limit, as $\mathcal{T} \nearrow R^d$, on the choice of approximating motion, such an independence may be proved for all concrete situations which are discussed below.

Observe that for the motion of a finite system of perticles in R^d determined by equations (10.35), the well-known conservation laws are valid:

$$|S_t x| = |x| \qquad (10.39)$$

(the law of conservation of the number of particles),

$$H(S_t x) = H(x) \qquad (10.40)$$

(the law of conservation of the energy) and, finally,

$$\sum_{(q,p) \in S_t x} p = \sum_{(q,p) \in x} p \qquad (10.41)$$

(the law of conservation of the momentum). For the particle motion in a domain $\mathcal{O}$ with the elastic reflections from the boundary $\partial \mathcal{O}$, the law of conservation of the momentum is not valid, and for motion in a potential field, one has to take into consideration, in the formulation of the law of conservation of the energy, the particle interaction with the field.

The conservation laws play a crucial role for constructing the limiting trajectory $x(t)$. In particular, the proof of compactness of the family $\{x^{\mathcal{T}}(t)\}$ for given x and t is based on these laws. The latter fact is the key point in the proof of existence of the limit. According to the definition of the topology in the space M, it is sufficient to control, as $\mathcal{T} \nearrow R^d$, the restriction $x^{\mathcal{T}}(t)_{\mathcal{O} \times R^d}$ on the fixed bounded domain $\mathcal{O} \subset R^d$. Possible violations of compactness are explained by two (mutually connected) reasons: a) individual particles gain infinite speed in finite time:

$$\max[\|p\|: p \in \mathbf{x}^{\mathcal{T}}(t)_{\mathcal{O} \times R^d}] \to \infty$$

and b) accumulations of particles ("collapses") arise:

$$|\mathbf{x}^{\mathcal{T}}(t)_{\mathcal{O} \times R^d}| \to \infty.$$

The estimations guaranteeing compactness are deduced by means of uniform in $\mathcal{T}$ estimates for $|\mathbf{x}^{\mathcal{T}}(t)_{\mathcal{O} \times R^d}|$ and $H(\mathbf{x}^{\mathcal{T}}(t)_{\mathcal{O} \times R^d})$ which are obtained by using the conservation laws for the number of particles and the energy. The fact is, that an increase in the number of particles and the energy within the domain $\mathcal{O}$ can occur only because of an "influx" through the boundary of the domain, and one is sometimes able to estimate such as influx by means of the corresponding values for a larger domain. This idea was first realized in the papers of Lanford [L1], [L2], [L4], initiating the mathematical study of the infinite particle dynamics. In these papers the one-dimensional case $(d = 1)$ was considered and $U(|q|)$ was supposed to be a smooth function of $q \in R^1$ with compact support. Under these conditions on the potential, the energy can be estimated by the number of particles. Hence, Lanford succeeded in producing necessary estimations by using the law of conservation of the number of particles only. He constructed the dynamics on the subset $\hat{M}^{(1)} \subset M$, characterized by the conditions

$$\sup_{(q,p) \in \mathbf{x}} \frac{|P|}{\ln_+ |q|} < \infty, \quad \sup_{y \in R^1} \frac{|\mathbf{x}_{\mathcal{T}(y, \ln_+|y| \times R^1)}|}{\ln_+ |y|} < \infty, \tag{10.42}$$

where $\ln_+ r = \max[1, \ln r]$, $r > 0$. It is not hard to show that this set is "massive" in the sense discussed in Section 3.1.

In the series of papers [DF], [F4], [FD], [MPP], the construction of dynamics was realized for a wide class of interaction potentials for the dimensions $d = 1$, 2. Here the law of conservation of the energy plays an essential role. The existence of the dynamics is proved for the subset $\hat{M}^{(2)} \subset M$, characterized by the conditions

$$\sup_{y \in R^d} \sup_{r < \ln_+|y|} r^{-d}[H(\mathbf{x}_{\mathcal{T}(y,r) \times R^d}) + A|\mathbf{x}_{\mathcal{T}(y,r) \times R^d}|] < \infty \tag{10.43}$$

where A is a constant depending on the potential U. The set $\hat{M}^{(2)}$ is massive in the previous sense.

Notice also the paper [GS3] concerning the one-dimensional case where the dynamics is constructed for potentials U which admit elastic collisions of particles (this case is excluded by the conditions in the papers cited above).

The fact that the results of such a type have not yet been extended to the physically realistic dimension $d = 3$ seems to indicate an essential difficulty. Let us explain, on an intuitive level, why the estimates based on the law of conservation of the energy are not applicable for $d = 3$. Let a particle $(q, p) \in \mathbf{x}^{\mathcal{T}}(t)$ be fixed. Furthermore, let a number $r(t)$ be such that in the time interval from 0 to t this particle interacts directly with the particles which are at the moment 0 within the cube $\mathcal{T}(q, (\frac{1}{2})r(t))$ only. Assuming that the full energy of particles within $\mathcal{T}(q,(\tfrac{1}{2})r(t))$ at the time moment 0 is proportional to $r(t)^d$ (i.e., to the volume of this cube) and taking, as an estimate from above, the case where the full energy is transmitted to the fixed particle, we see that the module of its momentum is bounded from above by the value $r(t)^{d/2}$. Assuming that this estimate holds for other particles too, we conclude that the rate with which $r(t)$ increases does not exceed the order $r(t)^{d/2}$. Thus we come to the differential inequality

$$\dot{r}(t) \leqslant \mathrm{const} \cdot r(t)^{d/2}.$$

The solutions of this inequality are bounded for $d \leqslant 2$ on any finite time interval, but may be unbounded for $d = 3$. This argument forms the basis of all proofs in the papers [DF], [F4], [FD], [MPP].

A somewhat different way of constructing the time evolution of a probability measure on the phase space was suggested recently by Siegmund-Schultze [SS]. He proved the following fact. Let the dimension d be arbitrary, U be a smooth nonnegative function with compact support, and P be an arbitrary translation invariant probability distribution on $(M, \mathcal{M})$ for which the mean energy of an individual particle is finite. Then, for initial conditions forming a subset of P-probability 1 in M, there exists a trajectory $\mathbf{x}(t) \in M$. It seems that the conditions on the potential listed above may be weakened. However, the methods of Siegmund-Schultze do not yet permit us to prove the uniqueness of a solution in some reasonable sense and moreover, to control either properties of the solution.

3.3. Hierarchy of the Bogolyubov Equations.
As was remarked in Sect. 3.1, the construction of a dynamics permits us to define the time evolution $\{P_t, t \in R^1\}$ of an initial measure $P_0 = P$ (cf (10.38)). For certain purposes (for instance, for deducing kinetic equations; cf Sect. 6) it is convenient to use another, more direct method of describing the time evolution which is especially popular in physical literature. This method is based on studying a hierarchy of equations for moment function (see Sect. 2.7) $\mathscr{k}_P^{(n)}$, $n = 1, 2, \ldots$, which are called the Bogolyubov equations [B1] (another frequently used term is the BBGKY hierarchy equations (Bogolyubov–Born–Green–Kirkwood–Yvon)). If one supposes that the interaction potential U "does not admit" collisions of particles, the *Bogolyubov hierarchy* takes the form

$$\frac{\partial}{\partial t}\mathscr{k}^{(n)}(t;(q,p_1),\ldots,(q_n,p_n))$$
$$= \{\mathscr{k}^{(n)}(t;(q_1,p_1),\ldots,(q_n,p_n)), H((q_1,p_1),\ldots,(q_n,p_n))\}$$
$$+ \int dq_0 dp_0 \{\mathscr{k}^{(n+1)}(t;(q_0,p_0),(q_1,p_1),\ldots,(q_n,p_n)), \sum_{j=1}^n U(\|q_0 - q_j\|)\} \quad (10.44)$$

(for $n = 0$ only the integral part remains in the equation). Here $\mathscr{k}^{(n)}(t;\cdot)$ is the n-moment function at the time moment t.

The connection between the two methods of description of time evolution has not yet been completely investigated. It is natural to expect that, for a large class of situations, the moment functions $\ell_{P_t}, t \in R^1$, give a solution of equations (10.44) (at least, in a weak sense). One can mention here the paper [GLL] and the note [Su3] where this fact is proved for the one-dimensional dynamics constructed, respectively, in [L1] and [GS3]. A series of papers is devoted to the proof of this fact for the case where the initial measure P is absolutely continuous or "almost" absolutely continuous with respect to a Gibbs distribution with the potential U and parameters (z, β, p_0) (cf [T1], [T2], [Z1] and also [Pu]).

Observe that another approach is simultaneously developed intensively and fruitfully. In this approach the problems of existence and uniqueness of a solution of the Bogolyubov hierarchy equations are treated using functional analysis. Under such an approach the Bogolyubov hierarchy equations are considered as an abstract evolution equation. In the first step, a solution is constructed in a Banach space of sequences of functions describing states of finite particle systems. Then one performs the thermodynamic limit passage to infinite systems. Here, as in Sect. 3.2, properties of the finite particle dynamics play an essential role. The functional analytic approach is developed in the papers of the Kiev specialists in mathematical physics [G], [GP1], [M], [Pet]. An explanatory review of this direction is contained in the recent paper of Petrina and Gerasimenko [GP2] to which we refer the reader for details. The consecutive exposition of all the circle of questions connected with the present state of the problem of solutions of the Bogolyubov hierarchy equations, is contained in the monograph of Petrina, Gerasimenko and Malyshev [GMP].

Notice also the paper [Sk] where the hierarchy of diffusion equations of the same type as the Bogolyubov diffusion equations is investigated from the functional-analytic positions.

§4. Equilibrium Dynamics

4.1. Definition and Construction of Equilibrium Dynamics. Let us consider the Hamiltonian system (10.35) in a bounded domain $\mathcal{O} \subset R^d$ with elastic reflection from the boundary $\partial \mathcal{O}$. The Liouville theorem and the laws of conservation of the number of particles (10.39) and energy (10.40) imply (under some weak additional assumptions) that the restriction of this dynamics to the hypersurface in the space $M_\mathcal{O}$, characterized by fixing the number of particles $|\mathbf{x}|$ and the energy $H(\mathbf{x})$, preserves the measure on this surface induced by the Lebesgue-Poisson measure $L_\mathcal{O}$. The corresponding normalized measure is called the microcanonical distribution. The Gibbs distribution in the volume $\mathcal{O}$ with the potential U and parameters z, β and $p_0 = 0$ is the result of averaging microcanonical distributions with some weight and hence is invariant, too. Analogously, the Gibbs distribution in the volume $\mathcal{O}$ with the boundary condition $\bar{\mathbf{x}}$, potential U and parameters

z, β and $p_0 = 0$ is invariant with respect to the motion in the potential field of the frozen particles from $\bar{\mathbf{x}}$.

Now let P be a probability measure on the phase space $(M, \mathcal{M})$. Let us consider a one-parameter automorphism group $\{S^t, t \in R^1\}$ of the measure space $(M, \mathcal{M}, P)$ which is generated by the infinite system of differential equations (10.35). Formally, this means that for any pair of smooth "local" functions f, g depending on positions and momenta of particles which are concentrated in a bounded domain of the space R^d, the following relation is valid

$$\frac{d}{dt} \int_M P(d\mathbf{x}) f(S^t \mathbf{x}) \bar{g}(\mathbf{x}) \bigg|_{t=0} = \int_M P(d\mathbf{x}) \mathscr{L} f(\mathbf{x}) \bar{g}(\mathbf{x}) \tag{10.45}$$

where

$$\mathscr{L} f(\mathbf{x}) = \sum_{(q,p) \in \mathbf{x}} \bigg((\mathrm{grad}_q f(\mathbf{x}), p) $$

$$- \bigg(\mathrm{grad}_p f(\mathbf{x}), \sum_{\substack{(q',p') \in \mathbf{x}: \\ q \neq q'}} \mathrm{grad}\, U(\|q' - q\|) \bigg) \bigg) \tag{10.46}$$

We say that the measure P is translation invariant in the direction of the vector p_0 if $P(T_{sp_0} A) = P(A)$ for any $A \in \mathcal{M}$ and $s \in R^1$. For Gibbs distributions, this property is known to be valid under the uniqueness conditions (cf Theorems 2.2 and 2.3).

Definition 4.1. *Let P be a Gibbs distribution with the interaction potential U and parameters (z, β, p_0) which is translation invariant in the direction of the vector p_0. Then $(M, \mathcal{M}, P, \{S_t\})$ is called the equilibrium dynamical system.*

The connection of Definition 4.1 with the constructions from Sect. 3 is the following. In the situation we discuss below where the equilibrium dynamics is constructed, for P-almost all $\mathbf{x}$, the trajectory $S^t \mathbf{x}$ gives a solution of the system of equations (10.35). On the other hand, if a dynamics is constructed, in the sense of Definition 3.1, on a set $\hat{M}$ which has full measure with respect to a Gibbs distribution P from the class we described just above, then P will be an invariant measure which determines the equilibrium dynamical system in the sense of Definition 4.1. The condition $P(\hat{M}) = 1$ may be easily verified in the situations discussed in Sect. 3.2. However, an equilibrium dynamical system may be constructed by other methods in a much more general situation (see papers [Z], [S3] treating the one-dimensional case and [S4], [PPT] devoted to the multidimensional case).

The main idea of these constructions which distinguishes them from the constructions of Sect. 3 is that one can get, for any fixed t, uniform estimates for trajectories $\mathbf{x}^{\mathcal{T}}(t)$. This holds due to the invariance of a Gibbs distribution with the parameter $p_0 = 0$ with respect to the approximating motion (cf Sect. 3.2). The passage to an arbitrary value p_0 is possible due to the obvious relation

$$S^t \Gamma_{p_0} \mathbf{x} = \Gamma_{p_0} T_{t p_0} S^t \mathbf{x} \tag{10.47}$$

where

$$\Gamma_{p_0}\mathbf{x} = \{(q,p): (q, p - p_0) \in \mathbf{x}\}.$$

In the case where the potential U has a finite range it is possible to verify a "cluster" character of an equilibrium dynamical system. More precisely, in dimension $d = 1$ [S3] and in dimension $d \geqslant 2$ for small values of z [S4] (cf Theorem 2.2) the typical (with respect to the corresponding Gibbs distribution) point $\mathbf{x}$ has the following property: one can divide the particles $(q, p) \in \mathbf{x}$ into finite pairwise disjoint groups (clusters) in such a way that the particles from different clusters will not interact in the course of motion within a given bounded interval of time. After that one introduces a new partition of particles into clusters which move again independently, etc.

The Gibbs distributions mentioned in Definition 4.1 are connected with the $(d + 2)$-parameter family of invariants $\mathbb{N}(P)$, $\mathbb{H}(P)$ and $\mathbb{J}(P)$ (cf Sect. 3.2). It is not hard to check that for a Gibbs distribution P with the potential U and parameters (z, β, p_0) the specific average momentum coinsides with p_0. The parameter z is responsible for the specific average density $\mathbb{N}(P)$ and the parameter β for the specific average energy $\mathbb{H}(P)$ in the sense that, for fixed U, β and p_0 (U, z and p_0), $\mathbb{N}(P)$ changes monotonically with z (correspondingly, $\mathbb{H}(P)$ changes monotonically with β). The Gibbs distribution in the uniqueness region is completely determined by the values $\mathbb{N}(P)$, $\mathbb{H}(P)$ and $\mathbb{J}(P)$.

The question about existence of an equilibrium dynamical system is closely related, at least at the level of main ideas, to the question about existence of a solution of the Bogolyubov hierarchy equations (10.44) for an initial moment function k_P corresponding to a "local perturbation" of an invariant Gibbs distribution (this means that the measure P on $(M, \mathcal{M})$ is absolutely continuous with respect to one of the invariant Gibbs distributions). Such results are obtained, by using the functional-analytic method, in the series of papers [G], [GP1], [M], [Pet], [GP2] previously mentioned. On the other hand, in the cycle of papers [T1], [T2], [Z1] this question is investigated by means of the properties of the corresponding equilibrium dynamics.

4.2. The Gibbs Postulate. The fundamental postulate of statistical mechanics asserts that systems with a large number of particles in thermodynamical equilibrium are described by Gibbs distributions. The question of distinguishing the class of Gibbs distributions by means of some a priori physically natural conditions is extremely important for the mathematical foundations of statistical mechanics.

In the traditional "finite-particle" approach (cf [AA]) one refers usually to the ergodic theorem of Birkhoff-Khinchin and the theorem of equivalence of ensembles, which asserts that a microcanonical distribution, considered as a probability measure on $M_{\mathcal{T}}$, converges, as $\mathcal{T} \nearrow R^d$, to a Gibbs distribution with some values of parameters z, β and p_0. This approach is based on the well-known *ergodic hypothesis* for the dynamical system described by equations (10.35) with a microcanonical distribution. The well-known result here is the theorem about ergodicity of the system of two hard balls with elastic collisions (cf Chap. 8, Sect. 1).[4] On the other hand (although there are as yet no explicitly formulated results), the method of constructing invariant measures provided by the theory of Kolmogorov–Arnold–Moser gives us little hope that such an ergodicity takes place in a general situation.

Within the limits of the infinite particle approach, developed in this paper, the analog of the assertion about ergodicity of a microcanonical distribution is the hypothesis that the class of all "good enough" invariant measures is exhausted by Gibbs distributions which figure in Definition 4.1. Of course, it is not hard to construct trivial counter examples. For instance, for an interaction potential of the finite range r_1, any measure such that with probability 1 all the particles are at a distance more than r_1 from each other and have zero momenta is invariant. It seems that a priori natural restrictions on a class of measures which exclude such examples, may consist of assumptions of the following types:

1) the restrictions of these measures onto $\mathcal{M}_\mathcal{O}$ for bounded $\mathcal{O} \subset R^d$ should be determined by "nice enough" densities with respect to the measure $\bar{L}_\mathcal{O}$,

2) certain conditions of "space mixing" must be fulfilled (cf (10.20), (10.21)) which mean that events from the σ-algebras $\mathcal{M}_{\mathcal{O}_1}$ and $\mathcal{M}_{\mathcal{O}_2}$ for "distant" regions $\mathcal{O}_1, \mathcal{O}_2 \subset R^d$ are "almost independent".

The next reduction of the a priori class of measures is related to the assumption that all these measures are Gibbs distributions in the sense of the general definition of Sect. 2.6, under reasonable restrictions on their potentials Φ. This problem has been studied in such a context in the cycle of papers [GS1], [GS2].

In the papers [GS1], [GS2] the interaction potential U is supposed to have a finite range and a hard core of diameter $r_0 > 0$ and to obey $\lim_{r \to r_0+} U(r) = \infty$. The problem of describing invariant measures is solved here in an a priori introduced class of Gibbs distributions corresponding to potentials $\Phi = (\Phi^{(1)}, \Phi^{(2)}, \ldots)$ which satisfy some qualitative restrictions. Two restrictions among these are of principal character: (a) there exists $n_0 = n_0(\Phi) \geq 2$ such that $\Phi^{(n)} \equiv 0$ for $n > n_0$, (b) for all $n = 2, \ldots, n_0$ and $(q_1, p_1), \ldots, (q_n, p_n) \in R^d \times R^d$

$$|\Phi^{(n)}((q_1, p_1), \ldots, (q_n, p_n))| \leq \psi\left(\max_{1 \leq j_1 < j_2 \leq n} \|q_{j_1} - q_{j_2}\|\right) \qquad (10.48)$$

where ψ is a monotone function which decreases rapidly enough.

The problem of finding invariant measures has been stated in [GS1], [GS2] as the problem of finding stationary-in-time solutions of the Bogolyubov hierarchy equations, i.e., as the problem of finding moment functions for which the left-hand side of the equations (10.44) is equal to zero. The main theorem of [GS1], [GS2] asserts that, inside the class of measures described above, every stationary solution is the moment function of one of the Gibbs distributions figuring in Definition 4.1.

---

[4] Recently, Kramli, Simany and Szasz have proved the same result for the system of three balls.

The proof of this theorem uses the following important notion. A function h^Φ of the form

$$h^\Phi(\mathbf{x}) = \sum_{n \geq 1} \sum_{\substack{\mathbf{x}' \subset \mathbf{x}: \\ |\mathbf{x}'|=n}} \Phi^{(n)}(\mathbf{x}'), \qquad \mathbf{x} \in M, |\mathbf{x}| < \infty, \qquad (10.49)$$

(cf (10.26)) is called the first integral of the motion in the space R^d if

$$h^\Phi(S^t \mathbf{x}) = h^\Phi(\mathbf{x}), \qquad t \in R^1, \qquad (10.50)$$

(cf (10.39)–(10.41)). In the first step one proves that h^Φ is a first integral of the motion whenever the moment function ℓ_P of a Gibbs distribution P with a potential Φ from the indicated class gives a stationary solution of the Bogolyubov hierarchy equations.

In the second step, one investigates the summatory first integrals of the motion of the form (10.49). It turns out that under the imposed conditions they are reduced to linear combinations of "canonical" first integrals (10.39)–(10.41). This leads to equalities (10.29a, b) according to which P will be one of the Gibbs distributions with the potential U.

4.3. Degenerate Models. The connection between the summatory first integrals and the class of invariant measures is illustrated by the example of "degenerate" models, where, in addition to the "canonical" first integrals (10.39)–(10.41), there exist other summatory first integrals. For instance, in the case of an ideal gas where the momenta of particles do not change during the motion, there exist first integrals of the form

$$h(\mathbf{x}) = \sum_{(q,p) \in \mathbf{x}} \varphi(p), \qquad \mathbf{x} \in M, \qquad |\mathbf{x}| < \infty, \qquad (10.51)$$

where φ is an arbitrary measurable function. Correspondingly, all Poisson measures $P^0_{z,\sigma}$ are invariant in this case (cf Sect. 2.5). An analogous situation takes place for the model of one-dimensional ($d=1$) hard rods of length $r_0 > 0$ where the particles change momenta under collisions. Here the measures $P^{r_0}_{z,\sigma}$ are invariant (see again Sect. 2.5). From the results of Section 5 (cf Sects. 5.2, 5.4) it follows that there are no other invariant measures for these two models inside a wide class of probability distributions on $(M, \mathcal{M})$.

The one-dimensional system (10.35) with the interaction potential

$$U(r) = (\mathrm{sh}(Ar))^{-2}, \qquad r > 0, \qquad (10.52)$$

with $A = \mathrm{const} > 0$, gives another interesting example of a degenerate system. Given any number of particles $N < \infty$, this system is integrated by the method of the inverse problem of scattering theory (cf [C]). Therefore, for this model there exists an infinite series of nontrivial summatory first integrals. In the paper [C] the simplest of these integrals is considered. It has the form (10.49) with some explicitly written Φ, and one proves the existence and invariance property of the Gibbs distribution with the potential $\beta\Phi, \beta > 0$.

In the note [Gu] a result is given which extends, in the one-dimensional case, the constructions of Section 4.2 to a new class of interaction potentials including the potential (10.52). It turns out that, within this class of interaction potentials, the potential (10.52) (and multiples of it) exhausts the set of potentials for which there exist additional invariant Gibbs measures. Recently, Gurevich investigated the multidimensional situation, too. Here, under quite general conditions on the interaction potential U, the only potential for which "non-canonical" invariant Gibbs distributions may appear is that of ideal gas.

4.4. Asymptotic Properties of the Measures P_t. The question about the asymptotic behaviour of the measures P_t, $t \in R^1$, giving the time evolution of an initial measure $P_0 = P$ (cf Definition 3.2) seems to be very important. A natural conjecture is that, for non-degenerate models of motion and "good enough" (in the sense of Sect. 4.2) initial measures P, the measures P_t converge, as $t \to \pm\infty$, to an invariant Gibbs distribution with the potential U for which the values of the specific average particle number, energy and momentum are the same as for P. This conjecture may be considered as an analog of the classical ergodic Boltzmann hypothesis.

A similar role in the equilibrium dynamics concept is played by the hypothesis that, for non-degenerate models of motion, the equilibrium dynamical system has mixing properties. From this hypothesis one can derive the convergence of the measures P_t for initial measures P which are absolutely continuous with respect to an invariant Gibbs distribution with the potential U. Physically, this fact corresponds to an assertion about asymptotic "dispersing" local fluctuations in equilibrium dynamical systems.

Both these problems are very difficult and results at the mathematical level are obtained here only for degenerate models of ideal gas and of one-dimensional hard rods which will be considered in the next sections.

§ 5. Ideal Gas and Related Systems

5.1. The Poisson Superstructure. We begin by studying the dynamical system on $(M, \mathcal{M})$ corresponding to the potential $U \equiv 0$. It is convenient to give a general definition including the example of the ideal gas as a particular case.

Let $(N^1, \mathcal{N}^1, \pi)$ be a metric space with a Borel σ-finite measure π, $\{\tau_t^1, t \in R^1\}$ be a one parameter measurable group of bijective transformations of N^1 which preserve the measure π. By analogy with the spaces $(Q^0, \mathcal{Q}^0)$ and $(M, \mathcal{M})$ (cf Sects. 2.1, 2.5) we introduce the measurable space $(N, \mathcal{N})$ the points of which are "locally finite" subsets of N^1. We shall fix $z > 0$ and, by analogy with Poisson measures P_z^0 and $P_{z,\sigma}^0$, introduce a probability measure $\mathbb{P}_{z,\pi}^0$ on $(N, \mathcal{N})$ defined by the same conditions 1), 2) (cf Sect. 2.2), replacing R^d by the space N^1 and the Lebesgue measure by the measure π. We shall define the flow $\{\tau_t, t \in R^1\}$ in

$(N, \mathcal{N}, \mathbb{P}^0_{z,\pi})$ by setting

$$\tau_t X = \{x \in N^1 : \tau^1_{-t} x \in X\}. \tag{10.53}$$

It is not hard to check that the measure $\mathbb{P}^0_{z,\pi}$ is invariant with respect to the flow $\{\tau_t\}$.

Definition 5.1. The dynamical system $(N, \mathcal{N}, \mathbb{P}^0_{z,\pi}, \{\tau_t\})$ is called the *Poisson superstructure* over $(N^1, \mathcal{N}^1, \pi, \{\tau^1_t\})$. The quadruple $(N^1, \mathcal{N}^1, \pi, \{\tau^1_t\})$ is called the one-particle dynamical system.

In the particular case where (i) $N^1 = R^d \times R^d$, (ii) π is a measure $l \times \sigma$ where l is the Lebesgue measure and σ is a probability distribution on R^d, and, finally, (iii) $\tau^1_t(q, p) = (q + tp, p)$, $(q, p) \in R^d \times R^d$, we get the dynamical system of ideal gas. Here $\mathbb{P}^0_{z,\pi}$ coincides with the probability measure $\mathbb{P}^0_{z,\sigma}$ (cf Sect. 2.5).

Clearly, all ergodic properties of the Poisson superstructure are completely determined by the one-particle dynamical system. However, generally the problem of obtaining necessary and sufficient conditions of ergodicity and mixing is difficult enough. The situation is simplified if one assumes in addition that a so-called trend to infinity takes place for the one-particle dynamical system $(N^1, \mathcal{N}^1, \pi, \{\tau^1_t\})$, i.e., there exists a set $C \in \mathcal{N}^1$ with $\pi(C) < \infty$ and a number $t_0 > 0$ such that $\pi(N^1 \setminus (\bigcup_{t \in R^1} \tau^1_t C)) = 0$ and the intersection $C \cap \tau^1_t C = \emptyset$ for all t with $|t| > t_0$. For ideal gas this condition is fulfilled provided the probability distribution σ has no atom at 0.

Theorem 5.1 (cf [CFS]). *If a trend to infinity takes place in a one-particle dynamical system, then the corresponding Poisson superstructure is a B-flow.*

The proof of this theorem is a verification of the B-property of the partition given by the intersection $X \cap C$, $X \in N$.

The physical explanation of the fact that the Poisson superstructure has such strong ergodic properties is connected with specific features of infinite particle systems. Any probability measure absolutely continuous with respect to $\mathbb{P}^0_{z,\pi}$ is determined by the corresponding density $f(X)$, $X \in N$. The function f is "local" (i.e., depends on the positions and momenta of the particles concentrated in a bounded domain of the space R^d) or may be approximated by local functions in the sense of L_1-convergence. The "shifted" function depends essentially on the positions and momenta of the particles which are at a distance of order $|t|$ from the origin. It follows from the definition of the measure $\mathbb{P}^0_{z,\pi}$ that the functions $f(\tau_{-t}X)$ and $f(X)$ are "almost" independent for large $|t|$, and this fact leads to good ergodic properties.

Another popular example of Poisson superstructure is the so-called *Lorentz gas* (cf Chap. 8). Let a countable set of points ("immovable scatterers") in the space R^d be thrown about. Every scatterer generates a potential field which decreases rapidly enough at a large distance from the scatterer. The one-particle system will be the dynamical system corresponding to the motion of a particle in the potential field given by the total potential of the scatterers. The Lorentz gas, in a general form, is the Poisson superstructure over such a system.

The particular case where the potentials of the scatterers are those of hard spheres (not necessarily of a constant radius) may be investigated at the mathematical level. This case corresponds to the motion of a particle with a constant velocity and with the elastic reflection from the spheres. The Lorentz gas is studied in this case by methods of the theory of dispersing billiards (cf [BS1], [BS2] and Chap. 8). The case of a periodic configuration of scatterers is directly reduced to the dispersing billiard on the torus. In the paper [BS2] it is proved that the Lorentz gas is a K-system under wide assumptions about the configuration of scatterers.

Notice the close connection of the models regarded in this section with infinite systems of independently moving particles (see Chap. 7 of the book [MKM] and the references therein). In particular, the invariance of the Poisson measure $\mathbb{P}^0_{z,\sigma}$ with respect to the dynamics of ideal gas, as well as the mixing property, follow from results which were already established in the 1950's (cf [D1], [Do]).

5.2. Asymptotic Behaviour of the Probability Distribution P_t as $t \to \infty$. In this section we discuss the question about the asymptotic behaviour of the probability measures P_t, $t \in R^1$, which describe the time evolution of an initial measure P (cf (10.38)) induced by the dynamics of ideal gas (10.37). Since we do not assume that the measure P is absolutely continuous with respect to one of the invariant measure $\mathbb{P}^0_{z,\sigma}$, this question is beyond the constructions of the preceding section.

The investigation of asymptotic properties of the probability measures P_t was initiated in the early paper [D1] as has been previously mentioned. A detailed analysis of this question is given in the more recent paper [DS2], the results of which are discussed below. Let us introduce the following coefficient of asymptotic mixing (comp. (10.20)):

$$\alpha_P(r,s) = \sup_{y \in R^d} \sup_{\substack{A_1 \in \mathcal{M}_{\mathcal{T}(y,r)}, \\ A_2 \in \mathcal{M}_{\mathcal{T}(y,r+s)^c}}} |P(A_1 \cap A_2) - P(A_1)P(A_2)|. \qquad (10.54)$$

The following conditions are imposed onto the initial measure P:

(I) for any $b > 0$

$$\lim_{s \to \infty} \alpha_P(bs, s) = 0,$$

(II) the 1-moment measure $K_P^{(1)}(dq \times dp)$ is absolutely continuous with respect to a measure $\hat{K}_P^{(1)}(dq) \times dp$ on $R^d \times R^d$ with $\sup_{y \in R^d} \hat{K}_P^{(1)}(\mathcal{T}(y, b)) < \infty$ for any $b > 0$ and has a bounded density with respect to this measure,

(III) the 2-moment measure $K_P^{(2)}(\prod_{i=1,2} dq_i \times dp_i)$ is absolutely continuous with respect to a measure $\hat{K}_P^{(2)}(dq_1 \times dq_2)\prod_{j=1,2} dp_j$ on $(R^d \times R^d)^2$ with $\sup_{y_1,y_2 \in R^d} \hat{K}_P^{(2)}(\times_{j=1,2} \mathcal{T}(y_j, b_j)) < \infty$ for any $b_1, b_2 > 0$ and has a bounded density with respect to this measure.

Theorem 5.2 (cf [DS2]). *Let the probability distribution P satisfy conditions (I)–(III). Then the probability distributions P_t converge, as $t \to \pm\infty$, to the Poisson* measure $\mathbb{P}_{z,\sigma}^0$ if and only if the 1-moment measures

$$K_{P_t}^{(1)}(A) = K_P^{(1)}(\{q, p\colon (q - tp, p) \in A\})$$

converge to the measure $z(l \times \sigma)$.

As to the convergence of measures $K_{P_t}^{(1)}$, it may be established by wide assumptions about the 1-moment measure $K_P^{(1)}$. For instance, if $K_P^{(1)}$ has the periodicity property with respect to the space translations $(q, p) \mapsto (q + y, p)$, then the measures $K_{P_t}^{(1)}$ converge to a measure of the form $z(l \times \sigma)$, which is obtained by space averaging the measure $K_P^{(1)}$.

The conditions of Theorem 5.2 may be verified for Gibbs distributions under the uniqueness conditions (cf Theorems 2.2 and 2.3). The proof of Theorem 5.2 is based on a variant of the Poisson limit theorem for sums of weakly dependent random variables.

Recently the result of Theorem 5.2 was extended by Willms who suggested a more general variant of the condition (I) (cf [W]). Notice also the paper [Su5] where an "abstract" theorem of convergence to a Poisson measure $\mathbb{P}_{z,\pi}^0$ on the space $(M, \mathcal{M})$ (cf Sect. 5.1) is established. From the results proved in [Su5] it follows that one can require property (I) not for P but for the conditional distribution $P(\cdot | \mathcal{M}^Q)$ with respect to the σ-algebra $\mathcal{M}^Q$ (cf Sect. 2.5) induced by the projection map Π. This allows us to extend essentially the class of initial probability distributions for which the assumptions of the convergence theorem are valid.

A result similar to Theorem 5.2 was obtained in [KS] for the Lorentz gas.

5.3. The Dynamical System of One-Dimensional Hard Rods. The dynamics of one-dimensional hard rods corresponding to the potential (10.14) is closely connected with the dynamics of ideal gas. The connection between the two dynamics is established by using special transformations of "dilatation" and "contraction" in the phase space $(M, \mathcal{M})$. Let $\mathbf{x} \in M$ and a particle $(q, p) \in \mathbf{x}$ be given. We enumerate the particles $(\tilde{q}, \tilde{p}) \in \mathbf{x}$ by integers in order of increasing coordinates $\tilde{q} \in R^1$; with the number 0 given to the distinguished particle (q, p). Under the "dilatation" transformation, $\mathbb{D}_q$, the particle (q_j, p_j) passes into $(q_j + jr_0, p_j)$, $j \in Z^1$, where r_0 is the hard rod length. The transformation of "contraction", $\mathbb{C}_q$, is inverse to $\mathbb{D}_q$ (it is correctly defined only if $q_{j+1} - q_j > r_0$ for all j). If one denotes the time shift in the dynamics of ideal gas by S_t^0 and the time shift in the dynamics of hard rods by $S_t^{r_0}$, then the following formula is valid:

$$S_t^{r_0}\mathbf{x} = T_{-r_0 n_t}\mathbb{D}_{q+tp}S_t^0\mathbb{C}_q\mathbf{x}, \quad t \in R^1, \tag{10.55}$$

where $n_t = n_t((q, p), \mathbb{C}_q\mathbf{x})$ is the total "algebraic" number of the intersections of the trajectory of the particle (q, p) by the trajectories of other particles $(\tilde{q}, \tilde{p}) \in \mathbb{C}_q\mathbf{x}$ under the free motion in the time interval from 0 to t.

Ergodic properties of the hard rods system with the invariant measure $P_{z,\sigma}^{r_0}$ (cf Sect. 2.5) have been investigated on the basis of this connection in the papers

[S2], [AGL1], [P]. The most general result is proved in the paper of Aizenman, Goldstein and Lebowitz [AGL1].

Theorem 5.3. *Let σ be a probability measure on R^1 with a finite first moment ($\int_{R^1} \sigma(dp)|p| < \infty$) such that $\sigma(\{0\}) < 1$. Then for any $z > 0$, the dynamical system $(M, \mathcal{M}, \mathbb{P}^{ro}_{z,\sigma}, \{S^{ro}_t\})$ is a K-flow. If z and σ satisfy the additional condition $\sigma((p^0_{z,\sigma} - \varepsilon, p^0_{z,\sigma} + \varepsilon)) = 0$ for some $\varepsilon > 0$ where*

$$p^0_{z,\sigma} = \frac{zr_0}{1 + zr_0} \int_{R^1} \sigma(dp) p, \qquad (10.56)$$

then the dynamical system $(M, \mathcal{M}, \mathbb{P}^{ro}_{z,\sigma}, \{S^{ro}_t\})$ is a B-flow.

The possibility of omitting this additional condition remains an open problem.

Formula (10.55) provides the basis for a generalization of the results of Sect. 5.2 to the dynamics of the hard rods given in [DS2]. Let us suppose that the measure P is translation invariant and concentrated on the set of points $\mathbf{x} \in M$ having the property: $|q - q'| > r_0$ for any pair of different particles $(q, p), (q', p') \in \mathbf{x}$. By means of the transformations $\mathbb{D}_q$ and $\mathbb{C}_q$, one can associate with P a corresponding "contracted" translation invariant measure $P^{(0)}$. If the measure $P^{(0)}$ satisfies conditions (I)–(III) (cf Sect. 5.2), then the measure $P^{(0)}_t$ obtained from $P^{(0)}$ in the course of the ideal gas dynamics converges, as $t \to \pm\infty$, to a Poisson measure $\mathbb{P}^0_{z,\sigma}$. Using formula (10.55) one can prove that the measures P_t obtained from P in the course of the hard rod dynamics converge to the corresponding measure $\mathbb{P}^{ro}_{z,\sigma}$. Notice that the conditions on the initial measure P are formulated here in terms of the measure $P^{(0)}$ whose connection with P is not so transparent. However, it is possible to verify such conditions for a wide class of Gibbs distributions (cf [DS2]).

§6. Kinetic Equations

6.1. Statement of the Problem. According to Definition 3.2, the time evolution of a probability measure P on the space $(M, \mathcal{M})$ is represented by the family of measures $P_t, t \in R^1$, determined by formula (10.38), or, equivalently, by the family of moment functions $\ell(t, \cdot)$ satisfying the hierarchy of the Bogolyubov equations (10.44). Kinetic equations of statistical mechanics are used for an approximate description of the time evolution in more simple terms. The purpose of rigorous mathematical investigations initiated here in recent years is to provide the mathematical background for the approximations used, and to give us a deeper understanding in this difficult area of problems. Meanwhile, these investigations are at an initial level, and a large part of what is said below must be treated only as preliminary conjectures.

Notice the profound review of the papers in this subject by Spohn [Sp1] where the majority of the themes we touch in this paragraph are discussed in detail.

Chapter 10. Dynamical Systems of Statistical Mechanics

The problems of kinetic theory for the Lorentz gas are considered in the review of van Bejeren [B].

In this section we shall try to describe, without pretending to be mathematically precise, some general features in stating problems of derivation of the basic kinetic equations: Boltzmann equation, Vlasov equation, Landau equation, and finally, hydrodynamic Euler equation, taking, as a starting point, the ideas developed in Sect. 3. After that we shall proceed to a consequent discussion of various equations and formulate the (non-numerous) mathematical results which are known here.

Let $F = F_P$ be a functional defined on some subset of the space of probability measures on $(M, \mathcal{M})$ with values in the space $\mathscr{F}$ of (generally speaking, vector-valued) functions depending on $q, p \in R^d$. For Boltzmann, Vlasov and Landau equations, $F_P = \mathscr{k}_P^{(1)}$ (the 1-moment function of a measure P). Suppose that a family of transformations $\mathbb{R}_t$, $t \in R^1$, of the space $\mathscr{F}$ is given, with the semi-group (or group) property

$$\mathbb{R}_{t_1+t_2} = \mathbb{R}_{t_1}\mathbb{R}_{t_2}, \ t_1, t_2 \geq 0 \quad \text{or } t_1, t_2 \leq 0 \ (t_1, t_2 \in R^1). \tag{10.57}$$

Suppose furthermore that for some classes of interaction potentials and initial measures P_0 on $(M, \mathcal{M})$ and for some interval $I \subset R^1$

1) an approximate equality takes place

$$F_{P_{t'}} \approx \mathbb{R}F_{P_0} \quad \text{where } t' = \kappa t, \ t \in I. \tag{10.58}$$

Here $\kappa > 0$ is a constant giving the "time-scaling";

2) the map $P \mapsto F_P$ is approximately invertible on some subset of the space of measures on $(M, \mathcal{M})$, including the measures $P_{t'}$, $t' \in R^1$, so that the measure $P_{\kappa t}$ may be "almost reconstructed" from the value of the functional $F_{P_{\kappa t}}$.

In addition, it is natural to assume that the transformations $\mathbb{R}_t$ may be described by means of a solution of a differential equation of the form

$$\frac{d}{dt} F(t) = AF(t) \tag{10.59}$$

where A is some operator in $\mathscr{F}$ which does not depend on t. Equations of this type are called kinetic equations.

In recent mathematical papers the approximate relations in conditions 1) and 2) are interpreted as asymptotic equalities with respect to an auxiliary parameter $\varepsilon \to 0_+$. One considers a family $\{P_0^\varepsilon, \varepsilon > 0\}$ of initial measures on $(M, \mathcal{M})$ onto which some restrictions are imposed. First, one assumes that the 1-moment function $\mathscr{k}_{P_0^\varepsilon}^{(1)}$ satisfies the relation

$$\mathscr{k}_{P_0^\varepsilon}^{(1)}(q, p) = \varepsilon^{\alpha_1} f_0(\varepsilon^{\alpha_2} q, p) \tag{10.60}$$

where f_0 is a fixed function $R^d \times R^d \to [0, \infty)$ and α_j are some constants (of course, one can consider a more general situation when equality (10.60) is fulfilled in the limit, as $\varepsilon \to 0_+$). Further, one supposes that $\kappa = \varepsilon^{-\alpha_3}$ (cf (10.58)) and finally, that the particle motion is determined by the interaction potential U^ε of the form

$$U^\varepsilon(r) = \varepsilon^{\alpha_4} U(\varepsilon^{\alpha_5} r), \qquad r \geqslant 0, \tag{10.61}$$

where U is a fixed function $[0, \infty) \to R^1 \cup \{\infty\}$ satisfying the conditions of Sect. 2.4.

Observe that the choice of constants α_j is not unique. This property is connected essentially with the fact that, under the simultaneous scale change $t = \varepsilon t'$, $q = \varepsilon q'$, $p = p'$, equations (10.35) are transformed into the equations of the same form with the new potential $U_\varepsilon(r) = U(\varepsilon^{-1} r)$ and the 1-moment function takes the form $k^{(1)\prime}(t'; (q', p')) = \varepsilon^{-d} k^{(1)}_{P^\varepsilon_{\varepsilon^{-1}t}}(\varepsilon^{-1} q, p)$.

For comparison, we present below a table indicating the choice of the parameters for the two most frequently used situations: $\alpha_2 = \alpha_3 = 1$ and $\alpha_2 = \alpha_3 = 0$. From the physical point of view, these situations correspond to the choice of the scale for measuring time and distance in "micro-" and "macro-world". We shall therefore speak about "micro"- and "macro-variables". The parameter ε gives the ratio of these scales and therefore may be considered as small in many applications.

The second part of the table is obtained from the first one by the change of variables $t = \varepsilon t'$, $q = \varepsilon q'$, $p = p'$.

Type of the limit passage	Equation	Microvariables: $t' = \varepsilon^{-1} t$		Macrovariables: $t' = t'$	
		1-moment function $k^{(1)}_{P^\varepsilon_0}(q, p)$	Potential	1-moment function $k^{(1)}_{P^\varepsilon_0}(q, p)$	Potential
Low density approximation	Boltzmann	$\varepsilon f_0(\varepsilon q, p)$	$U(r)$	$\varepsilon^{-(d-1)} f_0(q, p)$	$U(\varepsilon^{-1} r)$
Mean field approximation	Vlasov	$f_0(\varepsilon q, p)$	$\varepsilon^d U(r)$	$\varepsilon^{-d} f_0(q, p)$	$\varepsilon^d U(r)$
Weak coupling approximation	Landau	$f_0(\varepsilon q, p)$	$\varepsilon^{1/2} U(r)$	$\varepsilon^{-d} f_0(q, p)$	$\varepsilon^{1/2} U(\varepsilon^{-1} r)$
Hydrodynamic approximation	Euler for a compressible fluid	$f_0(\varepsilon q, p)$	$U(r)$	$\varepsilon^{-d} f_0(q, p)$	$U(\varepsilon^{-1} r)$

The conditions on the 1-moment function listed in the table are, of course, not sufficient for our purposes. Additional conditions on the whole measure P^ε_0 are formulated in a different form for different problems and it is too cumbersome to describe their form in detail at this stage of the development of the theory. It seems that, in the microvariables, these conditions must correspond, at a qualitative level, to the general conditions 1), 2) discussed in Sect. 4.2 and, in particular, they must be fulfilled for a large class of Gibbs measures.

In the macrovariables, the condition of the decay of correlations is sometimes formulated in terms of asymptotics of higher moment functions subject to the scaling transformations:

$$f_0^{(n)}((q_1,p_1),\ldots,(q_n,p_n)) = \varepsilon^{n\rho} k_{P_n^\varepsilon}^{(n)}((q_1,p_1),\ldots,(q_n,p_n)) \tag{10.62}$$

where $\rho = d - 1$, d, d and d for the corresponding lines of the table. This condition has the form

$$\lim_{\varepsilon \to 0} f_0^{(n)}((q_1,p_1),\ldots,(q_n,p_n)) = \prod_{j=1}^n f_0(q_j,p_j) \tag{10.63}$$

and is often called the "hypothesis of chaos" because it means (in macrovariables) the asymptotical "complete independence" of events which "occur" for different particles. For the case of the low density approximation, condition (10.63) leads to the well-known Boltzmann rule for calculating the number of collisions (Stoßzahlansatz). One can expect that in the situations we listed above, the fulfilment of conditions of the type (10.63) at the moment $t = 0$ involves the fulfilment of analogous conditions for all t.

The scheme described above leads to kinetic equations in the whole space R^d. Without special modifications of this construction one can get equations which describe the motion in a bounded domain with smooth or piecewise smooth boundary. In microvariables one can, for instance, consider a family $\{\mathcal{O}^\varepsilon, \varepsilon > 0\}$ of homotetic domains whose linear sizes increase like ε^{-1}. For any given ε one supposes that the probability measure P_0^ε is concentrated on the set $M_{\mathcal{O}^\varepsilon}$ and the time evolution P_t^ε is induced by the motion of particles in $\mathcal{O}^\varepsilon$ with boundary conditions on $\partial \mathcal{O}^\varepsilon$. For definiteness we consider below the boundary conditions of elastic reflection. Passing to microvariables, we get the motion in the fixed domain $\mathcal{O} \subset R^d$. The same is true for the limit kinetic equation as well.

6.2. The Boltzmann Equation. This equation is considered under the assumption that the dimension $d > 1$. It has the form

$$\frac{\partial}{\partial t} f_t(q,p) = -(p, \mathrm{grad}_q f_t(q,p)) + \int_{(S^{d-1} \times R^d)_p^+} |p - p_1|$$
$$\times B(p - p_1, e)[f_t(q, p')f_t(q, p_1') \tag{10.64}$$
$$- f_t(q,p)f_t(q,p_1)]\, de\, dp_1,\ q, p \in R^d.$$

Here

$$(S^{d-1} \times R^d)_p^+ = \{(e, p_1) \in S^{d-1} \times R^d : (p - p_1, e) > 0\},$$

S^{d-1} is a unit sphere in R^d. Vectors p', p_1' are connected with p, p_1 by the relations

$$p' = p - (p - p_1, e)e, \quad e \in S^{d-1},$$
$$p_1' = p_1 - (p_1 - p, e)e, \quad e \in S^{d-1},$$

and $B(\tilde{p}, e)$, $\tilde{p} \in R^d$, $e \in S^{d-1}$, is a function called differential cross section and is determined by the potential U. Physically, p, p_1 and p', p_1' correspond to the momenta of two colliding particles before their collision and after it, respectively.

The relation which connects them is equivalent to the natural conservation laws: $p + p_1 = p' + p_1'$, $\|p\|^2 + \|p_1\|^2 = \|p'\|^2 + \|p_1'\|^2$. For the function B one can write an explicit expression by integrating the equations of motion of two particles interacting via the potential U. In the case of the hard sphere potential (10.14) one has $B(\tilde{p}, e) = (\tilde{p}, e)$.

Problems of existence and uniqueness of a solution of the Boltzmann equation are extremely nontrivial. Chapter 11 of this volume is devoted to a review of results obtained here (see also the review papers in [No]).

As already mentioned above, the Boltzmann equation is obtained by the low density limit passage which is called also the *Boltzmann–Grad limit passage* (cf [Gr]). The physical meaning of this limit passage is most transparent under consideration in microvariables. The density of particles converges to 0 like ε and thereby (if we suppose for simplicity that the radius of interaction of the potential U is finite) particles move with constant momenta most of the time, changing their values now and then during relatively short intervals of collision interaction. The length of the time interval between subsequent collisions of a given particle with other particles (the time of free motion) has the order ε^{-1} and the time of each collision has the order of a constant. Most of these collisions are pairwise and asymptotically, as $\varepsilon \to 0$, one can neglect the multiple collisions. The normalization $t' = \varepsilon^{-1} t$ is natural because, during a time of the order ε^{-1}, each particle passes through a finite number of intervals of free motion and a finite number of collisions. The first term in the right hand side in (10.64) describes the free motion of particles while the second one corresponds to collisions.

In macrovariables, one considers particles of small diameter ε and their density, having the order $\varepsilon^{-(d-1)}$, diverges to ∞. The choice of macrovariables is more convenient for discussing details of the derivation of Boltzmann equation (see below).

An important result concerning the derivation of the Boltzmann equation was obtained by Lanford [L4]. He considered the motion generated by the potential of hard spheres of diameter $r_0 = \varepsilon$ (cf (10.14)) in a domain $\mathcal{O} \subset R^d$. It was assumed that for initial probability measures P_0^ε there exist moment functions $\ell_{P_0^\varepsilon}$ with the following properties:

1) for some positive constant z, β and all ε and n

$$\varepsilon^{(d-1)n} \ell_{P_0^\varepsilon}^{(n)}((q_1, p_1), \ldots, (q_n, p_n)) \leqslant z^n \exp\{-\beta(p_1^2 + \cdots + p_n^2)\}, \quad (10.65)$$

2) there exists $s > 0$ such that for any n, uniformly on every compact subset of $\Gamma_\mathcal{O}^{(n)}(s)$, the following convergence takes place

$$\lim_{\varepsilon \to 0} \varepsilon^{(d-1)n} \ell_{P_0^\varepsilon}^{(n)}((q_1, p_1), \ldots, (q_n, p_n)) = \prod_{j=1}^n f_0(q_j, p_j) \quad (10.66)$$

where f_0 is a smooth function $\mathcal{O} \times R^d \to [0, \infty)$.

Here and below $\Gamma_\mathcal{O}^{(n)}(s)$ denotes the set

$$\{(x_1, \ldots, x_n) \in (\mathcal{O} \times R^d)^n : q_\mathcal{O}^0(\tilde{s}, x_{j_1}) \neq q_\mathcal{O}^0(\tilde{s}, x_{j_2}) \text{ for any } \tilde{s} \in [-s, 0] \text{ and } j_1 \neq j_2\}$$
$$(10.67)$$

where $q_{\mathcal{O}}^{0}(\tilde{s},(q,p))$ is the position, at the time moment $\tilde{s}$ of the particle having, at time zero, the position $q \in \mathcal{O}$ and the momentum vector $p \in R^d$ under the free motion in the domain $\mathcal{O}$ (with the condition of the elastic reflection from the boundary).

Theorem 6.1 (cf [L4]). *Let condition 1) and 2) be fulfilled. Then there exists a constant $t_0(z,\beta) > 0$ such that for any $t \in [0, t_0(z,\beta))$ the moment functions $\ell_{P_t^\varepsilon}^{(n)}((q_1,p_1),\ldots,(q_n,p_n))$, $n \geq 1$, of the measure P_t^ε satisfy the relation*

$$\lim_{\varepsilon \to 0} \varepsilon^{(d-1)n} \ell_{P_t^\varepsilon}^{(n)}((q_1,p_1),\ldots,(q_n,p_n)) = \prod_{j=1}^{n} f_t(q_j, p_j)$$

for any n uniformly on every compact subset of $\Gamma_{\mathcal{O}}^{(n)}(t+s)$. Here $f_t(q,p)$ is the solution of the Boltzmann equation (10.64) in the domain $\mathcal{O}$ with the initial date $f_0(q,p)$.

An analogous assertion may be formulated as well for $t \in (-t_0(z,\beta), 0]$ (one assumes here that $s < 0$ in condition 2) and the interval $[-s, 0]$ in definition (10.67) is replaced by the interval $[0, -s]$ where $s < 0$). Instead of (10.64), the equation with the opposite sign in front of the integral term arises. This fact illustrates the well-known property of "irreversibility" of the Boltzmann equation.

Notice that, before passing to the limit, the system (10.35) has the reversibility property: if one replaces, at the moment zero, the values of momenta of all particles by the opposite values, then the motion in the "positive" direction will be the same as the motion in the "negative" direction for particles with the original values of momenta. The paradox of irreversibility which we encounter is explained by the fact that the sets $\Gamma_{\mathcal{O}}^{(n)}(s)$, $s > 0$, are not invariant under the change of time. Hence, condition 2) for the moment functions obtained by "reversing" the values of the particle momenta at time $t \in (0, t_0(z,\beta))$ does not follow from the assertion of Theorem 6.1 at all.

So, Theorem 6.1 guarantees a convergence to the solution of the Boltzmann equation on a finite interval of time. This restriction is closely connected with the fact that the existence of a solution of the Boltzmann equation itself, is proved in a general case only locally in time (cf Chap. 11). Conditions 1), 2) (reformulated in terms of microvariables) may be verified for a wide class of Gibbs measures with the value of activity $z(\varepsilon)$ depending on ε.

In the (unpublished) dissertation of King, the results of Lanford are extended onto a wider class of interaction potentials.

Lanford's method is based on the fact that the moment function $\ell_\varepsilon(t) = \ell_{P_t^\varepsilon}$ which gives the solution of Bogolyubov hierarchy equations (10.44) is expanded into the series of perturbation theory which converges for small enough t. The main term of the expansion is the term corresponding to the free motion of particles, and the terms corresponding to interaction between particles play the role of a "small" perturbation. After that, one passes separately to the limit for all terms, and thereby an analogous limiting series of perturbation theory arises which gives a solution of the Boltzmann equation.

In the recent paper [IP] the Boltzmann-Grad limit is performed for all times $t \in R^1$ for the two-dimensional case of a rarified gas.[5]

A problem which seems to be interesting is the study of the trajectory of a given particle under the conditions of the Boltzmann-Grad limit passage. It is natural to expect that this behaviour is described by some nonlinear Markov process whose infinitesimal generator at the time moment t is determined by the unconditional probability distribution for the state of the process at this moment. This generator is connected with an operator which describes the linearized Boltzmann equation. The trajectory of the process consists of intervals of deterministic motion with a constant speed along a straight line, interrupted by moments of random jump change of the velocity. Papers [Ta1], [Ta2], [U] are devoted to the study of such processes from a probabilistic point of view. As Spohn showed in [Sp1], under the conditions of Theorem 6.1, the probability distribution of the trajectory of a given particle converges weakly, for $0 \leq t \leq t_0(z, \beta)$, to the corresponding nonlinear Markov process. The problem of studying time fluctuations in the Boltzmann-Grad limit is of great interest as well. In terms of microvariables this problem in stated in the following way. Let $\mathbf{x}^\varepsilon(t)$, $t \in [0, \infty)$, be a trajectory of the motion in a domain $\mathcal{O} \subset R^d$ subjected to an interaction potential U_ε. An initial probability distribution P_0^ε for $\mathbf{x}^\varepsilon(0)$ determines the probability measure on the space of trajectories and hence defines the joint distribution of random variables

$$\zeta_g^\varepsilon(t) = \sum_{(q,p) \in \mathbf{X}^\varepsilon(t)} g(\varepsilon q, p) \qquad (10.68)$$

where g is a smooth function with a compact support. Change of the average value

$$E^\varepsilon(t, g) = \varepsilon^{d-1} E \zeta_g^\varepsilon(t) = \varepsilon^{d-1} \int_{R^d \times R^d} g(q,p) \ell_{P_{\varepsilon^{-1}t}^\varepsilon}(\varepsilon^{-1}q, p) \, dq \, dp \qquad (10.69)$$

is described in the limit, as $\varepsilon \to 0$, by the Boltzmann equation. The standard deviation of $\zeta_g^\varepsilon(t)$ is of the order $\varepsilon^{-(d-1)/2}$ and therefore it is natural to consider the normalized value

$$\eta_g^\varepsilon(t) = \varepsilon^{(d-1)/2}(\zeta_g^\varepsilon(t) - E^\varepsilon(t, g)). \qquad (10.70)$$

In papers [Sp1], [Sp2] it is proved that, under the conditions of Theorem 6.1 and the assumption that the initial probability measure P_0^ε is a Gibbs distribution in a volume $\mathcal{O}^\varepsilon$ with a potential U_ε, inverse temperature β and an appropriate value of the activity $z(\varepsilon)$, there exists a limit, as $\varepsilon \to 0$, of the covariance $\mathrm{Cov}(\eta_{g_1}^\varepsilon(t)\eta_{g_2}^\varepsilon(t))$. Evolution of limiting covariances may be described by the linearized Boltzmann equation. The natural conjecture that the values $\eta_g^\varepsilon(t)$ have in the limit a Gaussian distribution remains open.

In the review [Sp1] one noticed the possibility of extending Lanford's construction to the model of Lorentz gas (cf Sect. 5.1) where, in the low density limit,

---

[5] This result is now extended by Illner and Pulvirenti to the three-dimensional case, too.

a linear equation arises of the Kolmogorov type equation in the theory of Markov processes.

6.3. The Vlasov Equation. Under the assumption that $U(\|q\|)$ is a smooth function of the variable $q \in R^d$, $d \geq 1$, this equation takes the form

$$\frac{\partial}{\partial t} f_t(q,p) = -(p, \text{grad}_q f_t(q,p)) - (\text{grad}_p f_t(q,p)), \qquad (10.71)$$

$$\int_{R^d \times R^d} dq_1 \, dp_1 \, \text{grad}_q U(\|q - q_1\|) f_t(q_1, p_1)).$$

If one considers a motion in a bounded domain $\mathcal{O} \subset R^d$, then the integration is on the set $\mathcal{O} \times R^d$ and one introduces the condition of elastic reflection.

The physical meaning of the Vlasov equation is illustrated by the corresponding line of the table. In microvariables, in the course of the mean field limit passage, the radius of interaction of the potential increases, but the value of the interaction potential between fixed particles tends to 0 in such a way that the force of interaction between a given particle and all other particles in the system has the order ε. Therefore, a finite change of the particle position occurs during the time of the order ε. After passing to macrovariables the trajectory of every particle becomes deterministic in the mean field limit and is given by the equations

$$\dot{q}(t) = p(t),$$
$$\dot{p}(t) = -\int_{R^d \times R^d} f_t(q,p) \text{grad}\, U(\|q - q(t)\|) \, dq \, dp. \qquad (10.72)$$

The Vlasov equation describes the change of the density $f_t(q,p)$ induced by the dynamics of particles of form (10.72).

For the case of a smooth potential U considered in this section, the mathematically rigorous investigation of the limit is not difficult under the assumption that the total mass of the system is finite: one requires only weak additional conditions on the initial distribution (typically, the validity of a law of large numbers). For accuracy we shall consider the motion in a bounded domain $\mathcal{O} \subset R^d$ (analogous statements are valid for the motion in the space R^d).

Theorem 6.2 (cf [Ma], [BH]). *Let a smooth nonnegative integrable function f_0 on $\mathcal{O} \times R^d$ be fixed. Suppose that the initial probability distribution P_0^ε is concentrated on the set $M_\mathcal{O}$ and that for any smooth function $g: \mathcal{O} \times R^d \to R^1$ with compact support, the random variable $\varepsilon^d \xi_g^\varepsilon(0)$ (cf (10.68)) converges in the distribution, as $\varepsilon \to 0$, to the integral*

$$\int_{\mathcal{O} \times R^d} dq \, dp \, g(q,p) f_0(q,p).$$

Then for any $t \in [0, \infty)$ the random variable $\varepsilon^d \xi_g^\varepsilon(t)$ converges in distribution, as $\varepsilon \to 0$, to the integral

$$\int_{\mathcal{O} \times R^d} dq\, dp\, g(q,p) f_t(q,p)$$

where f_t is a solution of the Vlasov equation in the domain $\mathcal{O}$ with the initial function f_0.

Notice that the relations (10.60), (10.62) and (10.63) in the weak sense follow from the condition of Theorem 6.2.

Under the same conditions, one proves the convergence of the trajectory of a single particle to a solution of the equation (10.72).

Existence of the solution of the Vlasov equation (10.71) for all $t \in [0, \infty)$ follows from Theorem 6.2. Uniqueness of the solution was proved in papers [D7], [N] on the basis of the fact that equations (10.72) determine characteristics of the Vlasov equation and some probabilistic constructions.

The problem of the investigation of time fluctuations in the mean field limit is stated in the same way as for the low density limit, was described in Sect. 6.2. The difference is that in (10.70) and (10.71) one must replace $d-1$ by d and the linearized Vlasov equation arises instead of the linearized Boltzmann equation. The complete investigation of this problem including the proof of asymptotic normality of fluctuations was done in the paper of Braun and Hepp [BH]. Results connected with the derivation of the Vlasov equation may be extended to the Lorentz gas as well, where a linearized variant of this equation arises.

Notice however that the integrability of the initial data f_0 (finite total mass) as well as the smoothness of $U(\|q\|)$ are very essential in all the proofs. One of the difficulties arising when f_0 is assumed only to be bounded, is that for slowly decreasing potentials, the integral in the right hand side of (10.71) may diverge. However, only such potentials (for instance, the Coulomb potential $U(r) = 1/r$ in the dimension $d = 3$) are needed in most applications of the Vlasov equation. Another difficulty appears when one takes a potential which is singular near the origin.

6.4. The Landau Equation. This equation has the form

$$\frac{\partial}{\partial t} f_t(q,p) = -(p, \mathrm{grad}_q f_t(q,p))$$
$$+ \left[\int_{R^d} dp_1 f_t(q,p_1) \left(\sum_{j=1}^d \frac{\partial}{\partial p^j} a^j(p-p_1) \right) \right] f_t(q,p) \qquad (10.73)$$
$$+ \left[\int_{R^d} dp_1 f_t(q,p_1) \left(\sum_{j_1,j_2=1}^d \frac{\partial}{\partial p^{j_1}} D_{j_1,j_2}(p-p_1) \frac{\partial}{\partial p^{j_2}} \right) f_t(q,p). \right.$$

Here $a(p) = (a^1(p), \ldots, a^d(p))$ is the drift vector and $D(p) = (D_{j_1,j_2}(p), j_1, j_2 = 1, \ldots, d)$ is the diffusion matrix at a point $p \in R^d$ determined by the potential U.

The physical meaning of the weak coupling limit which leads to the Landau equation may be understood by considering the trajectory of a given particle under conditions which are written in the corresponding line of the table. In microvariables, the particle density has the order of a constant and the interaction radius does not depend on ε. Hence, it is natural to expect that, in time of the order ε^{-1}, approximately ε^{-1} "interaction acts" (collisions) of a given particle with other particles occur. The potential is multiplied by a small factor $\varepsilon^{1/2}$ and hence, during every collision act, the momentum of a particles changes a little and the dispersion of this change is proportional to ε. So, the dispersion of a general change of the momentum during time of the order ε^{-1} is of the order of a constant. It is natural to expect that the momentum of a particle will be determined in the limit by a non-linear Markov process of diffusion type. The Landau equation describes the change of the density of particle distribution whose motion is given by this non-linear Markov process.

At present, there are no mathematical results connected with the derivation of the Landau equation. As Spohn observed in [Sp1], some corollaries of results of Kesten and Papanicolau [KP] may be interpreted as a derivation of the linearized Landau equation for a Lorentz gas.

6.5. Hydrodynamic Equations. The special feature of the hydrodynamic limit passage is that, in microvariables, the radius of interaction of the potential and the density of particles (in the order of value) do not depend on ε. This fact is an imposing obstacle for rigorous investigation of the hydrodynamic limit passage. A large amount of literature is devoted to a discussion of physical and mathematical problems arising here, from which we mention the fundamental papers of Bogolyubov [B1], [B2] (see also the development of these ideas in [Zu1], [Zu2]).

We restrict ourselves to a brief discussion of the statement of the problem. As to the initial probability distribution P_0^ε it is natural to suppose, in addition to conditions of a general character (see conditions 1), 2) in Sect. 4.2), that it is locally translation invariant, i.e., "changes only a little" under space shifts of the order $o(\varepsilon^{-1})$; this agrees with the consideration of the 1-moment function of the form $f(\varepsilon q, p)$ (see the table drawn above).

One can expect that the property of local translation invariance is preserved in the course of the time evolution and moreover, in accordance with the conjectures discussed in Sect. 4.2, the probability distribution $P_{\varepsilon^{-1}t}^\varepsilon$ is approximated, as $\varepsilon \to 0$, in a neighborhood of the point $\varepsilon^{-1}q$ of the order $o(\varepsilon^{-1})$ by a Gibbs distribution with the potential U and parameters (z, β, p_0) depending on q and t. Therefore (at least in the domain of absence of phase transitions), the distribution $P_{\varepsilon^{-1}t}^\varepsilon$ may be approximately described by indicating "local" values of parameters (z, β, p_0) or corresponding "local" values of the invariants of the motion: the specific average density, specific average energy and specific average momentum (cf Sect. 4.1). Unlike the cases considered in the preceding sections, we now write kinetic equations for the set of the above listed parameters of the Gibbs distribution, not for the 1-moment function. More precisely, we connect with the probability measure P, the following family of functions

where
$$G_P(q) = (\bar{n}_P(q), \bar{E}_P(q), \bar{p}_P(q)) \tag{10.74}$$

where

$$\bar{n}_P(q) = \int_{R^d} k_P^{(1)}(q,p)\,dp, \tag{10.75a}$$

$$\bar{p}_P(q) = \int_{R^d} p k_P^{(1)}(q,p)\,dp, \tag{10.75b}$$

$$\bar{E}_P(q) = \int_{R^d} \frac{\|p\|^2}{2} k_P^{(1)}(q,p)\,dp$$
$$+ \frac{1}{2}\int_{(R^d)^3} U(\|\tilde{q}\|) k_P^{(2)}((q,p_1),(q+\tilde{q},p_2))\,d\tilde{q}\,dp_1\,dp_2) \tag{10.75c}$$

where $k_P^{(1)}$, $k_P^{(2)}$ are the 1- and 2-moment functions of measure P, respectively. Under the assumption that the limit for the initial family of probability distributions P_0^ε, $\varepsilon > 0$,

$$G_0(q) = \lim_{\varepsilon \to 0} G_{P_0^\varepsilon}(q), \tag{10.76}$$

exists, it is natural to expect the existence for any $t \in [0, \infty)$ of the limit

$$G_t(q) = \lim_{\varepsilon \to 0} G_{P_{\varepsilon^{-1}t}^\varepsilon}(q) \tag{10.77}$$

which gives a solution of the Euler equation for a compressible fluid with the initial date G_0.

Notice the interesting paper of Morrey [M], where an attempt to investigate the hydrodynamic limit was done in an approach closely connected with that described above. Unfortunately, Morrey introduced additional complicated conditions not only on the initial distribution P_0^ε, but on the distributions P_t^ε obtained in the course of the time evolution and the question of compatibility of these conditions remains open. One expects that, in the next asymptotic (in ε) approximation, the hydrodynamic equation contains additional terms of the order ε and describes the evolution of the system until the time of the order $\varepsilon^{-2}t$ (equations of *Navier-Stokes type*), but this circle of problems is even less clear.

We shall not discuss in detail the problems of existence and uniqueness of a solution of hydrodynamic equations, but we refer the reader to the wide literature existing here (see [Te] and the references therein).

It is worth mentioning an important direction of investigation where one takes the Boltzmann equation as the initial point for the derivation of hydrodynamic equations. These results are described in Sect. 5 of Chap. 11. Notice only that, although these results may be considered as an important stage in the foundation of the hydrodynamic limit passage for the case of a rarefied gas, there are some principal difficulties here from the point of view of the concepts developed in this section. Firstly, as was observed in Sect. 6.2, the derivation of the Boltzmann equation is performed now only for small t while, in the hydrodynamic limit passage, $t \to \infty$. Secondly, the possibility of changing the two limit passages: the low density and hydrodynamic is not clear.

Notice also an interesting attempt to derive hydrodynamic equations from Hamiltonian equations of the so-called motion of vortices [MP].

The situation is essentially simplified for the degenerate models of motion considered in Sect. 5. The simplest model where a non-trivial hydrodynamic equation arises is the model of one-dimensional hard rods. Since the invariant state $P_{z,\sigma}^{r_0}$ is determined here by its 1-moment function $\ell_{P_{z,\sigma}^{r_0}}^{(1)}$, the analog of the Euler equation is the following equation for the function $f_t(q,p) = \lim_{\varepsilon \to 0} \ell_{P_{\varepsilon^{-1}t}^{\varepsilon}}(\varepsilon^{-1}q, p)$:

$$\frac{\partial}{\partial t} f_t(q,p) = -p \frac{\partial}{\partial q} f_t(q,p) + r_0 \frac{\partial}{\partial q} \Big[f_t(q,p) \\ \times \int_{R^1} dp_1 (p - p_1) f_t(q, p_1) \\ \times \Big(1 - r_0 \int_{R^1} dp_2 f_t(q, p_2) \Big)^{-1} \Big]. \tag{10.78}$$

This equation was first written in the paper [Pe]. In [BDS2] it was proved that, under some sufficiently general assumptions on the initial state, the function $f_t(q,p)$ satisfies equation (10.78). In [BDS2] one proved as well the uniqueness of the solution of equation (10.78). The method used in this paper is based on the fact that the transformation described in Sect. 5.3, which reduces the one-dimensional hard rod system to the ideal gas, may also be extended to the limit equation (10.78).

Bibliography

In the monograph [B1] the hierarchy of equations is written and investigated which describes the change in the time of the moment functions of a probability measure in the course of the motion of interacting particles. A new method of derivation of kinetic equations (Boltzmann, Vlasov and Landau) from the hierarchy equations for moment functions is developed, on the basis of profound general considerations. The series of fundamental facts characterizing the process of convergence to an equilibrium state was formulated for the first time in this literature. In the paper [B2] the first derivation of hydrodynamic equations (Euler equations for a compressible ideal fluid) from the hierarchy equations for moment functions is given. Ideas of the book [B1] and paper [B2] compose the ground of modern conceptions about the connection between kinetic equations and equations which describe the motion of a large system of particles.

The paper [GO] contains a review of mathematical papers concerning the themes of the present chapter which are published mainly from 1968 to 1975. In the monograph [Zu1] the results developing the approach proposed in [B1], [B2] are presented. A review of further results in this direction is contained in [Zu2]. Papers [Ma], [MT] are devoted to results which connect kinetic equations with different types of hierarchy equations (similar to the Bogolyubov hierarchy equations) arising in the description of motion of various physical systems.

The paper [GP2] contains a systematized analysis of results on existence of a solution of the Bogolyubov hierarchy equations obtained by means of functional-analytic approach.

Papers [AGL2], [L3], [L4], which are close to this chapter in methods and style, are devoted to detailed and consequent exposition of results which are discussed in Sections 3, 4, 5.

In the paper [Gr] a limiting procedure was formulated explicitly (the low density, or Boltzmann-Grad limit) by means of which one must get the Boltzmann equation from equations describing the motion of particles (in particular, from the Bogolyubov hierarchy equations). A rigorous derivation of the Boltzmann equation from the Bogolyubov hierarchy equations in the course of the low density limit is given in the paper [L4].

Different aspects of the material presented in Chapter 10 are also discussed in reviews and papers by the authors of this chapter which partly have the character of a review ([DS], [DS1], [DS2], [S1]).

Finally, in connection with the last paragraph of this chapter, we mention papers [BDS2], [H] and the important review [Sp1] which contains the first systematized analysis of the problem of derivation of kinetic equations on the basis of a unified and precisely formulated approach.

For the convenience of the reader, references to reviews in Zentralblatt für Mathematik (Zbl.), compiled using the MATH database, have, as far as possible, been included in this bibliography.

[AGL1] Aizenman, M., Goldstein, S., Lebowitz, J.L.: Ergodic properties of an infinite one-dimensional hard rod system. Commun. Math. Phys. *39*, 289–301 (1975). Zbl. 352.60073

[AGL2] Aizenman, M., Goldstein, S., Lebowitz, J.L.: Ergodic properties of infinite systems. Lect. Notes Phys. *38*, 112–143 (1975). Zbl. 316.28008

[A] Alexander, R.: Time evolution for infinitely many hard spheres. Commun. Math. Phys. *49*, 217–232 (1976)

[AA] Arnold, V.I., Avez, A.: Ergodic problems of classical mechanics. New York—Amsterdam: Benjamin 1968, 286 p. Zbl. 167.229

[B] Beieren, H. van: Transport properties of stochastic Lorentz models. Rev. Mod. Phys. *54*, No. 1, 195–234 (1982)

[B1] Bogolyubov, N.N.: Problems of Dynamical Theory in Statistical Physics [Russian]. Moscow—Leningrad: OGIZ, Gostechnizdat 1946; see also Bogolyubov, N.N., Selected Papers, Vol. 2, 99–196. Kiev: Naukova Dumka 1970 [Russian]. Zbl. 197.536

[B2] Bogolyubov, N.N.: Equations of hydrodynamics in statistical mechanics [Ukrainian]. Sb. Tr. Inst. Mat. Akad. Nauk USSR *10*, 41–59 (1948); see also Bogolyubov, N.N., Selected Papers, Vol. 2, 258–276. Kiev: Naukova Dumka 1970 [Russian]. Zbl. 197.536

[BK] Bogolyubov, N.N., Khatset, B.I.: On some mathematical questions in the theory of statistical equilibrium. Dokl. Akad Nauk SSSR *66*, 321–324 (1949) [Russian] Zbl. 39.217, see also Bogolyubov, N.N., Selected Papers, Vol. 2, 494–498. Kiev: Naukova Dumka 1970 [Russian]. Zbl. 197.536

[BPK] Bogolyubov, N.N., Petrina, D.Ya., Khazet, B.I.: Mathematical description of an equilibrium state of classical systems on the basis of the canonical ensemble formalism. Teor. Mat. Fiz. *1*, No. 2, 251–274 (1969) [Russian]

[BDS1] Boldrighini, C., Dobrushin, R.L., Sukhov, Yu.M.: Hydrodynamics of one-dimensional hard rods. Usp. Mat. Nauk. *35*, No. 5, 252–253 (1980) [Russian]

[BDS2] Boldrighini, C., Dobrushin, R.L., Sukhov, Yu.M.: One dimensional hard rod caricature of hydrodynamics. J. Stat. Phys. *31*, 577–615 (1983)

[BPT] Boldrighini, C., Pellgrinotti, A., Triolo, L.: Convergence to stationary states for infinite harmonic systems. J. Stat. Phys. *30*, 123–155 (1983)

[BM] Botvich, D.A., Malyshev, V.A.: Unitary equivalence of temperature dynamics for ideal and locally perturbed Fermi-gas. Commun. Math. Phys. *91*, 301–312 (1983). Zbl. 547.46053

[BR1] Bratteli, O., Robinson, D.: Operator Algebras and Quantum Statistical Mechanics. I. New York–Heidelberg–Berlin, Springer-Verlag 1979. Zbl. 421.46048

Bibliography

[BR2] Bratteli, O., Robinson, D.: Operator Algebras and Quantum Statistical Mechanics. II. New York–Heidelberg–Berlin, Springer-Verlag 1981. Zbl. 463.46052

[BH] Braun, W., Hepp, K.: The Vlasov dynamics and its fluctuations in the 1/N limit of interacting classical particles. Commun. Math. Phys. 56, 101–113 (1977)

[Bu] Bunimovich, L.A.: Dynamical systems with elastic reflections. Usp. Mat. Nauk. 39, No. 1, 184–185 (1984) [Russian]

[BS1] Bunimovich, L.A., Sinai, Ya.G.: Markov partitions for dispersed billiards. Commun. Math. Phys. 78, 247–280 (1980). Zbl. 453.60098

[BS2] Bunimovich, L.A., Sinai, Ya.G.: Statistical properties of Lorentz gas with a periodic configuration of scatterers. Commun. Math. Phys. 78, 479–497 (1981). Zbl. 459.60099

[C] Chulaevsky, V.A.: The method of the inverse problem of scattering theory in statistical physics. Funkts. Anal. Prilozh. 17, No. 1, 53–62 (1983) [Russian]

[CFS] Cornfeld, I.P., Fomin, S.V., Sinai, Ya.G.: Ergodic theory. New York–Heidelberg–Berlin, Springer-Verlag 1982. 486 p. Zbl. 493.28007

[DIPP] De Masi, A., Ianiro, N., Pellegrinotti, A., Presutti, E.: A survey of the hydrodynamical behaviour of many-particle systems. In: Nonequilibrium phenomena. II, pp. 123–294. Stud. Stat. Mech. 11. Amsterdam–Oxford–New York, North-Holland 1984. Zbl. 567.76006

[D1] Dobrushin, R.L.: On the Poisson law for the distribution of particles in space. Ukr. Mat. Z. 8, No. 2, 127–134 (1956) [Russian]. Zbl. 73.352

[D2] Dobrushin, R.L.: Gibbsian random fields for lattice system with pair interaction. Funkts. Anal. Prilozh. 2, No. 4, 31–43 (1968) [Russian]

[D3] Dobrushin, R.L.: The uniqueness problem for a Gibbsian random field and the problem of phase transitions. Funkts. Anal. Prilozh. 2, No. 4, 44–57 (1968) [Russian]. Zbl. 192.617. English transl.: Funct. Anal. Appl. 2, 302–312 (1968)

[D4] Dobrushin, R.L.: Gibbsian random fields. General case. Funkts. Anal. Prilozh. 3, No. 1, 27–35 (1969) [Russian]. Zbl. 192.618. English transl.: Funct. Anal. Appl. 3, 22–28 (1969)

[D5] Dobrushin, R.L.: Gibbsian random fields for particles without hard core. Teor. Mat. Fiz. 4, No. 1, 101–118 (1970) [Russian]

[D6] Dobrushin, R.L.: Conditions for the absence of phase transitions in one-dimensional classical systems. Mat. Sb., Nov. Ser. 93, No. 1, 29–49 (1974) [Russian]. Zbl. 307.60081. English transl.: Math. USSR, Sb. 22 (1974) 28–48 (1975)

[D7] Dobrushin, R.L.: Vlasov equations. Funkts. Anal. Prilozh. 13, No. 2, 48–58 (1979) [Russian]. Zbl. 405.35069. English. transl.: Funct. Anal. Appl. 13, 115–123 (1979)

[DF] Dobrushin, R.L., Fritz, J.: Nonequilibrium dynamics of one-dimensional infinite particle systems with a singular interaction. Commun. Math. Phys. 55, No. 3, 275–292 (1977)

[DSS] Dobrushin, R.L., Siegmund-Schultze, R.: The hydrodynamic limit for systems of particles with independent evolution. Math. Nachr. 105, No. 1, 199–224 (1982)

[DS] Dobrushin, R.L., Sinai, Ya.G.: Mathematical problems in statistical mechanics. Math. Phys. Rev. 1, 55–106 (1980). Zbl. 538.60094

[DS1] Dobrushin, R.L., Sukhov, Yu.M.: On the problem of the mathematical foundation of the Gibbs postulate in classical statistical mechanics. Lect. Notes Phys. 80, 325–340 (1978). Zbl. 439.70016

[DS2] Dobrushin, R.L., Sukhov, Yu.M.: Time asymptotics for some degenerated models of evolution of systems with infinitely many particles. In: Itogi Nauki Tekh., Ser. Sovrem. Probl. Mat. 14, 147–254. (1979) [Russian]. Zbl. 424.60096. English transl.: J. Sov. Math. 16, 1277–1340 (1981)

[Do] Doob, J.: Stochastic processes. Chichester–New York–Brisbane–Toronto, Wiley 1953. Zbl. 53.268

[Du] Durret, R.: An introduction to infinite particle systems. Stochast. Process Appl. 11, No. 2, 109–150 (1981)

[FF] Fichtner, K.H., Freudenberg, W.: Asymptotic behaviour of time evolutions of infinite

	particle systems. Z. Wahrscheinlichkeitstheorie Verw. Gebiete. *54*, 141–159 (1980). Zbl. 438.60088
[F1]	Fritz, J.: An ergodic theorem for the stochastic dynamics of quasi-harmonic crystals. In: Random fields. Rigorous results in statistical mechanics and quantum field theory. Coll. Math. Soc. J; Bolyai, 27 vol. 373–386. Amsterdam–Oxford–New York, North-Holland 1981. Zbl. 489.60066
[F2]	Fritz, J.: Infinite lattice systems of interacting diffusion processes. Existence and regularity properties. Z. Wahrscheinlichkeitstheor. Verw. Gebiete. *59*, 291–309 (1982). Zbl. 477.60097
[F3]	Fritz, J.: Local stability and hydrodynamical limit of Spitzer's one-dimensional lattice model. Commun. Math. Phys. *86*, 363–373 (1982). Zbl. 526.60092
[F4]	Fritz, J.: Some remarks on nonequilibrium dynamics of infinite particle systems. J. Stat. Phys. *34*, 539–556 (1984). Zbl. 591.60098
[FD]	Fritz, J., Dobrushin, R.L.: Nonequilibrium dynamics of two-dimensional infinite particle systems with a singular interaction. Commun. Math. Phys. *57*, 67–81 (1977)
[GLL]	Gallavotti, G., Lanford, O.E., Lebovitz, J.L.: Thermodynamic limit of time dependent correlation functions for one dimensional systems. J. Math. Phys. *13*, No. 11, 2898–2905 (1972)
[GMS]	Gallovotti, G., Miracle-Sole, S.: Absence of phase transition in hard-core one-dimensional systems with long-range interaction. J. Math. Phys. *11*, No. 1, 147–155 (1970)
[G]	Gerasimenko, V.I.: The evolution of infinite system of particles interacting with nearest neighbours. Dokl. Akad. Nauk Ukr. SSR, Ser. A. No. 5, 10–13 (1982) [Russian]. Zbl. 486.45009
[GMP]	Gerasimenko, V.I., Malyshev, P.V., Petrina, D.Ya.: Mathematical foundations of classical statistical mechanics. Kiev. Naukova Dumka 1985 [Russian]
[GP1]	Gerasimenko, V.I., Petrina, D.Ya.: Statistical mechanics of quantum-classical systems. Nonequilibrium systems. Teor. Mat. Fiz. *42*, No. 1, 88–100 (1980) [Russian]
[GP2]	Gerasimenko, V.I., Petrina, D.Ya.: Mathematical description of the evolution of the state of infinite systems in classical statistical mechanics. Usp. Mat. Nauk. *38*, No. 5, 3–58 (1983) [Russian]. Zbl. 552.35069. English transl.: Russ. Math. Surv. *38*, No. 5, 1–61 (1983)
[Gr]	Grad, H.: Principles of the kinetic theory of gases. In: Handbuch der Physik, B. XII, Sekt. 26, pp. 205–294. Berlin, Springer-Verlag, 1958
[Gri]	Griffeath, D.: Additive and cancellative interacting particle systems. Lect. Notes Math. *724* (1979). Zbl. 412.60095
[Gu]	Gurevich, B.M.: Additive integrals of motion of point particles in one dimension. Math. Res. *12*, 59–63 (1982). Zbl. 515.70017
[GO]	Gurevich, B.M., Oseledets, V.I.: Some mathematical problems connected with nonequilibrium statistical mechanics of an infinite number of particles. In: Probab. Teor. Mat. Stat., Teor. Kybern. *14*, 5–40. 1977 [Russian]. Zbl. 405.60093
[GSS]	Gurevich, B.M., Sinai, Ya.G., Sukhov, Yu.M.: On invariant measures of dynamical systems of one dimensional statistical mechanics. Usp. Mat. Nauk. *28*, No. 5, 45–82 (1973) [Russian]. Zbl. 321.28011. English transl.: Russ. Math. Surv. *28*, No. 5, 49–86 (1973)
[GS1]	Gurevich, B.M., Sukhov, Yu.M.: Stationary solutions of the Bogolyubov hierarchy equations in classical statistical mechanics. 1–4. Commun. Math. Phys. *49*, No. 1, 69–96 (1976); *54*, No. 1, 81–96 (1977); *56*, No. 3, 225–236 (1977); *84*, No. 4, 333–376 (1982)
[GS2]	Gurevich, B.M., Sukhov, Yu.M.: Stationary solutions of the Bogolyubov hierarchy in classical statistical mechanics. Dokl. Akad. Nauk SSSR *223*, No. 2, 276–279 (1975) [Russian]. Zbl. 361.60090. English transl.: Sov. Math., Dokl. *16*, 857–861 (1976)
[GS3]	Gurevich, B.M., Sukhov, Yu.M.: Time evolution of the Gibbs states in one-dimensional classical statistical mechanics. Dokl. Akad. Nauk SSSR. *242*, No. 2, 276–279 (1978) [Russian]. Zbl. 431.46054. English transl.: Sov. Math., Dokl. *19*, 1088–1091 (1978)
[H]	Hauge, E.G.: What can one learn from Lorentz models? Lect. Notes Phys. *31*, 337–353 (1974)

Bibliography

[IP]   Illner, R., Pulvirenti, M.: Global validity of the Boltzmann equation for a two-dimensional rare gas in vacuum. Commun. Math. Phys. *105*, No. 2, 189–203 (1986). Zbl. 609.76083

[K1]   Kallenberg, O.: Random measures: London–New York–San Francisco, Academic Press., 104 p. 1976. Zbl. 345.60032; Akademie-Verlag 1975. Zbl. 345.60031

[K2]   Kallenberg, O.: On the asymptotic behaviour of line processes and systems of non-interacting particles. Z. Wahrscheinlichkeitstheorie Verw. Gebiete *43*, N 1, 65–95 (1978). Zbl. 366.60086 (Zbl. 376.60059)

[KP]   Kesten, H., Papanicolau, G.: A limit theorem for turbulent diffusion. Commun. Math. Phys. *65*, No. 2, 97–128 (1979). Zbl. 399.60049

[K]   Kinetic theories and the Boltzmann equation. ed. C. Cercignani Lect. Notes Math. *1048*, 1984. Zbl. 536.00019

[Ko]   Kozlov, O.K.: Gibbsian description of point random fields. Teor. Veroyatn. Primen. *21*, No. 2, 348–365 (1976) [Russian]. Zbl. 364.60086. English transl.: Theory Probab. Appl. *21* (1976), 339–356 (1977)

[KS]   Kramli, A., Szász, D.: On the convergence of equilibrium of the Lorentz gas. In: Functions, series, operators. Colloq. Math. Soc. János Bolyai *35*, vol. II. 757–766. Amsterdam–Oxford–New York, North-Holland 1983. Zbl. 538.60099

[L1]   Lanford, O.E.: The classical mechanics of one-dimensional systems of infinitely many particles. I. An existence theorem. Commun. Math. Phys. *9*, 176–191 (1968). Zbl. 164.254

[L2]   Lanford, O.E.: The classical mechanics of one-dimensional systems of infinitely many particles. II. Kinetic theory. Commun. Math. Phys. *11*, 257–292 (1969). Zbl. 175.214

[L3]   Lanford, O.E.: Ergodic theory and approach to equilibrium for finite and infinite systems. Acta Phys. Aust., Suppl. *10*, 619–639 (1973)

[L4]   Lanford, O.E.: Time evolution of large classical systems. Lect. Notes Phys. *38*, 1–111 (1975). Zbl. 329.70011

[LL]   Lanford, O.E., Lebowitz, J.L.: Time evolution and ergodic properties of harmonic systems. Lect. Notes Phys. *38*, 144–177 (1975). Zbl. 338.28011

[LLL] Lanford, O.E., Lebowitz, J.L., Lieb, E.: Time evolution of infinite anharmonic systems. J. Stat. Phys. *16*, No. 6, 453–461 (1977)

[LR]   Lanford, O.E., Ruelle, D.: Observable at infinitely and states with short range correlations in statistical mechanics. Commun. Math. Phys. *13*, 194–215 (1969)

[L]   Lang, R.: On the asymptotic behaviour of infinite gradient systems. Commun. Math. Phys. *65*, 129–149 (1979). Zbl. 394.60098

[LS]   Lebowitz, J.L., Spohn, H.: Steady self-diffusion at low density. J. Stat. Phys. *29*, 39–55 (1982)

[Le1] Lenard, A.: Correlation functions and the uniqueness of the state in classical statistical mechanics. Commun. Math. Phys. *30*, 35–44 (1973)

[Le2] Lenard, A.: States of classical statistical mechanical systems of infinitely many particles. I, II. Arch. Rat. Mech. Anal. *50*, 219–239; 241–256 (1975)

[Li]   Liggett, T.M.: The stochastic evolution of infinite systems of interacting particles. Lect. Notes Math. *598*, 187–248 (1977). Zbl. 363.60109

[M]   Malyshev, P.V.: Mathematical description of evolution of classical infinite system. Teor. Mat. Fiz. *44*, No. 1, 63–74 (1980) [Russian]

[MPP] Marchioro, C., Pellegrinotti, A., Pulvirenti, M.: Remarks on the existence of non-equilibrium dynamics. In: Random fields. Rigorous results in statistical mechanics and quantum field theory, Colloq. Math. Soc. Janos Bolyai *27*, vol. II, 733–746. Amsterdam–Oxford–New York, North-Holland 1981. Zbl. 496.60100

[MP]   Marchioro, C., Pulvirenti, M.: Vortex methods in two-dimensional fluid dynamics. (Roma, Edizione Klim), 1984. Zbl. 545.76027. Lect. Notes Phys. *203*. Berlin etc.: Springer-Verlag III, 137 p. (1984)

[Ma]   Maslov, V.P.: Equations of the self-consisted field. In: Itogi Nauki Tekh., Ser. Sovrem. Probl. Mat. *11*, 153–234. (1978) [Russian]. Zbl. 404.45008. English transl.: J. Sov. Math. *11*, 123–195 (1979)

[MT] Maslov, V.P., Tariverdiev, C.E.: The asymptotics of the Kolmogorov-Feller equation for a system of a large number of particles. In: Itogi Nauki Tekh., Ser. Teor. Veroyatn. Mat. Stat. Teor. Kibern. *19*, 85–125 (1982) [Russian]. Zbl. 517.60100. English transl.: J. Sov. Math. *23*, 2553–2579 (1983)

[MKM] Matthes, K., Kerstan, J., Mecke, J.: Infinitely divisible point processes. Chichester–New York—Brisbane–Toronto, Wiley 1978. 532 p. Zbl. 383.60001

[M1] Minlos, R.A.: The limit Gibbs distribution. Funkts. Anal. Prilozh. *1*, No. 2, 60–73 (1967) [Russian]. Zbl. 245.60080. English transl.: Funct. Anal. Appl. *1*, 140–150 (1968)

[M2] Minlos, R.A.: Regularity of the limit Gibbs distribution. Funkts. Anal. Prilozh. *1*, No. 3, 40–53 (1967) [Russian]. Zbl. 245.60081. English transl.: Funct. Anal. Appl. *1*, 206–217 (1968)

[M] Morrey, C.B.: On the derivation of the equations of hydrodynamics from statistical mechanics. Commun. Pure Appl. Math. *8*, 279–326 (1955)

[N] Neunzert, H.: The Vlasov equation as a limit of Hamiltonian classical mechanical systems of interacting particles. Trans. Fluid Dynamics *18*, 663–678 (1977)

[No] Nonequilibrium phenomena I. The Boltzmann equation. Studies in Statistical Mechanics, Z (J.L. Lebowitz, E.W. Montroll, Eds.) Amsterdam–New York–Oxford, North-Holland 1983. Zbl. 583.76004

[P] de Pazzis, O.: Ergodic properties of a semi-infinite hard rods systems. Commun. Math. Phys. *22*, 121–132 (1971). Zbl. 236.60071

[Pe] Perkus, J.K.: Exact solution of kinetics of a model of classical fluid. Phys. Fluids. *12*, 1560–1563 (1969)

[Pet] Petrina, D.Ya.: Mathematical description for the evolution of infinite systems in classical statistical physics. Locally perturbed one dimensional systems. Teor. Mat. Fiz. *38*, No. 2, 230–250 (1979) [Russian]

[PPT] Presutti, E., Pulvirenti, M., Tirozzi, B.: Time evolution of infinite classical systems with singular, long range, two body interactions. Commun. Math. Phys. *47*, 85–91 (1976)

[Pu] Pulvirenti, M.: On the time evolution of the states of infinitely extended particles systems. J. Stat. Phys. *27*, 693–709 (1982). Zbl. 511.60096

[R1] Ruelle, D.: Correlation functions of classical gases. Ann. Phys. *25*, 109–120 (1963)

[R2] Ruelle, D.: Statistical Mechanics. Rigorous results. New York–Amsterdam, Benjamin 1969. Zbl. 177.573

[R3] Ruelle, D.: Superstable interactions in classical statistical mechanics. Commun. Math. Phys. *18*, 127–159 (1970)

[SS] Siegmund-Schultze, R.: On non-equilibrium dynamics of multidimensional infinite particle systems in the translation invariant case. Commun. Math. Phys. *100*, 245–265 (1985). Zbl. 607.60097

[S1] Sinai, Ya.G.: Ergodic theory. Acta Phys. Aust., Suppl. *10*, 575–606 (1973)

[S2] Sinai, Ya.G.: Ergodic properties of a gas of one dimensional hard rods with an infinite number of degrees of freedom. Funkts. Anal. Prilozh. *6*, No. 1, 41–50 (1972) [Russian]. Zbl. 257.60036. English. transl.: Funct. Anal. Appl. *6*, 35–43 (1972)

[S3] Sinai, Ya.G.: Construction of dynamics in one dimensional systems of statistical mechanics. Teor. Mat. Fiz. *11*, No. 2, 248–258 (1972) [Russian]

[S4] Sinai, Ya.G.: Construction of cluster dynamics for dynamical systems of statistical mechanics. Vestn. Mosk. Gos. Univ. Mat. Mekh. *29*, No. 3, 152–158 (1974) [Russian]

[S5] Sinai, Ya.G.: Ergodic properties of the Lorentz gas. Funkt. Anal. Prilozh. *13*, No. 3, 46–59 (1979) [Russian]. Zbl. 414.28015. English transl.: Funct. Anal. Appl. *13*, 192–202 (1980)

[S6] Sinai, Ya.G.: The theory of phase transitions. M.: Nauka 1980 [Russian]. Zbl. 508.60084

[Sk] Skripnik, V.I.: On generalized solutions of Gibbs type for the Bogolyubov-Streltsova diffusion hierarchy. Teor. Mat. Fiz. *58*, No. 3, 398–420 (1984) [Russian]

[Sp1] Spohn, H.: Kinetic equations from Hamiltonian dynamics: Markovian limits. Rev. Mod. Phys. *52*, No. 3, 569–615. (1980)

[Sp2]	Spohn, H.: Fluctuation theory for the Boltzmann equation. In: Nonequilibrium phenomena I. The Boltzmann equation. Studies in statistical mechanics; pp. 418–439. (J.L. Lebowitz, E.W. Montroll, Eds.). Amsterdam–Oxford–New York, North-Holland 1983
[Sp3]	Spohn, H.: Hydrodynamical theory for equilibrium time correlation functions of hard rods. Ann. Phys. *141*, No. 2, 353–364 (1982)
[Su1]	Sukhov, Yu.M.: Random point processes and DLR equations. Commun. Math. Phys. *50*, No. 2, 113–132 (1976). Zbl. 357.60056
[Su2]	Sukhov, Yu.M.: Matrix method for continuous systems of classical statistical mechanics. Tr. Mosk. Mat. O.-va. *24*, 175–200 (1971) [Russian]. Zbl. 367.60117. English transl.: Trans. Mosc. Math. Soc. *24*, 185–212 (1974)
[Su3]	Sukhov, Yu.M.: The strong solution of the Bogolyubov hierarchy in one dimensional classical statistical mechanics. Dokl. Akad. Nauk SSSR *244*, No. 5, 1081–1084 (1979) [Russian]. Zbl. 422.45007. English transl.: Sov. Math., Dokl. *20*, 179–182 (1979)
[Su4]	Sukhov, Yu.M.: Stationary solutions of the Bogolyubov hierarchy and first integrals of the motion for the system of classical particles. Teor. Mat. Fiz. *55*, No. 1, 78–87 (1983) [Russian]
[Su5]	Sukhov, Yu.M.: Convergence to a Poissonian distribution for certain models of particle motion. Izv. Akad. Nauk SSSR, Ser. Mat. *46*, No. 1, 135–154 (1982) [Russian]. Zbl. 521.60063. English transl.: Math. USSR, Izv. *20*, 137–155 (1983)
[T1]	Takahashi, Y.: A class of solutions of the Bogolyubov system of equations for classical statistical mechanics of hard core particles. Sci. Papers Coll. Gen. Educ., Univ. Tokyo. *26*, No. 1, 15–26 (1976). Zbl. 402.70013
[T2]	Takahashi, Y.: On a class of Bogolyubov equations and the time evolution in classical statistical mechanics. In: Random fields. Rigorous results in statistical mechanics and quantum field theory, vol. II. Amsterdam–Oxford–New York. Colloq. Math. Soc. János Bolyai, *27*, 1033–1056 (1981). Zbl. 495.28017
[Ta1]	Tanaka, H.: On Markov processes corresponding to Boltzmann's equation of Maxwellian gas. In: Proc. Second Japan–USSR Sympos. Prob. Theory, Kyoto. 1972, Lect. Notes Math. *330*, 478–489 (1973). Zbl. 265.60095
[Ta2]	Tanaka, H.: Stochastic differential equation associated with the Boltzmann equation of Maxwellian molecules in several dimensions. In: Stochastic analysis. London–New York–San Francisco, Acad. Press 1978. 301–314. Zbl. 445.76056. Proc. int. Conf., Evanston. Ill.
[Te]	Temam, R.: Navier-Stokes equations. Theory and numerical analysis. Stud. Math. Appl. 2 Amsterdam–New York–Oxford, North-Holland P.C., 1977. 500 p. Zbl. 383.35057
[U]	Ueno, T.: A stochastic model associated with the Boltzmann equation. Second Japan–USSR Symp. Prob. Theory. v. 2, 183–195, Kyoto 1972
[VS]	Volkovysskij, K.L., Sinai, Ya.G.: Ergodic properties of an ideal gas with an infinite number of degrees of freedom. Funkt. Anal. Prilozh. *5*, No. 3, 19–21 (1971) [Russian]. Zbl. 307.76005. English transl.: Funct. Anal. Appl. *5* (1971), 185–187 (1972)
[W]	Willms, J.: Convergence of infinite particle systems to the Poisson process under the action of the free dynamics. Z. Wahrscheinlichkeitstheor. Verw. Gebiete *60*, 69–74 (1982). Zbl. 583.60096. Infinite dimensional analysis and stochastic processes, Semin. Meet., Bielefeld 1983, Res. Notes Math. *124*, 161–170 (1985)
[Z]	Zemlyakov, A.N.: The construction of the dynamics in one dimensional systems of statistical physics in the case of infinite range potentials. Usp. Mat. Nauk. *28*, No. 1, 239–240 (1973) [Russian]
[Z1]	Zessin, H.: Stability of equilibria of infinite particle systems. III. International Vilnius Conf. Prob. Theory Math. Stat. vol. III, 371–372, Vilnius 1981
[Z2]	Zessin, H.: The method of moments for random measures. Z. Wahrscheinlichkeitstheor. Verw. Gebiete *62*, 395–409 (1983). Zbl. 503.60060

[Z3]   Zessin, H.: BBGKY hierarchy equations for Newtonian and gradient systems. In: Infinite dimensional analysis and stochastic processes Semin. Meet., Bielefeld 1983, Res. Notes Math. *124* (S. Albeverio, Ed.), pp. 171–196. Boston–London–Melbourne: Pitman Adv. Publ. Program, 1985. Zbl. 583.60097

[Zu1] Zubarev, D.N.: Nonequilibrium statistical thermodynamics. Moscow, Nauka 1973 [Russian]

[Zu2] Zubarev, D.: Modern methods of statistical theory of nonequilibrium processes. Itogi Nauki Tekh. Ser. Sovrem. Probl. Mat. *15*, 131–226 (1980) [Russian]. Zbl. 441.60099. English transl.: J. Sov. Math. *16*, 1509–1572 (1981)

Chapter 11
Existence and Uniqueness Theorems for the Boltzmann Equation

N.B. Maslova

§ 1. Formulation of Boundary Problems. Properties of Integral Operators

1.1. The Boltzmann Equation. In general, the situation which arises when mathematically studying the Boltzmann equation may be outlined as follows. There are two well-studied limiting regimes. The first of them is the so-called free-particle flow, where the particles do not interact. The second one corresponds to thermodynamic equilibrium which is described by the Maxwell distribution. Almost all theorems which are known at present guarantee existence of a solution of boundary problems in situations which are close to one of the mentioned regimes. The only problem for which one succeeds to prove global eixtence without serious restrictions on initial deta, is the Cauchy problem for a homogeneous gas.

Flows which are close to free-particle flows are described by the non-linear Boltzmann equation with a small parameters ε in front of the collision integral (the value ε^{-1} is called the Knudsen number). Local existence theorems for non-stationary problems (such theorems are now proven under very general conditions) may be regarded as statements about solutions of the Boltzmann equation for large Knudsen numbers.

More difficult and deeper results are related to the study of those flows which are close to equilibrium. The first approximation for describing these flows is given by the linearized Boltzmann equation. Boundary problems for this equation turn out to be hard enough. However, these problems have now been studied quite well so that one can effectively use them for investigating solutions of

Chapter 11. Uniqueness Theorems for the Boltzmann Equation

non-linear equations. In this way one has obtained theorems which guarantee existence for stationary problems and global existence for non-stationary problems.

Below, only the simplest variant of the Boltzmann equation is considered. This is a kinetic equation which describes the motion in R^3 of a system of identical particles which interact via a pair potential depending on the distance between particles (cf Chap. 10, Sect. 2). One may include into consideration, without difficult mixtures of particles of different types which are moving, subject to extra forces.

As noted in Sect. 6 of Chap. 10, the main characteristic of a gas in kinetic theory is the 1-moment function (*Boltzmann distribution density*) which is denoted in the sequel[1] as $F(t, \xi, x)$ and describes the distribution of particles in coordinates x ($x \in \Omega, \Omega \subseteq R^m, m = 1, 2, 3$) and momenta ξ ($\xi \in R^3$). More precisely, this function describes the density of a distribution of particles in the phase space $R^3 \times \Omega$ at the time moment t. In the limit of low density, the function $F(t, \xi, x)$ must satisfy the Boltzmann equation

$$\frac{\partial}{\partial t} F + DF = J(F, F).$$

Here $D = \sum_{\alpha=1}^{m} \xi_\alpha \frac{\partial}{\partial x_\alpha}$, J is the *collision integral* given, at the point t, ξ, x, by the formulae:

$$J(F, F) = \int_{R^3 \times \Sigma} \left[F(t, \xi', x) F(t, \xi_1', x) \right.$$
$$\left. - F(t, \xi, x) F(t, \xi_1, x) \right] B(|\xi - \xi_1|, |\xi - \xi_1|^{-1} |\langle \xi - \xi_1, \alpha \rangle|) \, d\alpha \, d\xi,$$

$$\Sigma = \{\alpha \in R^3 | |\alpha| = 1\},$$

$$\xi' = \xi - \alpha \langle \xi - \xi_1, \alpha \rangle, \quad \xi_1' = \xi_1 + \alpha \langle \xi - \xi_1, \alpha \rangle.$$

The non-negative function B, continuous on $(0, \infty) \times (0, 1)$, is uniquely determined by the interaction potential between particles. The main mathematical results are obtained for the class of "hard" potentials, introduced by Grad [G2]. This class is determined by the following conditions on the function B:

$$B(v, z) \leq b^1 z(v + v^{\beta-1}), \qquad (11.1)$$

$$\int_0^1 B(v, z) \, dz \geq b^0 v(v + 1)^{-1}, \qquad (11.2)$$

where $\beta \in (0, 1)$ and $b^i (i = 0, 1)$ are positive constants. The potential of hard spheres (balls) (cf (10.14)) for which $B = b^1 vz$ is, of course, in this class. For the case of power potentials $U(r) = cr^{-s}$, classical mechanics gives the relation

---

[1] The system of notations adopted in this chapter differs somewhat from that of Chap. 10. This fact is explained by traditions established in literature in this area.

$$B(v, z) = v^p z b(z) b_1(z), \ p = 1 - 4/s, \qquad (11.3)$$

where the function b is continuous on $[0, 1]$ and $b_1(z) = z^{p-3}$. From the physical point of view, the singularity of the function b_1 at the origin is connected with the interaction at large distances. If this interaction is essential, then the Boltzmann equation gives a bad approximation. Replacing the function b_1 by a constant (which is called the "angular cut-off") guarantees the fulfilment of Grad's conditions for $s \geqslant 4$.

For slowly decreasing potentials ($s < 4$), the second Grad condition which guarantees boundedness from below of the frequency of collisions is violated. Such potentials are included in a class of "*soft*" *potentials*, which was introduced by Grad, and is determined by the relation (1.1) and condition

$$\int_0^1 B(v, z) \, dz \leqslant b^1 (1 + v\beta^{-1}), \ \beta \in (0, 1) \qquad (11.4)$$

which guarantees boundedness of the frequency of collisions from above. A specific position in kinetic theory is occupied by the "Maxwell molecules" determined by the relation (11.3) for $s = 4$. This class was introduced by Maxwell because of the simplicity of the structure of corresponding moment equations. Within this class one has found nontrivial exact solutions of the Boltzmann equation (cf [B1]).

For most interaction potentials for particles in a neutral gas (cf Chap. 10, Sect. 2.4), conditions (11.1), (11.2) are fulfilled. In particular, the functions B, obtained in the course of quantum mechanical computations of collisions, satisfy these conditions. Below we assume that conditions (11.1), (11.2) are fulfilled, whenever a contrary assumption is not explicity specified.

The *Maxwell distribution*

$$\omega(\xi) \equiv \omega(\rho, h, V, \xi) = \rho(h/2\pi)^{3/2} \exp(-h/2|\xi - V|^2)$$

$$(\rho > 0, h > 0, V \in R^3)$$

for any constant values of parameters β, h, V gives an exact solution of the Boltzmann equation. A large area of physical applications and mathematical investigations is connected with the study of perturbations of the state $\omega = \omega(1, 1, 0, \xi)$. The perturbation f determined by the equality $F = \omega(1 + f)$ should be found from the equation

$$\frac{\partial}{\partial t} f + Df = Lf + \Gamma(f, f). \qquad (11.5)$$

Here, the *linearized collision operator*, L, and the operator Γ are given by the formulae

$$\omega Lf = J(\omega(1 + f), \omega(1 + f)) - J(\omega f, \omega f),$$

$$\Gamma(f, f) = J(\omega f, \omega f). \qquad (11.6)$$

Chapter 11. Uniqueness Theorems for the Boltzmann Equation

Omitting from (11.5) the nonlinear part, Γ leads to the linearized Boltzmann equation.

Let S be the boundary of a domain Ω in R^m, $n(x)$ be the internal normal vector to S at a point x. Let us denote by F^- and F^+ the densities of distributions of falling and reflected molecules on S, respectively:

$$F^+ = F\chi(\langle \xi, n(x) \rangle), \quad F^- = F - F^+, \qquad x \in S;$$

χ is the indicator function of the interval $(0, \infty)$.

Boundary conditions on the surface S lead to a connection between the densities of distributions of falling and reflected molecules:

$$F^+ = \mathscr{R}F^- + \Phi^+, \qquad x \in S,$$

$$(\mathscr{R}F^-)(t, \xi, x) = \int_{R^3} r(t, x, \xi, \eta) F^-(t, \eta, x) \, d\eta,$$

with some given functions Φ^+, r.

In the investigation of linear problems, one usually assumes that the operator $\mathscr{R}$ admits a representation $\mathscr{R} = \mathscr{R}_0 + \mathscr{R}_1$, such that $\omega^+ = \mathscr{R}_0 \omega^-$ and the function $\mathscr{R}_1 \omega^- F^-$ is negligibly small. Then the boundary condition takes the form

$$F^+ = RF^- + \Phi^+$$

with $R = (\omega^+)^{-1} \mathscr{R}_0 \omega^-$ and some given function Φ^+. If the domain Ω is unbounded, one adds to the boundary conditions on S the condition at infinity

$$F \to F_\infty \qquad \text{as } |x| \to \infty. \tag{11.7}$$

Hence, the following list of boundary problems may be connected with the Boltzmann equation:

1) the *internal stationary problem*

$$DF = J(F, F), \, x \in \Omega; \, F^+ = \mathscr{R}F^- + \Phi, \, x \in S; \tag{11.8}$$

considered for a bounded domain Ω;

2) the *external stationary problem*: to find functions satisfying the relations (11.8), (11.7) where Ω is the complement of the closure of a bounded domain in R^m;

3) the *nonstationary boundary problem*

$$\frac{\partial}{\partial t} F + DF = J(F, F), \, x \in \Omega, \, t \in (0, T), \tag{11.9}$$

$$F^+ = \mathscr{R}F^- + \Phi^+, \, x \in S, \, t \in (0, T), \tag{11.10}$$

$$F = F_0, \, x \in \Omega, \, t = 0; \tag{11.11}$$

4) the *Cauchy problem* (11.9), (11.11) with $\Omega = R^m$.

One formulates in an analogous way the corresponding problems for the linearized equation.

A specific role is played by problems on initial and boundary Knudsen strata (layers) for applications of the above. The first of them consists in finding a solution of the Cauchy problem independent of the coordinates x. Solutions of this problem describe a "fast" process of approaching local thermodynamical equilibrium (i.e., Maxwell distribution with variable parameters ρ, h, V) and give the possibility to find asymptotical initial data for hydrodynamic equations. The second problem is a generalization of the *Kramers problem* and consists of the description of a gas flow over a plane under the action of a given momentum flow $B_\alpha (\alpha = 1, 2, 3)$ and energy flow B_4. In the linearized stationary version, this problem is reduced to finding a function $f(\xi, x)$ satisfying the conditions:

$$\xi_1 \frac{\partial}{\partial x} f = Lf + g, \, x > 0, \, f^+ = Rf^- + \Phi^+, \, x = 0, \tag{11.12}$$

$$\lim_{x \to \infty} \int_{R^3} (\psi_j - \delta_{j4}(2/3)^{1/2} \psi_0) f \omega \, d\xi = B_j, \, j = 1, \ldots, 4; \tag{11.13}$$

$$\psi_0 = 1, \, \psi_\alpha = \xi_\alpha, \, \alpha = 1, 2, 3, \, \psi_4 = (|\xi|^2 - 3) 6^{-1/2}. \tag{11.14}$$

In the asymptotical approach, a solution of the problem (11.12), (11.13) determines the velocity of sliding and temperature jump on the boundary of a streamlined body and thereby, the correct boundary conditions for hydrodynamic equations.

1.2. Formulation of Boundary Problems. A solution of boundary problems and properties of the collision integral are described below in terms of spaces $L_p(X)$ and $L_p(X, \varphi) = \{f \mid \varphi f \in L_p(X)\}$ where φ is a nonnegative weight function. The spaces $H = L_2(R^3, \omega^{1/2})$, $\mathcal{H} = L_2(R^3 \times \Omega, \omega^{1/2})$ and

$$\mathcal{H}(S) = L_2(R^3 \times S, \omega^{1/2} \varphi), \qquad \varphi = |\langle \xi, n(x) \rangle|^{1/2},$$

play the main role in the investigation of linear problems.

Notations $\langle f, g | X \rangle$ and $\|f | X\|$ are used below for the scalar product and norm in a space X.

Let us write the definition of a solution of the above-formulated boundary problems in a more precise form. For fixed ξ, x we set

$$\hat{F}(\tau) = F(\xi, x + \xi^{(m)}\tau), \, \tau \in R^1, \, \xi^{(m)} = \{\xi_1, \ldots, \xi_m\}.$$

Associated with the space $X = L_p(R^3 \times \Omega, \varphi)$ is the functional class $\mathcal{B}(X)$ consisting of the functions F which satisfy the following conditions: 1) $F \in X$, 2) $F^+ \in L_p(R^3 \times S, \varphi_1)$, $\varphi_1 = \varphi |\langle \xi, n \rangle|^{1/p}$, 3) for almost all ξ, x the function $\hat{F}$ has the generalized first order derivative and $(1 + |\xi|)^{-1} \frac{\partial}{\partial \tau} \hat{F} \in X$. Functions F from $\mathcal{B}(X)$ do not need to have all the derivatives figuring in the Boltzmann equation. However, for these functions, the value of the operator $D \left(DF = \frac{\partial}{\partial \tau} \hat{F} \right)$ and

Chapter 11. Uniqueness Theorems for the Boltzmann Equation 259

boundary value F^- are correctly defined. A solution of the stationary problem (11.8) in $L_p(R^3 \times \Omega, \varphi)$ is a function F from a corresponding class $\mathscr{B}$ satisfying relations (11.8) almost everywhere. In an analogous way, one defines a solution of the non-stationary problems in $L_p([0, T] \times R^3 \times \Omega, \varphi)$. In general, such solutions have no derivatives in t and x_α, but the action of the operator $\frac{\partial}{\partial \tau} + D$ is correctly defined for them.

In constructing solutions of boundary problems, one uses essentially integral forms of kinetic equations. For their description, consider an auxiliary problem

$$Df = g_1 - fg_2, \quad x \in \Omega, \quad f^+ = \Phi^+, \quad x \in S, \quad (11.15)$$

where $g_i (i = 1, 2)$, Φ^+ are given functions and $g_2 \geq 0$. If the ray $\{y \in \Omega | y = x - \xi^{(m)}\tau, \tau > 0\}$ does not intersect the boundary S, then one adds to relation (11.15) the following condition

$$f(\xi, x - \xi^{(m)}\tau) \to 0, \quad \tau \to \infty.$$

A solution, $f = U(g_1, g_2) + E(\Phi^+, g_2)$, of the problem (11.15) leads to the following integral form of the problem (11.6):

$$f = VF, \quad x \in \Omega, \quad F^+ = \mathscr{R}F^- + \Phi^+, \quad x \in S,$$

$$VF = U(\Phi(F, F), Q(F)) + E(F^+, Q(F)),$$

$$\Phi(F, F) = J(F, F) + FQ(F), \quad (11.16)$$

$$Q(F) = 2\pi \int_{R^3} F(\eta, x) \left[\int_0^1 B(|\xi - \eta|, z) \, dz \right] d\eta.$$

By setting

$$v = Q(\omega), \quad Kf = Lf + vf,$$

we obtain the integral form of the stationary linear problem

$$f = UKf + Ef^+, x \in \Omega, \quad f^+ = Rf^- + \Phi^+, \quad x \in S. \quad (11.17)$$

One constructs analogously integral kinetic equations connected with non-stationary problems.

The problem of constructing a solution of an integro-differential equation in $\mathscr{B}(X)$ for an appropriate space X is equivalent to the problem of constructing a solution of the corresponding integral equation in X.

1.3. Properties of the Collision Integral. Investigation of properties of the collision integral J for the gas of elastic balls was done in the papers of Carleman. Below we give some bounds generalizing Carleman's results.

We set

$$\varphi = \exp\{s|\xi|^2\}(1 + |\xi|^2)^{r/2} \quad (11.18)$$

If $r = 0, s > 0$, then the bound for $\Phi(F, F)$ is obtained in quite a simple way: due to the equality $\Phi(\varphi^{-1}, \varphi^{-1}) = \varphi^{-1} Q(\varphi^{-1})$, one gets

$$|\varphi \Phi(F,F)| \leqslant Q(\varphi^{-1}) \| F | L_\infty(R^3, \varphi) \|^2. \tag{11.19}$$

For soft potentials, one deduces from this estimate the boundedness of Φ in $L_\infty(R^3, \varphi)$. For hard potentials this estimate is not too useful because of unboundedness of $Q(\varphi^{-1})$ for large $|\xi|$.

Lemma 1.1. *If $s > 0$, $r > 2$, then the operator Φ is bounded in $L_\infty(R^3, \varphi)$* (cf [M1]).

Lemma 1.1 permits to prove the existence of a local solution for initial-boundary values problems. However, one does not succeed to prove the existence of a global solution in the class of exponentially decreasing (in ξ) functions. Therefore, following Carleman, we consider distribution functions with inverse-power decreasing.

Lemma 1.2 (cf [MC1], [MC2]). *Let φ be defined by equality* (11.18) *with $s = 0$, $r > 5$ and $B(v, z) \leqslant z(b_0 v^{1-\gamma} + b_2 v^{\varepsilon - 1})$ with $\gamma \in [0, 1]$, $\varepsilon \in (0, 1)$. Then there exists a positive constant C such that, for all F from $L_\infty(R^3, \varphi)$, the following inequality is valid*:

$$|\varphi \Phi(F,F)| \leqslant C \| F | L_\infty(R^3, \varphi) \|^2$$
$$+ 4\pi b_0 (r-2)^{-1} (1 + |\xi|)^{1-\gamma} \| F | L_\infty(R^3, \varphi) \| \, \| F | L_1(R^3) \|.$$

The bound stated in lemma cannot be improved, and therefore, the operator Φ is unbounded in $L_\infty(R^3, \varphi)$. However, if $\int_0^1 B(v, z) dz \geqslant b_1(v^{1-\gamma} + 1)$, then the operator $\Phi[Q + \alpha]^{-1}$, for large enough α and r, is bounded and moreover, has a small norm. This fact provides a basis for the proof of the existence of a global solution of the Cauchy problem for a space-homogeneous gas and local solution of some initial-boundary value problems.

Functions ψ_j given by equalities (11.14) are called *invariants of collisions*. The conservation laws for the mass, momentum and energy guarantee the fulfilment of the relations

$$\int_{R^3} \psi_j J(F,F) d\xi = 0 \, (j = 0, 1, \ldots, 4) \tag{11.20}$$

for all good enough functions F. We set

$$P_0 f = \sum_{j=0}^{4} \psi_j \langle \psi_j, f | H \rangle, \qquad Pf = f - P_0 f, \tag{11.21}$$

and denote by $P_0 H$ and PH the corresponding subspaces of the space H. Let be $\varphi_r = \omega^{1/2} (1 + |\xi|)^r$. The following lemma contains a description of properties of the collision integral which guarantee the "extinguishing" of the small perturbations.

Lemma 1.3. 1) *For any real r, the operator $K = L + \nu I$ is continuous as an operator from $L_2(R^3, \varphi_r)$ to $L_2(R^3, \varphi_{r+1/2})$ and is completely continuous in H.*

2) *There exists a constant* $l > 0$ *such that*

$$\langle f, Lf | H \rangle \leq -l \| v^{1/2} Pf | H \|^2$$

for all f from $L_2(R^3, (\omega v)^{1/2})$. 3) $\langle f, Lg | H \rangle = \langle g, Lf | H \rangle$, $P_0 L = L P_0 = 0$.

§2. Linear Stationary Problems

2.1. Asymptotics. In the investigation of boundary problems one essentially uses information about solutions of the non-homogeneous linearized equation

$$Df = \varepsilon^{-1} Lf + \varepsilon g, \quad x \in R^m, \, \varepsilon \in (0, 1]. \tag{11.22}$$

The investigation of asymptotics of f, as $|x| \to \infty$ and $\varepsilon \to 0$, is closely connected with the investigation of the connection of the Boltzmann equation with hydrodynamic equations. The pressure, velocity and temperature of a gas are determined by functions $A_j = \langle \psi_j, f | H \rangle$ and so, the subspace $P_0 H$ contains, in a certain sense, the whole hydrodynamic information about the gas. A traditional method of obtaining such information from kinetic equations is to cut (in quite an arbitrary way) the infinite hierarchy equations for moments. The first equations of this hierarchy (the conservation laws) are obtained by projecting (11.22) onto the subspace $P_0 H$. Together with the functions A_j, the higher moments figure in the system of the conservation laws: the strain (tension) tensor $\tau_{\alpha\beta} = \langle P\psi_\alpha \psi_\beta, f | H \rangle$ and the vector of the thermal flow $q_\alpha = \langle P\psi_\alpha \psi_4, fH \rangle$. We define functions φ_{α_j} by the relations

$$\varphi_{\alpha_j} = L^{-1}(P\psi_\alpha \psi_j), \quad P_0 \varphi_{\alpha_j} = 0. \tag{11.23}$$

By scalar multiplying (11.22) in H by the functions φ_{α_j} we get

$$\varepsilon^{-1} \tau_{\alpha\beta} = -\mu \left(\frac{\partial}{\partial x_\alpha} A_\beta + \frac{\partial}{\partial x_\beta} A_\alpha \right) + 2/3 \mu \delta_{\alpha\beta} \sum_{\gamma=1}^{m} \frac{\partial}{\partial x_\gamma} A_\gamma$$

$$+ \sum_{\gamma=1}^{m} \frac{\partial}{\partial x_\gamma} \langle \varphi_{\alpha\beta} \xi_\gamma, Pf | H \rangle - \varepsilon \langle \varphi_{\alpha\beta}, g | H \rangle, \tag{11.24}$$

$$\varepsilon^{-1} q_\alpha = -\lambda \frac{\partial}{\partial x_\alpha} A_4 + \sum_{\gamma=1}^{m} \frac{\partial}{\partial x_\gamma} \langle \varphi_{\alpha 4} \xi_\gamma, Pf | H \rangle$$

$$- \varepsilon \langle \varphi_{\alpha 4}, g | H \rangle, \tag{11.25}$$

where μ, λ are positive coefficients which have the sense of coefficients of viscosity and thermal conductivity.

If one throws away the two last terms in the right hand side of (11.25), (11.24), these formulae give Newton and Fourier laws, correspondingly. The system of conservation laws completed by formulae (11.25), (11.24) makes it possible to get an exact estimate for $P_0 f$ in terms of initial and boundary data of the problem under consideration and to describe asymptotic properties of f. (Notice that the functions $\varphi_{\alpha j}$ are polynomials in ξ only for Maxwell molecules. In general, the system of moment equations does not include the relations (11.25), (11.24).)

It is convenient to formulate the main result in terms of the Fourier transform $\tilde{f}$ (in the variables x) of the function $\tilde{f}$. The function $\tilde{f}$ has to satisfy the equation

$$\tilde{D}\tilde{f} = \varepsilon^{-1} L\tilde{f} + \varepsilon \tilde{g}, \qquad \tilde{D} = i \sum_{\alpha=1}^{m} \mathscr{k}_\alpha \xi_\alpha, \, \varepsilon \in (0,1]. \tag{11.26}$$

The structure of the power decomposition in ε

$$\tilde{f} = \sum_{n \geq 0} \varepsilon^n \tilde{f}_n \tag{11.27}$$

was investigated, for the nonlinear nonstationary problem, by Hilbert. Formal substitution of the Hilbert series (11.27) into (11.22) gives the equations

$$L\tilde{f}_0 = 0, \qquad L\tilde{f}_{n+1} = D\tilde{f}_n - g\delta_{n1}. \tag{11.28}$$

The condition $P_0(D\tilde{f}_n - g\delta_{n1}) = 0$ of existence of a solution of equation (11.28) is equivalent to the equations of linear hydrodynamics. It is not difficult to check that $\tilde{f}_n = 0(|\mathscr{k}|^{n-2})$ for fixed ξ and $\mathscr{k} \to 0$.

Theorem 2.1 (cf [M9], [M10]). *If $\tilde{g} \in H$, then the equation (11.26) has, for $\mathscr{k} \neq 0$, a unique solution in H. If $(1 + |\xi|)^{n-1} g \in H$, then for $n \geq 1$ this solution admits the representation*

$$\tilde{f} = \sum_{l=0}^{n} \varepsilon^l \tilde{f}_l + \varepsilon^{n+1} \tilde{F}_n,$$

$$\|\tilde{F}_n | H\| \leq C(1 + |\mathscr{k}|)^{n-1} \|(1 + |\xi|)^{n-1} \tilde{g} | H\|$$

where C is a positive constant independent of $\mathscr{k}$ and ε.

2.2. Internal Problems. Consider the problem of finding a function f satisfying the conditions

$$Df = Lf, \qquad x \in \Omega, \qquad f^+ = Rf^- + \Phi^+, \qquad x \in S, \tag{11.29}$$

in a bounded domain Ω. The first result concerning a solution of problem (11.29) was obtained by Grad [G1]. Grad proved the existence and uniqueness of a solution of problem (11.29) for $m = 1$, $R = 0$, $\Omega = (0, d)$ and small enough d. Guiraud investigated conditions of existence of a solution of problem (11.29) for the gas of elastic balls (cf [Gu1]).

Lemma 1.3 and Theorem 2.1 give a possibility of obtaining the following a priori estimates for a solution of problem (11.29) in $\mathscr{H}$:

$$\tfrac{1}{2}\|f^- | \mathscr{H}(S)\| + l\|v^{1/2} Pf | \mathscr{H}\|^2 \leq \tfrac{1}{2}\|f^+ | \mathscr{H}(S)\|^2,$$

$$\|P_0 f | \mathscr{H}\| \leq C[\|Pf | \mathscr{H}\| + \|f | \mathscr{H}(S)\|].$$

These estimates guarantee the uniqueness of a solution of the problem in $\mathcal{H}$ provided $\|R|\mathcal{H}(S)\| < 1$. To prove its existence, we consider the system (11.17).

Lemma 2.1. *The operator UK is completely continuous in $\mathcal{H}$.*

From Lemma 2.1, one easily deduces the Grad result and even the following, much more general theorem:

Theorem 2.2. *If $\Phi^+ \in \mathcal{H}(S)$, $\|R|\mathcal{H}(S)\| < 1$, then problem (11.29) has a unique solution in $\mathcal{H}$.*

The lack of conservation laws for the interaction of molecules with the boundary is guaranteed by the condition $\|R|\mathcal{H}(S)\| < 1$. However, in applications, one often encounters situations where this condition is not fulfilled. Let us set for $x \in S$, $\varphi = |(\xi, n(x))|$,

$$\Pi_0 f^\pm = (\varphi, f^\pm|H)(\varphi, 1^\pm|H)^{-1}, \qquad \Pi f^\pm = f^\pm - 1^\pm \Pi_0 f^\pm.$$

Suppose the operator R satisfies the following conditions (cf [M1], [Gu1], [Gu3]):

$$R1^- = 1^+, \qquad R\Pi = \Pi R, \qquad \|R\Pi|\mathcal{H}(S)\| < 1. \tag{11.30}$$

From the physical point of view, these conditions mean that, for the interaction with the boundary, the law of conservation of mass is fulfilled but the others are not valid.

Theorem 2.3 (cf [M1]). *Let $\Phi^+ \in \mathcal{H}(S)$ and conditions (11.30) be fulfilled. Then the fulfilment of the condition $\langle \Phi^+, 1|\mathcal{H}(S)\rangle = 0$ is necessary and sufficient for the existence of a solution of problem (11.29) in $\mathcal{H}$. Moreover, the difference of any two solution is constant.*

2.3. External Problems. Usually, the external problem is formulated as being the problem of finding a function f satisfying the conditions (11.29) and the relation

$$f \to f_\infty, \qquad |x| \to \infty. \tag{11.31}$$

One supposes that the domain $R^m \backslash \bar{\Omega}$ is bounded, $m \geq 2$. Experts in classical mechanics are usually unanimous about the questions of existence and uniqueness of a solution of problems they formulated. This is no longer the situation for external stationary problems for the Boltzmann equation. The review [Gu3] shows how contradictory opinions were announced in connection with these problems, on the basis of "physical" considerations.

To start off, it is convenient to leave the traditional statement of the problem and to consider the problem of constructing a function f satisfying the conditions (11.29) and the relation $f \in \mathcal{H}_\ell$, where

$$\mathcal{H}_\ell = \{f | (1 + |x|)^{-\ell} f \in \mathcal{H}\}.$$

The condition $f \in \mathcal{H}_\ell$ gives a restriction on the rate of increase of the function f, as $|x| \to \infty$. In particular, for $\ell \leq m/2$, this condition means that f converges, in some sense, to zero, as $|x| \to \infty$. We set $\hat{\psi}_j = \psi_j - \delta_{j4}(2/3)^{1/2}$ and

$$Q_j = \langle \hat{\psi}_j \operatorname{sgn}\langle \xi, n(x)\rangle, f | \mathscr{H}(S)\rangle,$$

$$p = \langle \psi_0 + \psi_4(2/3)^{1/2}, f | H \rangle,$$

$$f_0 = \sum_{j=0}^{4} a_j(x)\hat{\psi}_j, \qquad a_0 = C_0,$$

$$a_\alpha = C_\alpha + \sum_{\beta=1}^{3} Q_\beta H_{\alpha\beta}, \qquad a_4 = C_4 + Q_4 H,$$

where C_j are constants, $\{H_{\alpha\beta}\}$ is a matrix of fundamental solutions of the Stokes system, H is a fundamental solution of the Laplace equation. Constants Q_j determine the force and thermal flow acting on the "body" $R^m \setminus \Omega$, function p determines the pressure corresponding to the distribution f.

Under sufficiently general assumptions about the operator R, the following theorem holds.

Theorem 2.4 (cf [M5], [M10]). *If $\Phi^+ \in \mathscr{H}(S)$, then for any given constants B_j, the problem (11.29) has a unique solution satisfying the condition*

$$p - B_0 \in \mathscr{H}_1, \qquad Q_j = B_j, \qquad j \geq 1.$$

This solution admits the representation $f = f_0 + f_1$ where $f_1 \in \mathscr{H}_\varepsilon$ for any positive ε.

Theorem 2.4 gives a possibility of examining conditions of existence of a solution of the external problems in the traditional statement. If $m = 3$, then, due to the theorem, one has $f - f_0 \in \mathscr{H}_1$ for any Q_j. Hence, one gets the unique possible condition at infinity:

$$f_\infty = \sum_{j=0}^{4} C_j \hat{\psi}_j. \tag{11.32}$$

The arbitrariness in the selecting of constants B_j prescribed by Theorem 2.4 provides the possibility of determining in an arbitrary way constants C_j (i.e. the pressure, velocity and temperature of the gas at infinity). This fact allows us to prove the following theorem.

Theorem 2.5 (cf [M5], [M9]). *If $m = 3$, $\Phi^+ \in \mathscr{H}(S)$, then for any constants C_j the problem (11.29) has a unique solution in $\mathscr{H}_2$.*

The situation is changed for the plane case. If the force and thermal flow differ from zero, then the solution of the problem (11.29) increases like $\ln|x|$, as $|x| \to \infty$. In order to fulfil condition (11.32) or even only a condition of boundedness of the solution, one needs to put $Q_j = 0$. Hence, due to Theorem 2.4, one obtains the result that a phenomenon which is completely analogous to Stokes' paradox in hydrodynamics takes place for the plane problem: it is impossible to define in an arbitrary way the velocity and temperature of a gas at infinity, because these

Chapter 11. Uniqueness Theorems for the Boltzmann Equation

parameters are uniquely determined by the requirement of boundedness of the solution.

2.4. Kramers' Problem. The problem (11.12)–(11.13) was widely discussed in connection with the applications we mentioned above (cf [G3], [Gu3]). In particular, Grad [G3] formulated conjectures about conditions of existence and uniqueness of a solution of this problem and asymptotic properties of solutions. Set

$$f_0 = \sum_{j=0}^{4} C_j \hat{\psi}_j + \sum_{j=2}^{4} B_j(\varphi_{1j} + x\hat{\psi}_j) < \varphi_{1j}, L\varphi_{1j}|H>^{-1}, \qquad (11.33)$$

where φ_{1j} are the functions determined by the relations (11.23).

Theorem 2.6 (cf [M4], [M8]). *If $g = 0$, $R = 0$, $|\xi_1|^{1/2}\Phi^+ \in H$, then for any k from $[2, +\infty)$ the problem (11.12)–(11.13) has a unique solution in $\mathscr{H}_k$. This solution admits the representation $f = f_0 + f_1$ where $f_1 \in \mathscr{H}$ and f_0 is determined by formula (11.33) with $B_0 = C_0$.*

Constants $C_j(j \geq 2)$ determine the velocity of sliding and temperature jump. Theorem 2.6 remains true under sufficiently general restrictions onto the function g and operator R (cf [M4], [M8]). It is natural to consider this theorem as a confirmation and precision of the Grad conjecture.

§3. Nonlinear Stationary Problems

In situations near equilibrium, it is possible to construct a solution of the stationary problem (11.8) in a bounded domain Ω by methods of iterations taking, as first approximation, the corresponding solution of the linearized equation. Guiraud [Gu1] proved this way the existence and uniqueness of a solution of the problem (11.8) for the gas of elastic balls. Theorems 2.2 and 2.3 give a possibility to obtain an analogous result for all hard potentials under conditions about the form of the boundary and the law of interaction between molecules and the boundary which are more general than in the Guiraud paper (cf [G]).

The external stationary problem, in particular the problem of describing a flow around a body, is more difficult. Let us consider the problem (11.7), (11.8) with $F_\infty = \omega(1, 1, V, \xi)$, $V = \{|V|, 0, 0\}$. The Mach number M is connected with V by the equality $M = (3/5)^{1/2}|V|$. By setting $f = \omega^{-1}(F - \omega)$, $\omega = F_\infty$, $v = \xi - V$ and assuming that the function F^+ is given, we obtain the following reformulation of the problem (11.7), (11.8):

$$\sum_{\alpha=1}^{3} (|V|\delta_{\alpha 1} + v_\alpha) \frac{\partial}{\partial x_\alpha} f = Lf + \Gamma(f, f), \qquad x \in \Omega, \qquad (11.34)$$

$$f^+ = \Phi^+, \qquad x \in S, \qquad f \to 0, \qquad |x| \to \infty. \qquad (11.35)$$

In situations near equilibrium the velocity V and the nonlinear term in (11.34) are small. By omitting these terms one gets the problem, whose solution has Stokes' asymptotics, as $|x| \to \infty$, due to Theorem 2.4. However, as in hydrodynamic problems, the corresponding method of iterations leads to the appearance of secular terms for large $|x|$.

Set

$$\|f|B_p(\Omega)\| = \sup(1 + |\xi|)^3 \omega^{1/2} \|f|L_p(\Omega)\|,$$
$$\|f|B_p(S)\| = \sup(1 + |\xi|)^4 \omega^{1/2} \|f|L_p(S)\|$$

and denote by $B_p(\Omega)$, $B_p(S)$ the corresponding Banach function spaces.

Let be $\Phi^+ \in B_\infty(S)$, $B_{p,\infty} = B_p(\Omega) \cap B_\infty(\Omega)$. Since functions f from $B_{p,\infty}$ are bounded, the index p measures the velocity of decrease of f, as $|x| \to \infty$ (the smaller p is, the faster the decrease).

Theorem 3.1 (cf [M6]). *Let f be a solution of the problem (11.34), (11.35) in $B_{p,\infty}$, $p < 12/5$. Then f admits the representation*

$$f(\xi, x) = \sum_{j=0}^{4} a_j(x)\hat{\psi}_j(v) + \hat{f}(\xi, x), \qquad (v = \xi - V),$$

$$\hat{f} \in B_2(\Omega), \qquad a_j \in L_\beta(\Omega), \qquad \beta > 2.$$

Functions a_j give the main terms for the asymptotics of the hydrodynamic parameters (pressure, velocity, temperature) at large distances from the boundary. Exact formulae for the Fourier transform of a_j show that these functions decrease, as $|x| \to \infty$, more slowly, of course, than functions from L_2. For computing the functions a_j, one needs only information about the flows Q_j. In particular, for $M < 1$,

$$a_\alpha(x) = \sum_{\beta=1}^{3} Q_\beta \hat{H}_{\alpha\beta}(x), \qquad \alpha = 1, 2, 3,$$

where $\{\hat{H}_{\alpha\beta}\}$ is the matrix of fundamental solutions of Oseen's system of equations. For subsonic flows (below the chock barrier) the functions a_j decrease like $|x|^{-2}$ outside the trace behind the body. For $M > 1$, along with the trace some new domains of slow decreasing of the functions a_j appear which are connected with Mach's cone.

Theorem 3.1 has a conditional character: one supposes there the existence of a solution of the problem (11.34), (11.35) from $B_{p,\infty}$. The following theorem shows that the problem (11.34), (11.35) may in fact be solved for small Φ^+.

Theorem 3.2 (cf [M6]). *There exists a positive number δ with the following property: if $\|\Phi^+|B_\infty(S)\| < \delta$, then the problem (11.34), (11.35) has a unique solution in $B_{p,\infty}$, $p \in (2, 12/5)$ satisfying the condition $\|f|B_{p,\infty}\| \leq C\delta$.*

From Theorem 3.2 it follows that the problem (11.34), (11.35) has a unique solution for the diffusive reflection if Mach's number is small.

The uniqueness of the solution of stationary problems may be proven also for another limiting case, for the case of a strongly rarefied gas (cf [M2], [M3]). In this case the integral operator determined by the equality (11.16) is a contraction in a suitable functional space. The rate of convergence of the iterations $F^{(n)} = VF^{(n-1)}$ depends on the dimension of the space. In particular, for $m = 3$, $\|F - F^{(n)}\| = 0(\varepsilon^n)$ and for $m = 1$, $\|F - F^{(n)}\| = O\left(\left(\varepsilon \ln \frac{1}{\varepsilon}\right)^n\right)$.

§4. Non-Stationary Problems

4.1. Relaxation in a Homogeneous Gas.
The problem of describing the relaxation is to find a function $F(t, \xi)$ satisfying the following conditions

$$\frac{\partial}{\partial t} F = J(F, F), \quad t \in (0, T), \quad F = F_0, \quad t = 0. \tag{11.36}$$

The first results concerning a solution of the problem (11.36) were obtained by Carleman for the gas of elastic balls.

The theorems we formulate below (cf [MC1], [MC3]) are the generalizations of Carleman's theorems to the class of potentials satisfying the following conditions

$$B(v, z) \leqslant b^1 z(v^{1-\gamma} + 1), \quad \int_0^1 B(v, z) \, dz \geqslant b^0 v^{1-\gamma}, \gamma \in [0, 1). \tag{11.37}$$

Let

$$\varphi = (1 + |\xi|)^r, \quad r \geqslant r_0, \quad r_0 = 2 + \max\{3, 8b^1(b^0)^{-1}\}. \tag{11.38}$$

Theorem 4.1. *If $F_0 \in L_\infty(R^3, \varphi)$, then for all T the problem (11.36) has a unique solution in $L_\infty([0, T] \times R^3, \varphi)$. This solution satisfies the condition*

$$\sup_t \|F|L_\infty(R^3, \varphi)\| < \infty.$$

Denote by ω the Maxwell distribution with the parameters satisfying the condition $\langle \psi_j, F_0 - \omega | L_2(R^3) \rangle = 0$.

Theorem 4.2. *Let the condition of Theorem 4.1 be fulfilled and $F_0 \in C(R^3)$. The solution of the problem (11.36) is continuous in ξ uniformly with respect to t and satisfies the relation $\sup_\xi |F(t, \xi) - \omega| \to 0$, $t \to \infty$.*

For applications, it is important to have an estimate of the rate of approach to the equilibrium state ω. For sufficiently general initial conditions, so far one has such estimates only for the case of the Maxwell molecules (cf [B2], [V]). By reducing drastically the class of initial distributions, Grad proved the exponentially rapid convergence of solutions of the problem (11.36) to the equilibrium. Let $F = \omega(1 + f)$, and

$$\varphi = (1+|\xi|)^3 \omega^{1/2}, \qquad N(f) = \|f|L_\infty(R^3,\varphi)\|.$$

Grad's theorem asserts that there exists a positive constant β for which $\sup_t N(f)\exp(\beta t) < \infty$ provided the norm $N(f)$ is sufficiently small for $t=0$. However, there is no reason to hope to drop the condition on $N(f)$ for $t=0$. Moreover, Bobylev's results show that for the Maxwell molecules and compact support initial data the perturbation may leave the space H in finite time (cf [B2], [B3]).

Notice also that from Povzner's results one can deduce that the global uniqueness for the problem (11.36) is preserved for a much wider class of initial distributions (cf [P], [M1]). However, for conditions of Povzner's theorem one does not even succeed to show the boundedness of F for large t.

4.2. The Cauchy Problem. Now let us consider the Cauchy problem in a general case:

$$\frac{\partial}{\partial t}F + DF = J(F,F), \qquad t \in (0,T), \qquad F = F_0, \qquad t = 0. \qquad (11.39)$$

Under weak restrictions on the initial distribution it is possible to prove the existence of a unique solution of the problem (11.39) on the time interval $(0,T)$ the length of which depends on initial data (cf [M1]). In particular, the following theorem holds.

Theorem 4.3 (cf [MC4]). *If $F_0 \in L_\infty(R^6,\varphi)$, $F_0 \geq 0$, and conditions (11.38), (11.37) are fulfilled, then there exists a positive number T_1 such that for $T \leq T_1$ the problem (11.39) has a unique solution in $L_\infty([0,T] \times R^6, \varphi)$.*

If φ is determined by the formula (11.18), then for $s > 0$, $r > 2$ the local uniqueness of a solution of the problem (11.39) in $L_\infty(R^6,\varphi)$ takes place for all potentials satisfying the first Grad condition (11.1) (cf [M1]). For the potentials satisfying the condition (11.4) the problem (11.39) has a locally unique solution in

$$L_\infty([0,T] \times R^6, \varphi) \quad \text{for } s > 0, r = 0.$$

For each case mentioned the solution may be obtained as the limit of a sequence of iterations in an integral kinetic equation. The proof of convergence of the iterations is based on the estimates contained in Lemmas 1.1, 1.2. It is not hard to prove that the solution is nonnegative for nonnegative F_0 and has the same smoothness in x, ξ as the initial distribution.

Now consider the problem (11.39), supposing that the initial distribution is close to the Maxwell distribution $\omega = \omega(1,1,0,\xi)$. In this situation it is possible to prove the existence of a unique solution of the problem on the infinite time interval. Let us set

$$\|f|B_{p,\infty}\| = \|f|B_p(R^3)\| + \|f|B_\infty(R^3)\|,$$

$$\|f|\hat{B}_p\| = \sup_{t \in [0,T]} \|f|B_{p,\infty}\|.$$

Theorem 4.4 (cf [M7], [MF1], [MF2]). *There exists a positive number δ such that, if $\|f_0|B_{2,\infty}\| < \delta$, then for all T the problem (11.39) has a unique solution satisfying the condition $\omega^{-1}(F - \omega) \in \hat{B}_2$. Moreover, the following estimate is valid:*

$$\sup_{t \geq 0} \|\omega^{-1}(F - \omega)|B_{2,\infty}\| < \infty.$$

The function F has the same smoothness in x as the initial distribution. Under additional assumptions about smoothness of F_0 and velocity of decrease of f_0, as $|x| \to \infty$, it is possible to obtain estimates of the rate of approach to the equilibrium state ω. In particular, if f_0 and its derivatives in x up to the second order are contained in $B_{1,\infty}$, then

$$\|f|B_{2,\infty}\| \leq C(1 + t)^{-3/4}, \qquad \|f|H\| \leq C(1 + t)^{-3/2} \text{ (cf [M1])}.$$

The difference h between the solutions of the nonlinear and linearized Boltzmann equations satisfies the condition $\|h|B_{2,\infty}\| \leq C(1 + t)^{-5/4}$. In the paper [KMN] similar estimates for the difference between the moments (ψ_j, f) and solutions of Navier-Stokes equations were obtained. An analogous theorem for soft potentials was obtained in the paper [UA1].

4.3. Boundary Problems. The existence of unique solutions of nonstationary boundary problems may be proven under sufficiently general assumptions about initial and boundary conditions (cf [M1], [MC4]).

As well as for the Cauchy problem, it is possible to obtain global theorems only in situations which are close to equilibrium ones. The first results in this direction were obtained in the papers [F], [U] where a domain Ω is a cube with the normally reflecting boundary. For this case the boundary problem is reduced to the Cauchy problem with periodic initial conditions.

In a general case such a reduction is, of course, impossible but for homogeneous boundary conditions one reduces the solution of the nonstationary problem, as well as the solution of the Cauchy problem, to the investigation of the equation

$$f(t) = T(t)f_0 + \int_0^t T(t - \tau)\Gamma(f(\tau), f(\tau))\,d\tau,$$

where $T(t)f_0$ is the solution of the nonstationary boundary problem for the linearized Boltzmann equation.

A solution of this equation requires information about properties of the semi-group $T(t)$. Guiraud [Gu2] investigated its properties for the gas of elastic balls contained in a bounded domain with a smooth boundary. Using some additional assumptions of a technical character, he proved exponential bounds for the decrease of $T(t)$ for growing t and the existence of a unique solution of the initial-boundary value problem (11.9)–(11.11). In Guiraud's proof, his unpublished result is essentially used for the complete continuity of the tenth degree of the operator UK figuring in the integral Boltzmann equation (11.17). From Lemma 2.1 the compactness of the first degree of this operator follows for all hard potentials. This fact and the above formulated theorems about solutions of linear stationary problems allow us to obtain estimates for the semi-group $T(t)$ and furthermore to use them for the investigation of nonlinear problems. As an example of such an application, we consider the problem (11.9)–(11.11) with $\Phi^+ = \omega(1,1,0,\xi)$, $\mathscr{R} = 0$. Assuming $f = \omega^{-1}(F - \omega)$ we obtain the following formulation of this problem:

$$\frac{\partial}{\partial t} f = -Df + Lf + \Gamma(f,f), \qquad t \in (0,T), \qquad x \in \Omega, \qquad (11.40)$$

$$f^+ = 0, \qquad t \in (0,T), \qquad x \in S, \qquad f = f_0, \qquad t = 0, \qquad x \in \Omega. \qquad (11.41)$$

If Ω is a bounded domain with a smooth boundary S, then the following theorem is valid for all potentials satisfying the conditions (11.1), (11.2).

Theorem 4.5 (cf [M7]). *There exists a positive constant δ such that if $\|f_0|B_\infty(\Omega)\| < \delta$, then for all T the problem (11.40), (11.41) has a unique solution in $\hat{B}_\infty$. This solution satisfies the condition*

$$\sup_t \|f|B_\infty(\Omega)\| \exp\{\sigma t\} \leqslant C \|f_0|B_\infty(\Omega)\|$$

with some positive constant σ.

§5. On a Connection of the Boltzmann Equation with Hydrodynamic Equations

5.1. Statement of the Problem. Investigation of the connection of the Boltzmann equation with the equations of hydrodynamics is one of the classical problems of statistical physics. In papers of Maxwell, for the first time in the literature, an infinite chain of equations for the moments of the Boltzmann distribution function appeared and the problem of the foundation of hydrodynamics was formulated as a problem of "closing" this chain. From the Boltzmann equation and the equality (11.20) it follows that the moments $M_j = \int_{R^3} \psi_j F \, d\xi$ ($j = 0, 1, \ldots 4$) must satisfy the system of equations

$$\frac{\partial}{\partial t} M_j = -\sum_{\alpha=1}^{3} \frac{\partial}{\partial x_\alpha} T_{\alpha_j}, \qquad T_{\alpha_j} = \int_{R^3} \xi_\alpha \psi_j F \, d\xi. \qquad (11.42)$$

Equations (11.42) are the first five equations of the above-mentioned chain. In hydrodynamics this chain is closed by means of the Newton and Fourier laws which express the moments T_{α_j} in terms of the functions M_j and their space derivatives. One should expect that, in situations close to equilibrium, a "synchronisation" of higher moments is rapidly taking the place which leads at a

Chapter 11. Uniqueness Theorems for the Boltzmann Equation

possibility of a simplified description. However, the physical and mathematical nature of this process is not clear so far.

One of the directions of investigations is related to the study of the asymptotics of a solution of the Cauchy problem for the Boltzmann equation

$$\frac{\partial}{\partial t}F^\varepsilon + DF^\varepsilon = \varepsilon^{-1}J(F^\varepsilon, F^\varepsilon), \qquad F^\varepsilon|_{t=0} = F^\varepsilon(0) \qquad (x \in \Omega,\ t \in [0, T]) \quad (11.43)$$

with the large parameter ε^{-1} in front of the collision integral, which corresponds to the hydrodynamic limit passage discussed in Chap. 10, Sect. 6. This direction has its own meaningful history which is connected with the names of Hilbert and Carleman. Hilbert described a structure of formal power expansions $F^\varepsilon = \sum_{n \geq 0} \varepsilon^n F_n$. The main term of the Hilbert series, F_0, is a Maxwell distribution whose moments satisfy the Euler equations. The Hilbert paradox discussed in 20-th is that, to construct all terms of the Hilbert series, one needs only information on hydrodynamic moments $M_j(j = 0, \ldots, 4)$ of an initial distribution. The resolution of the paradox is that the Hilbert series (whenever it converges) guarantees the validity of initial conditions only for initial data of a very special form.

The existence theorems formulated in Sect. 4 do not give any background for discussing the hydrodynamic limit passage since local theorems (in particular Theorem 4.3) guarantee the existence of a solution on a time interval of order ε only, and global theorems allow us to consider only small perturbations (of order ε).

In the papers of Japanese mathematicians [N], [UA2], theorems are obtained which give a foundation for the main term of the Hilbert series for initial data which are close to a Maxwell distribution with constant parameters and are analytic in x. The time of existence of a solution of the Cauchy problem guaranteed by these theorems depends on a norm of the initial perturbation in the corresponding Banach space of analytic functions and decreases when this norm is growing.

In fact, solutions of the Cauchy problem (11.43) exist at least on an interval where Kato's theorem guarantees the existence of solutions of Euler equations in Sobolev spaces. The precise formulation of this statement is given below, in Theorem 5.1. The only essential condition for an application of this theorem is a closeness of $F^\varepsilon(0)$ to a local Maxwell distribution. However, it is sufficient to suppose a closeness in L_∞ without introducing any restriction on gradients of the hydrodynamic momenta M_j.

One of the corollaries of Theorem 5.1 is the possibility to give the complete *foundation* for the Hilbert method: for smooth initial data the partial sum of the Hilbert series differs from the exact solution by a quantity of order $\varepsilon^n + e^{-t/\varepsilon}$ (cf Theorem 5.2).

As to global solutions of the problem (11.43), one succeeds in constructing them only for small initial perturbations (of order ε in W_2^2) of hydrodynamic moments. The principal term of the asymptotics is determined by a solution of Navier-Stokes equations.

5.2. Local Solutions. Reduction to Euler Equations.

Before passing to the discussion of the problem (11.43), we consider the Cauchy problem for the Euler equation. Let $M^{(F)} = (M_0^{(F)}, \ldots, M_4^{(F)})$ be the vector of the hydrodynamic momenta of a function F. The density ρ, velocity V and temperature θ are connected with $M^{(F)}$ by the relations

$$M_0^{(F)} = \rho, \quad M_j^{(F)} = \rho V_j (j = 1, 2, 3), \quad \sqrt{6} M_4^{(F)} = \rho V^2 + 3\rho(\theta - 1). \quad (11.44)$$

Euler equations admit the following representation

$$\frac{\partial}{\partial t} M(t) + \sum_{j=1}^{3} A_j(M(t)) \frac{\partial}{\partial x_j} M(t) = 0 \quad (11.45)$$

where A_j are smooth matrix functions. If, for $t = 0$, the condition

$$M_j(0) - \delta_{j0} \in W_2^l(\Omega), \, l \geqslant 2, \, \inf \rho > 0, \, \inf \theta > 0. \quad (11.46)$$

holds, then there exists a time interval $[0, T]$ such that the problem (11.45) has a unique solution in $L_\infty([0, T], W_2^l(\Omega))$ satisfying the condition: $\inf \rho > 0$, $\inf \theta > 0$.

By $\omega = \omega(M, \xi)$ we denote the Maxwell distribution with the momenta M. If the condition (11.46) is valid, then there exists a Maxwell distribution $\bar{\omega}$ with constant parameters such that

$$\omega(M(t), \xi) \leqslant \bar{\omega}. \quad (11.47)$$

Given a distribution function F for which the condition (11.46) holds, we denote by $M^{(F)}(t)$, $t \in [0, T]$, the local solution of (11.45) with the initial data $M^{(F)}$.

Now let us turn to the problem (11.43), assuming for simplicity that $\Omega = R^3$ and the initial distribution does not depend on ε. Principal statements which are given below remain true under the assumption that the initial function F^ε is a polynomial as well as for the case of where Ω is a torus. Denote by $H_m(\omega)$ the Hilbert space of functions F with the norm

$$\|F | H_m(\omega)\| = \| \|F | W_2^m(\Omega)\| |L_2(R^3, \omega^{-1/2})\|. \quad (11.48)$$

We shall consider initial data which belong to the set

$$\mathbb{B}(m, \alpha, c) = \{F | M_j^{(F)} - \delta_{j0} \in W_2^m(\Omega), \inf \rho > 0, \inf \theta > 0,$$

$$(1 + |\xi|)^\alpha (F - \omega) \in H_m(\omega), \|F - \omega | L_2(R^3, \omega^{-1/2})\| \leqslant C\} \quad (11.49)$$

Assume that the function B which characterizes the interaction potential satisfies, in addition to (11.1), (11.2), the following condition uniformly in v:

$$b_2 |v|^p \leqslant \left(\int_0^1 B^2(v, z) \, dz \right)^{1/2} \leqslant b_3 (1 + |v|^p), \quad p \geqslant 0, \quad b > 0. \quad (11.50)$$

This inequality holds, in particular, for the hard spheres as well as for power potentials $U(r) = cr^{-s}$, $s \geqslant 4$, with an "angular cut-off".

It follows from the inequality (11.50) that, in a gas with the distribution ω, for the collision frequency v (cf Sect. 1) and the operator Γ, the following estimates

Chapter 11. Uniqueness Theorems for the Boltzmann Equation

hold uniformly in x, t

$$v_1 \leq v(1 + |\xi|)^{-p} \leq v_2 \quad (0 < v_1 < v_2),$$

$$\|h_\beta v^{-1} \Gamma(f_1, f_2)|H\| \leq \gamma(\|h_\beta f_1|H\| \|f_2|H\|$$
$$+ \|f_1|H\| \|h_\beta f_2|H\|), \quad H = L_2(R^3, \omega^{1/2}),$$
$$h_\beta = (1 + |\xi|)^\beta, \quad h_\beta f_j \in H, \quad \beta \geq 0, \quad \gamma > 0.$$

The bounds from Lemma 1.3 are valid, too, uniformly in x, t.

Theorem 5.1. *Let $F \in \mathbb{B}(m, \alpha, c)$, $m \geq 5$, $\alpha \geq 4$. Then there exists a positive constant $c^* = c^*(l, \gamma)$ such that for $c \leq c^*$ the problem (11.43) has for all ε a unique solution on an interval $[0, T]$ satisfying the conditions: $h_\beta(F - \omega) \in L_\infty([0, T], H_{m-3}(\bar{\omega}))$, and*

$$\sup_{x, t} \|F - \omega|L_2(R^3, \bar{\omega}^{-1/2})\| \leq (2\gamma)^{-1}.$$

A sketch of the proof is as follows. Let $S_0 = S_0(n, \varepsilon)$ be a partial sum of the Hilbert series. For smooth initial data the method of Hilbert allows us to reduce equation (11.43) to the non-homogeneous Boltzmann equation

$$\left(\frac{\partial}{\partial t} + D\right)\Phi = 2\varepsilon^{-1} J(S_0, \Phi) + \varepsilon^{-1} J(\Phi, \Phi) + R_0 \tag{11.51}$$

with the function R_0 of order ε^n. The initial value of Φ under assumptions of the theorem does not need to be small. We introduce into equation (11.51) a "rapid" time $\tau = \varepsilon^{-1} t$ and consider a formal power expansion $\Phi(\tau) = \sum_{n \geq 0} \varepsilon^n \Pi_n(\tau)$. The function $\Pi_0 + F_0(0)$ must be a solution of the Carleman problem (11.36). For $n > 0$, the functions Π_n are solutions of linear equations corresponding to the problem (11.36). The Hilbert series terms and functions Π_n are uniquely determined by the following conditions

$$\Pi_n(0) + F_n(0) = F(0)\delta_{n0}, \quad \Pi_n(\tau) \to 0, \quad \tau \to \infty. \tag{11.52}$$

By setting $S = \sum_{j=0}^n \varepsilon^j (\Pi_j + F_j)$, $G = F^\varepsilon - S$, we obtain the following problem for the function G

$$\left(\frac{\partial}{\partial t} + D\right) G = 2\varepsilon^{-1} J(S, G) + \varepsilon^{-1} J(G, G) + R, \quad G\bigg|_{t=0} = 0,$$

with the function R satisfying the condition $\|R|L_2([0, T], H_s(\omega))\| \leq C\varepsilon^n$, $s \leq m - 2n - 1$, uniformly in ε.

To prove a possibility of applying a perturbation theory to the problem (11.51), one necessarily needs uniform in ε estimates for solutions of the linear problem

$$\left(\frac{\partial}{\partial t} + D\right) G = 2\varepsilon^{-1} J(S, G) + X, \quad G\bigg|_{t=0} = 0, \tag{11.53}$$

where X is a given function from $L_2([0, T], H_s(\bar{\omega}))$. Therefore, a key role in the proof of Theorem 5.1 is played by the following lemma.

Lemma 5.1. *The problem* (11.52) *has a unique solution in* $L_\infty([0, T], H_s(\bar{\omega}))$. *This solution satisfies the inequality*

$$\|G\|_\infty \leq C(T)\|v^{-1/2}X\|_2$$

uniformly in ε *where* $C(T)$ *is a positive constant and*

$$\|G\|_p = \|G|L_p([0, T], H_0(\bar{\omega}))\|, \qquad p \in [1, \infty].$$

The main role in the proof of this lemma is played by the bounds for the operator L which are given by Lemma 1.3, together with a rapid convergence of solutions of the Carleman problem (11.36) to a Maxwell distribution. In the course of the proof one uses as well some ideas of the paper [C].

Lemma 5.1 allows to obtain the bound

$$\|F^\varepsilon - S\|_\infty \leq C(T)\varepsilon^{n+1/2}, \alpha \geq \max\{n, 4\}, \quad m \geq 2n + 3. \tag{11.54}$$

From the bound (11.54) and Theorem 5.1 the following theorem is easily deduced.

Theorem 5.2. *Let the assumptions of Theorem 5.1 be valid with* $\alpha \geq n$ *and* $m \geq 2n + 3$. *Then the solution of the problem* (11.43) *admits the representation*

$$F^\varepsilon = \sum_{j=0}^{n-1} \varepsilon^j F_j + R(n, \varepsilon).$$

The function $R(n, \varepsilon)$ *obeys*

$$\|R(n, \varepsilon)|H_0(\bar{\omega})\| \leq C(\varepsilon^h + e^{-t/\varepsilon})$$

uniformly in ε, t.

5.3. A Global Theorem. Reduction to Navier-Stokes Equations. Let us return to the problem of evolution of small perturbations of the state $\omega = \omega(1, 1, 0, \xi)$ which are described by equation (11.5). The hydrodynamic parameters are uniquely determined by the function $P_0 f$: $M_j = \delta_{j0} + (\psi_j, P_0 f)$. Hence, the problem of closing the moment equations (11.42) may be formulated as a problem of describing the dependence of Pf on $P_0 f$. Equation (11.5) is equivalent to the system of equations

$$\frac{\partial}{\partial t} P_0 f = -P_0 Df, \quad f = P_0 f + Pf, \tag{11.55}$$

$$\frac{\partial}{\partial t} PS = -PDf + \varepsilon^{-1} LPf + \varepsilon^{-1} \Gamma(f, f). \tag{11.56}$$

Let us denote by Y Banach space of functions φ with the norm

$$\|\varphi\|_Y = \|\varphi|L_\infty([0, T], H_m)\| + \|v^{1/2}\varphi|L_2([0, T], H_m)\|$$

where $H_m = H_m(\omega)$ is the space introduced in Sect. 5.2, and $m \geq 2$. We shall seek a solution of the Cauchy problem for equation (11.56) in the ball $S_1 = \{\varphi:$

$\|\varphi\|_Y \leq a_1\}$ for given values $Pf(0)$ and $P_0 f$ from the ball $S_0 = \{(Pf(0), P_0 f) \in H_m \times Y : \|Pf(0)|H_m\|^2 + \|P_0 f|L_\infty([0, T], H_m)\|^2 \leq a_0^2, m \geq 2\}$.

We denote by V the operator which associates with the point $(Pf(0), P_0 f)$ from S_0 a solution of equation (11.56).

Theorem 5.3. *There exist positive constants a_0, a_1 such that for any ε and T the Cauchy problem for equation (11.56) has a unique solution in the ball S_1. The operator V satisfies uniformly in x and T the condition*

$$\|V(P\varphi_1, P_0 f) - V(P\varphi_2, P_0 f)|H_m\|$$
$$\leq C \exp\{-ct/\varepsilon\} \|P(\varphi_1 - \varphi_2)|H_m\|, \quad C, c > 0, (P\varphi_j, P_0 f) \in S_0.$$

From Theorem 5.3 it follows that the Boltzmann equation is reduced to a (non-local in x, t) system of equations for hydrodynamic moments. When $t\varepsilon^{-1} \to \infty$, the operator V is well-approximated by a local one. More precisely, for smooth and bounded uniformly in ε initial data and the moments M, it follows from Theorem 5.3 that the solution of the problem (11.43) satisfies the condition

$$F^\varepsilon = \omega(M, \xi) + O(\varepsilon) + O(e^{-ct/\varepsilon})$$

where, as above, $\omega(M, \xi)$ is the Maxwell distribution with the momenta M. If perturbations of the moments $(M_j - \delta_{j0})$ have an order ε, then the function Pf admits the representation

$$Pf = \varepsilon L^{-1} PDP_0 f + \Phi, \tag{11.57}$$

$$\Phi = P\omega^{-1}(\omega(M, \xi) - \omega) + \varphi_0 e^{-ct/\varepsilon} + O(\varepsilon^3) + O(\varepsilon^2 e^{-t/\varepsilon}), \tag{11.58}$$

where φ_0 is a function bounded in H_m uniformly in ε, t and depending on the initial date f only. This is the only situation where one succeeds to construct a global solution for the moments.

Denote by $E(t)$ the group in H generated by the operator $P_0 DP_0$ and by $N_\varepsilon(t)$ the semi-group in H generated by the operator $P_0 D(P_0 + \varepsilon L^{-1} PDP_0)$. The operators $E(t)$ and $N_\varepsilon(t)$ determine solutions of linear Euler and Navier-Stokes equations. The problem of constructing the moments is reduced, due to Theorem 5.3 and the relations (11.57), (11.58), to solving the equation

$$E(-t)P_0 f = U(P_0 f) + \bar{N}_\varepsilon(t) P_0 f(0) \tag{11.59}$$

where $\bar{N}_\varepsilon(t) = E(-t)N_\varepsilon(t)$ is the semigroup introduced in [EP] and the operator U is defined by

$$Uf = \int_0^t \bar{N}_\varepsilon(t - \tau) \Phi(\varphi(\tau), \tau) d\tau.$$

Let

$$\|\varphi\|_{Y_0}^2 = \|\varphi|L_\infty([0, T], H_m)\|^2 + \varepsilon \|\nabla_x \varphi|L_2([0, T], H_m)\|^2.$$

We shall consider initial data satisfying the condition

$$\varepsilon^{-1}\|P_0 f(0)|H_m\| + \|(1 + |\xi|)^2 Pf(0)|H_{m+2}\| \leqslant C_0 \qquad (m \geqslant 2).$$

If $P_0\varphi \in Y_0$, then, uniformly in T, ε,

$$\|U(P_0\varphi)|Y_0\| \leqslant C(\varepsilon + \varepsilon^{-1}\|E(-t)P_0\varphi|Y_0\|^2)$$

where C is a constant depending on $\|P_0\varphi|Y_0\|$.

Theorem 5.4. *There exist constants c_0, c_1 such that the Cauchy problem for the system (11.55), (11.56) has, for any ε and T, a unique solution satisfying the condition*

$$\varepsilon^{-1}\|P_0\varphi|Y_0\| + \|Pf|Y\| \leqslant C_1.$$

Replacing the function Φ by the sum of the first two terms from the right hand side of (11.58), corresponds to passing from the Boltzmann equation to non-linear Navier-Stokes equations. An error generated by such a replacement is of order ε. More precisely, let $\varphi = \varepsilon^{-1}E(-t)P_0 f$ and denote by φ_N the described above Navier-Stokes approximation of this function. Under assumptions of Theorem 5.4 the following relations holds true

$$\sup_T \|\varphi - \varphi_N|Y_0\| = O(\varepsilon).$$

Bibliography

Results on existence, uniqueness and properties of solutions of various problems for the Boltzmann equation which were obtained before 1978 are presented in the paper [M1]. The new results are contained in the papers of Soviet [B2], [G], [M4]–[M10], [MC4], [V] and foreign [A1], [AEP], [U], [UA1] authors. See also the review papers in the issue [No] and references therein. We particularly mention the classical papers of Grad [G1]–[G3] which have been the starting points for many contemporary investigations.

The papers [C], [EP], [KMN], [N], [UA2] contain results of foreign authors about the connection between the Boltzmann equation and equations of hydrodynamics (see also [No]). In [L] bounds for the Boltzmann collision integral are obtained. The papers [BMT], [MT] concern the derivation of the Boltzmann equation from stochastic dynamical models.

For the convenience of the reader, references to reviews in Zentralblatt für Mathematik (Zbl.), compiled using the MATH database, have, as far as possible, been included in this bibliography.

[A1] Arkeryd, L.: On the Boltzmann equation in unbounded space far from equilibrium, and the limit of zero-mean free path. Commun. Math. Phys. *105*, 205–219 (1986). Zbl. 606.76094

[AEP] Arkeryd, L., Esposito, R., Pulvirenti, M.: The Boltzmann equation for weakly inhomogeneous data. Preprint 1986

[BMT] Belavkin, V.P., Maslov, V.P., Tariverdiev, S.E.: Asymptotic dynamics of a system with a large number of particles described by the Kolmogorov—Feller equations. Teor. Mat. Fiz. *49*, No. 3, 298–306 (1981) [Russian]

[B1] Bobylev, A.V.: Exact solutions of the Boltzmann equation. Dokl. Akad. Nauk SSSR *225*,

No. 6, 1296–1299 (1975) [Russian]. English transl.: Sov. Phys., Dokl. *20* (1975), 822–824 (1976). Zbl. 361.76082

[B2] Bobylev, A.V.: Asymptotic properties of solutions of the Boltzmann equation. Dokl. Akad. Nauk SSSR. *261*, No. 5, 1099–1104 (1981) [Russian]

[B3] Bobylev, A.V.: Exact solutions of the nonlinear Boltzmann equation and the theory of relaxation of a Maxwell gas. Teor. Mat. Fiz. *60*, No. 2, 280–310 (1984) [Russian]. Zbl. 565.76074. English transl.: Theor. Math. Phys. *60*, 820–841 (1984)

[C] Caflish, R.: The fluid dynamical limit of the nonlinear Boltzmann equation. Commun. Pure Appl. Math. *33*, 651–666 (1980). Zbl. 424.76060

[EP] Ellis, R., Pinsky, M.: The first and second fluid approximations to the linearized Boltzmann equation. J. Math. Pures Appl., IX. Ser. *54*, 125–156 (1975). Zbl. 297.35066

[F] Firsov, A.N.: On a Cauchy problem for the nonlinear Boltzmann equation. Aehrodinamika Razr. Gasov. *8*, 22–36 (1976) [Russian]

[G] Gejnts, A.G.: On the solvability of the boundary value problem for the nonlinear Boltzmann equation. Aehrodin. Razr. Gazov. *10*, 16–24 (1980) [Russian]. Zbl. 493.76073

[G1] Grad, H.: High frequency sound according to the Boltzmann equation. SIAM J. Appl. Math. *14*, 935–955 (1966). Zbl. 163.232

[G2] Grad, H.: Asymptotic theory of the Boltzmann equation. II. Rarefield Gas Dynamics *1*, 25–59 (1969)

[G3] Grad, H.: Singular and non-uniform limits of solutions of the Boltzmann equation. Transport Theory, New York 1967. SIAM AMS Proc. *1*, 269–308 (1969). Zbl. 181.285

[Gu1] Guiraud, J.P.: Problème aux limites intérieur pour l'équation de Boltzmann en régime stationnaire, faiblement non-lineaire. J. Méc, Paris *11*, No. 2, 183–231 (1972). Zbl. 245.76061

[Gu2] Guiraud, J.P.: An H-theorem for a gas of rigid spheres in a bounded domain. Theor. Cinet, class. relativ., Colloq. int. CNRS 236, Paris 1974, 29–58 (1975). Zbl. 364.76067

[Gu3] Guiraud, J.P.: The Boltzmann equation in kinetic theory. A survey of mathematical results. Fluid Dynamics Trans. *7*, Part II, 37–84 (1976)

[KMN] Kawashima, S., Matsumura, A., Nishida, T.: On the fluid dynamical approximation to the Boltzmann equation at the level of the Navier-Stokes equation. Commun. Math. Phys. *70*, 97–124 (1979). Zbl. 449.76053

[L] Lukshin, A.V.: On a property of the collision integral. Zh. Vychisl. Mat. Mat. Fiz. *25*, No. 1, 151–153 (1985) [Russian]. Zbl. 586.76140. English transl.: USSR Comput. Math. Math. Phys. *25*, No. 1, 102–104 (1985)

[MT] Maslov, V.P., Tariverdiev, S.E.: The asymptotics of the Kolmogorov-Feller equation for a system of a large number of particles. In: Itogi Nauki Tekhn., Ser. Teor. Veroyatn. Mat. Stat. Teor. Kibern. *19*, 85–125. (1982) [Russian]. Zbl. 517.60100. English transl.: J. Sov. Math. *23*, 2553–2579 (1983)

[M1] Maslova, N.B.: Theorems on the solvability of the nonlinear Boltzmann equation. Complement II to the Russian translation of the book: Cercignani C.: Theory and applications of the Boltzmann equation. Edinburgh-London, Scottish Academic Press 1975. Zbl. 403.76065

[M2] Maslova, N.B.: Stationary problems for the Boltzmann equation for large Knudsen numbers. Dokl. Akad. Nauk SSSR *229*, No. 3, 593–596 (1976) [Russian]. Zbl. 355.45012. English translation: Sov. Phys., Dokl. *21*, 378–380 (1976)

[M3] Maslova, N.B.: The solvability of stationary problems for the Boltzmann equation for large Knudsen numbers. Zh. Vychisl. Mat. Mat. Fiz. *17*, 1020–1030 (1977) [Russian]. Zbl. 358.35067. English transl.: USSR Comput. Math. Math. Phys. *17* (1977), No. 4, 194–204 (1978)

[M4] Maslova, N.B.: Stationary solutions of the Boltzmann equation and the Knudsen boundary layer. [Aehrodinamika Razr. gazov *10*, 5–15 (1980)] [Russian]. Zbl. 493.76074. Molecular gas dynamics, Interuniv. Collect., Aerodyn. Rarefied Gases *10*, Leningrad 1980, 5–15 (1980)

[M5] Maslova, N.B.: Stationary solutions of the Boltzmann equations in unbounded domains. Dokl. Akad. Nauk SSSR 260, (M 4), 844–848 (1981) [Russian]. Zbl. 522.76073. English transl.: Sov. Phys., Dokl. 26, 948–950 (1981)

[M6] Maslova, N.B.: Stationary boundary value problems for the nonlinear Boltzmann equation. Zap. Nauchn. Semin. Leningr. Otd. Mat. Inst. Steklova 110. 115, 100–104 (1981) [Russian]. Zbl. 482.76073. English transl.: J. Sov. Math. 25, 869–872 (1984)

[M7] Maslova, N.B.: Global solutions of nonstationary kinetic equations. Zap. Nauchn. Semin. Leningr. Otd. Mat. Inst. Steklova 115, 169–177 (1982) [Russian]. Zbl. 493.35027. English transl.: J. Sov. Math. 28, 735–741 (1985)

[M8] Maslova, N.B.: The Kramers problem in kinetic theory of gases. Zh. Vychisl. Mat. Mat. Fiz. 22, No. 3, 700–704 (1982) [Russian]. Zbl. 502.76089. English transl.: USSR Comput. Math. Math. Phys. 22, No. 3, 208–209 (1982)

[M9] Maslova, N.B.: Stationary solutions of the linearized Boltzmann equation. Tr. Mat. Inst. Steklova 159, 41–60 (1983) [Russian]. Zbl. 538.76069. English transl.: Proc. Steklov Inst. Math. 159, 41–60 (1984)

[M10] Maslova, N.B.: Exterior stationary problems for the linearized Boltzmann equation. Aehrodinamika Razr. Gasov. 11, 143–165 (1983) [Russian]

[MC1] Maslova, N.B., Chubenko, R.P.: Limiting properties of solutions of the Boltzmann equation. Dokl. Akad. Nauk SSSR 202, No. 4, 800–803 (1972) [Russian]

[MC2] Maslova, N.B., Chubenko, R.P.: Estimations of the integral of collisions. Vestn. Leningr. Univ. No. 13, 130–137 (1973) [Russian]

[MC3] Maslova, N.B., Chubenko, R.P.: Relaxation in a mono-atomic space-homogeneous gas. Vestn. Leningr. Univ. No. 13, 90–97 (1976) [Russian]. Zbl. 373.45008

[MC4] Maslova, N.B., Chubenko, R.P.: On solutions of the nonstationary Boltzmann equation. Vestn. Leningr. Univ. No. 19, 100–105 (1973) [Russian]. Zbl. 278.35075

[MF1] Maslova, N.B., Firsov, A.I.: The solution of the Cauchy problem for the Boltzman equation. I. The theorem of existence and uniqueness. Vestn. Leningr. Univ. No. 19, 83–88 (1975) [Russian]. Zbl. 325.76105

[MF2] Maslova, N.B., Firsov, A.I.: The solution of the Cauchy problem for the Boltzmann equation. II. Estimations of solutions of the inhomogeneous linearized equation. Vestn. Leningr. Univ. No. 1, 97–103 (1976) [Russian]. Zbl. 371.76062

[N] Nishida, T.: Fluid dynamic limit of the nonlinear Boltzmann equation to the level of the compressible Euler equation. Commun. Math. Phys. 61, 119–148 (1978). Zbl. 381.76060

[No] Nonequilibrium phenomena I. The Boltzmann equation. Studies in Statistical Mechanics, X (J.L. Lebowitz, E.W. Montroll, Eds.). Amsterdam–New York–Oxford, North-Holland 1983. Zbl. 583.76004

[P] Povzner, A.Ya.: On the Boltzmann equation in kinetic theory of gases. Mat. Sb., Nov. Ser. 58, No. 1, 65–86 (1962) [Russian]. Zbl. 128.225

[U] Ukai, S.: On the existence of global solutions of mixed problem for non-linear Boltzmann equation. Proc. Japan Acad., Ser. A 50, 179–184 (1974). Zbl. 312.35061

[UA1] Ukai, S., Asano, K.: On the Cauchy problem of the Boltzmann equation with a soft potential. Publ. Res. Inst. Math. Sci., Kyoto Univ. 18, 477–519 (1982). Zbl. 538.45011

[UA2] Ukai, S., Asano, K.: The Euler limit and initial layer of the non-linear Boltzmann equation. Hokkaido Math. J. 12, 311–332 (1983). Zbl. 525.76062

[V] Vedenyapin, V.V.: Anisotropic solutions of the nonlinear Boltzmann equation for the Maxwell molecules. Dokl. Akad. Nauk SSSR 256, No. 2, 338–342 (1981) [Russian]

Subject Index

Absolute 140
— continuity 124
Activity (fugacity) 213
Approximation of the first type by periodic transformations 59
— of the second type by periodic transformations 60
Asymptotic geodesics 140
Attractor 119
—, Belykh 178
—, Henon 178
—, hyperbolic 119
—, Lorenz 173
—, Lozi 178
—, stochastic 173
—, uniformly partially hyperbolic 120
Automorphism 4
—, aperiodic 59
—, Bernoulli 7
—, Gauss 35
—, induced 25
—, integral 25
—, Markov 8
— with pure point spectrum 30
— with quasi-discrete spectrum 32
B-automorphism 45
K-automorphism 20

Billiards 153
—, semidispersing 164
—, Sinai (dispersing) 157
Bifurcation period-doubling 190
Boltzmann distribution density 255
Boltzmann-Grad limit passage 240

Capacity, lower 149
—, upper 149
Cauchy problem 257
Caustic 155
Characteristic exponent 15
Cocycle 24
—, cohomological 24
—, measurable multiplicative 16
Collision integral 255

— operator linearized 256
Condition of finite range 216
Conditions DRL 214
Configuration with a minimal energy 106
Conjugate points 141
Curve, convex 157
—, decreasing 162
—, increasing 162
Cycle, attracting 196
—, irrational neutral 196
—, rational neutral 196
—, repelling 196

Diffeomorphism, Anosov 113
—, locally transitive 137
Dimension, Hausdorff 147
—, Lyapunov 147
— of measure 149
Direct product of automorphisms 22
Distribution, Gibbs 219
—, Maxwell 256
Doubling transformation 191
Dynamical system, Anosov 113
— —, equilibrium 228
— —, ergodic 18
— —, Gauss 35
— —, Hamiltonian 5
— —, uniformly partial hyperbolic 120
Dynamics determined by the interaction potential 223

Endomorphism of the measurable space 4
—, Bernoulli 7
—, exact 27
—, Markov 8
Entropy, conditional 37
— of a flow 39
— of a partition 36
— of automorphism 39
— of the action of group 75
— per unit time 38
—, topological 130
— formula 182
Equations of Navier-Stokes type 246

Subject Index

Equilibrium, dynamical system 228
— state 130
Equivalence in the sense of Kakutani 51
Ergodic components 19
Expanding property 179

Feigenbaum universality law 190
Finite horizon 171
Flow 4
—, Anosov 113
—, Gauss 35
—, geodesic 5
B-flow 49
K-flow 20
Foliations locally transitive 137
Foundation for the Hilbert method 271

Groups of automorphisms orbitally isomorphic 79

Hopf chain 136
Horosphere, stable 139
—, unstable 139

Information dimension upper 150
— —, lower 150
Interval exchange transformation 64
Interaction potential 213
Invariants of collisions 260
Isomorphism finitary 50
—, metrical 4
—, weak 44

Julia set 196

K-action of the group 77

Lebesgue space 10
Ledrappier capacity of measure lower 150
— — — upper 150
Limit sphere 140
Linear measurable bundle 15
Linearized collision operator 256
Lorentz gas 170

Manifold, local stable 112
—, — unstable 112
—, global stable 113
—, — unstable 113
—, — weakly stable 113
—, — — unstable 113
Manifolds of Anosov type 141
— of hyperbolic type 142

Mather spectrum 121
Maxwell molecules 256
Measurable action of the group 3
Measure, Gibbs 56
—, invariant 4
—, quasi-invariant 4
—, Sinai 134
—, Sinai-Ruelle-Bowen 131
—, translation invariant 211
— with maximal entropy 131
— with nonzero exponents 134
—, u-Gibbs 125
Mixing 20
—, r-fold 20
—, weak 20
Modular subgroup 145
Moment measure 220
Motions in Lobachevsky geometry 144

Natural extension of an endomorphism 27

Ornstein distance 46

Particle configuration 210
Partition, Bernoulli 45
—, exhaustive 42
—, extremal 42
—, finitely determined 46
—, loosely Bernoulli 53
—, Markov 126
—, perfect 42
—, tame 81
—, very weak Bernoulli 48
—, weak Bernoulli 49
Perron-Frobenius operator 180
Point, forward regular 16
—, heteroclinic 118
—, Lyapunov regular (biregular) 17
—, regular 152
—, singular 152
—, transversal homoclinic 117
Poisson measure 212
— super structure 233
Potential "hard" 255
— "soft" 256
Principal curvatures 143
Problem, external stationary 257
—, internal stationary 257
—, nonstationary boundary 257

Quadrilateral 162

Rectangle 126

Subject Index

Regular component of the boundary 152
Return time function 24

Schwarzian derivative 185
Set, hyperbolic 115
—, invariant 18
—, uniformly partially hyperbolic 120
—, locally maximal compact invariant 118
Skew product 23
Smale horseshoe 116
Special representation of a flow 26
Spectral equivalence of dynamical systems 28
Stable and unstable manifolds of hyperbolic set 116
State conditional Gibbs 56
Submanifold, concave 159
—, convex 159
Symbolic representation 127

System of conditional measures 10
— of variational equations 109
— with nonzero Lyapunov exponents 124
—, nonuniformly completely hyperbolic 123
—, — partially hyperbolic 123

Theorem, Bogolubov-Krylov 8
—, Oseledets 14
Topological transitivity 8
Trajectory, nonuniformly completely hyperbolic 110
—, — partially hyperbolic 111
—, uniformly completely hyperbolic 109

Uniquely ergodic homeomorphism 8

Variational principle for Gibbs measures 57
— — for topological entropy 130

Encyclopaedia of Mathematical Sciences
Editor-in-chief: R. V. Gamkrelidze

Springer-Verlag – synonymous with quality in publishing

- Our **Encyclopaedia of Mathematical Sciences** is much more than just a dictionary. It gives complete and representative coverage of relevant contemporary knowledge in mathematics.
- The volumes are monographs in themselves and contain the principal ideas of the underlying proofs.
- The authors and editors are all distinguished researchers.
- The average volume price makes **EMS** an affordable acquisition for both personal and professional libraries.

Dynamical Systems

Volume 1: **D. V. Anosov, V. I. Arnold** (Eds.)
Dynamical Systems I
Ordinary Differential Equations and Smooth Dynamical Systems
1988. IX, 233 pp. ISBN 3-540-17000-6

Volume 2: **Ya. G. Sinai** (Ed.)
Dynamical Systems II
Ergodic Theory with Applications to Dynamical Systems and Statistical Mechanics
1989. IX, 281 pp. 25 figs.
ISBN 3-540-17001-4

Volume 3: **V. I. Arnold** (Ed.)
Dynamical Systems III
1988. XIV, 291 pp. 81 figs.
ISBN 3-540-17002-2

Volume 4: **V. I. Arnold, S. P. Novikov** (Eds.)
Dynamical Systems IV
Symplectic Geometry and its Applications
1989. VII, 283 pp. 62 figs.
ISBN 3-540-17003-0

Volume 5: **V. I. Arnold** (Ed.)
Dynamical Systems V
Theory of Birfurcations and Catastrophes
1990. Approx. 280 pp. ISBN 3-540-18173-3

Volume 6: **V. I. Arnold** (Ed.)
Dynamical Systems VI
1990. ISBN 3-540-50583-0

Volume 16: **V. I. Arnold, S. P. Novikov** (Eds.)
Dynamical Systems VII
1990. ISBN 3-540-18176-8

Several Complex Variables

Volume 7: **A. G. Vitushkin** (Ed.)
Several Complex Variables I
Introduction to Complex Analysis
1989. VII, 248 pp. ISBN 3-540-17004-9

Volume 8: **A. G. Vitushkin, G. M. Khenkin** (Eds.)
Several Complex Variables II
Function Theory in Classical Domains. Complex Potential Theory
1990. ISBN 3-540-18175-X

Volume 9: **G. M. Khenkin** (Ed.)
Several Complex Variables III
Geometric Function Theory
1989. VII, 261 pp. ISBN 3-540-17005-7

Volume 10: **S. G. Gindikin, G. M. Khenkin** (Eds.)
Several Complex Variables IV
Algebraic Aspects of Complex Analysis
1989. Approx. 265 pp. ISBN 3-540-18174-1

Springer-Verlag
Berlin Heidelberg New York London
Paris Tokyo Hong Kong

Encyclopaedia of Mathematical Sciences
Editor-in-chief: R. V. Gamkrelidze

Algebra

Volume 11: **A. I. Kostrikin, I. R. Shafarevich** (Eds.)
Algebra I
Basic Notions of Algebra
1989. Approx. 272 pp. 45 figs.
ISBN 3-540-17006-5

Volume 18: **A. I. Kostrikin, I. R. Shafarevich** (Eds.)
Algebra II
1990. ISBN 3-540-18177-6

Topology

Volume 12: **D. B. Fuks, S. P. Novikov** (Eds.)
Topology I
1990. ISBN 3-540-17007-3

Volume 17: **A. V. Arkhangelskij, L. S. Pontryagin** (Eds.)
General Topology I
1990. ISBN 3-540-18178-4

Analysis

Volume 13: **R. V. Gamkrelidze** (Ed.)
Analysis I
Integral Representations and Asymptotic Methods
1989. VII, 238 pp. ISBN 3-540-17008-1

Volume 14: **R. V. Gamkrelidze** (Ed.)
Analysis II
1990. Approx. 270 pp. 21 figs.
ISBN 3-540-18179-2

Volume 15: **V. P. Khavin, N. K. Nikolskij** (Eds.)
Commutative Harmonic Analysis I
1990. ISBN 3-540-18180-6

Volume 19: **N. K. Nikolskij** (Ed.)
Functional Analysis I
1990. ISBN 3-540-50584-9

Volume 20: **A. L. Onishchik** (Ed.)
Lie Groups and Lie Algebras I
1990. ISBN 3-540-18697-2

Volume 21: **A. L. Onishchik, E. B. Vinberg** (Eds.)
Lie Groups and Lie Algebras II
1990. ISBN 3-540-50585-7

Volume 22: **A. A. Kirillov** (Ed.)
Representation Theory and Non-Commutative Harmonic Analysis I
1990. ISBN 3-540-18698-0

Springer-Verlag
Berlin Heidelberg New York London
Paris Tokyo Hong Kong